Software Methodologies
A Quantitative Guide

Software Methodologies
A Quantitative Guide

By
Capers Jones

CRC Press is an imprint of the
Taylor & Francis Group, an **informa** business
AN AUERBACH BOOK

CRC Press
Taylor & Francis Group
6000 Broken Sound Parkway NW, Suite 300
Boca Raton, FL 33487-2742

Printed on acid-free paper

International Standard Book Number-13: 978-1-138-03308-5 (Hardback)

Library of Congress Cataloging-in-Publication Data

Names: Jones, Capers, author.
Title: Software methodologies : a quantitative guide / Capers Jones.
Description: Boca Raton : Taylor & Francis, a CRC title, part of the Taylor & Francis imprint, a member of the Taylor & Francis Group, the academic division of T&F Informa, plc, [2017] | Includes bibliographical references.
Identifiers: LCCN 2016058829 | ISBN 9781138033085 (hb : alk. paper)
Subjects: LCSH: Software engineering--Statistical methods. | Computer software--Development. | Computer programming--Technique--Evaluation.
Classification: LCC QA76.758 .J6455 2017 | DDC 005.1--dc23
LC record available at https://lccn.loc.gov/2016058829

Visit the Taylor & Francis Web site at
http://www.taylorandfrancis.com

and the CRC Press Web site at
http://www.crcpress.com

Contents

Preface

This is my 17th book overall and my second book on software methodologies. My first methodology book was *Software Engineering Best Practices*, which was published by McGraw-Hill in 2010.

I decided on a new book on methodologies instead of a second edition of my older book because there are now so many methodologies (60 in this book) that an update to my first book might have pushed the page count close to 1000 pages, which is too large for easy reading and far too large for most publishers of trade books.

Since I'm age 78 writing this book, I've lived through the development of essentially every methodology in the book. As a former practicing programmer, I've actually used quite a few, although nobody I know has used all 60 of the methods discussed.

As a young IBM 1401 programmer in the 1960s, my first applications were small (converting older IBM 650 programs to run on the 1401). I worked solo and used cowboy development because that was the norm in the early 1960s.

Later at IBM, when I developed IBM's first parametric estimation tool in 1973, I had a colleague, Dr. Charles Turk. We used waterfall development for the first two versions. This IBM parametric estimation tool was called Development Planning System and it grew to have about 200 users in the IBM laboratories in the United States and Europe.

Then, at the first company I founded in 1984, Software Productivity Research (SPR), our development team used iterative development for our estimating tools of SPQR/20, Checkpoint, and KnowledgePlan.

As a friend and colleague of the late Watts Humphrey, I also experimented with his personal software process and team software process.

Our current parametric estimation tool at Namcook Analytics LLC, Software Risk Master (SRM), also used iterative development. Agile would have been inconvenient because we are not located in the same building or even in the same state.

Also, Agile assumes the existence of a user population. This does not occur for software such as SRM that is brand new and uses new inventions with patentable features that have never been seen before. Some of the SRM features, such as early sizing prior to requirements, needed to be kept secret until the application was released, so as not to give away valuable trade secrets to competitors. Although we had Beta test users, they had to sign non-disclosure agreements.

A basic question that the book will address is why software has at least 60 methodologies in 2016 and keeps developing new ones about every 10 months. This seems to be a sociological issue rather than a technical issue.

Several methods such as team software process (TSP), the rational unified process (RUP), and disciplined Agile development (DAD) could probably be used for every known software application in the world. Unfortunately, the software industry is not very literate and does not measure quality or productivity very well. The lack of effective measurements slows the adoption of technically successful methods such as TSP, DAD, and RUP. It is also slowing the use of the software engineering method and theory (SEMAT) approach, which is proving to be valuable but not at all well known in 2016.

For example, the Agile methodology is very difficult to measure due to the use of sprints and non-standard metrics such as story points, velocity, burn down, and so on. None of these Agile metrics have International Standards Organization (ISO) standards and they vary widely from project to project and company to company. Story points vary by over 400% from group to group. In this book, Agile productivity and quality data are converted into International Function Point Users' Group (IFPUG) function points to allow side-by-side comparisons with the other methodologies.

Agile is not a panacea and it has its own problems (consider all the thousands of Agile coaches). Agile sometimes fails or falls into traditional problems such as schedule delays and cost overruns. While Agile quality and productivity are a bit better than waterfall, Agile is not really as quality strong as TSP and RUP, although DAD is in the same category as those two.

A common failing of all current methodologies, including Agile, is that they view every application as novel and unique. This is not the case in 2016 when about 80% of software applications are replacements for aging legacy software or enhancements to legacy software.

Other modern software projects are those that involve modifying commercial off-the-shelf (COTS) packages; modifying and porting legacy applications to enterprise resource planning (ERP) packages; and even modifying open-source packages. None of these existed in the 1960s when methodologies were first developed and formalized.

The proper starting place for a new wave of software development methodologies will not be looking at requirements but instead extracting and examining the design and architecture patterns of similar applications already built. Once you know the feature sets and construction plans for similar legacy applications then new requirements are much easier and simpler than today.

It is much easier to build a tract home that has already been built dozens of times and has readily available blueprints and construction parts lists than it is to hire an architect and design a brand new home from scratch.

Unfortunately, software engineering ignores similar existing applications and assumes every application is new, novel, and in need of brand new requirements, architecture, and designs even though there may be hundreds or even thousands of almost identical applications already in existence.

Another question the book will address is why some methods, such as Agile, suddenly become cults and rapidly expand to millions of applications and users. It would be nice to say that the expansion is due to the success of the method, but that is not really the case.

Software methodologies come and go at fairly frequent intervals. Here, too, sociological issues seem to be at play. If you consider past methods that surged in popularity (rapid application development [RAD], computer-aided software engineering [CASE]), and Spiral development), they remained popular for about 5–7 years and then dropped out of use.

This situation reminds me of changes in clothing styles. Men's clothes change about every 5–7 years; ladies clothing styles seem to change more often at intervals of 2–3 years. Software development methodologies seem to cycle at intervals that are closer to men's clothing styles than women's clothing styles. This is probably because methodologies need to be used for several years before any results can be known.

The book will also address the topic of what might come next after Agile has run its course. It is obvious that custom designs and manual coding, which are part of essentially all current methodologies except mashups, are intrinsically expensive and buggy. This includes Agile, which definitely assumes new and novel applications.

The post-Agile wave of methodologies may well feature construction from standard reusable components instead of custom designs and manual coding. In order for reuse to become an integral part of a new generation of software methodologies, it is obvious that we need software application parts lists and software application construction blueprints.

Parts lists and blueprints will need forensic analysis of legacy applications in order to ascertain all of the common generic features used by many applications. It is possible that the essence of the SEMAT approach will prove useful in constructing these blueprints and parts lists to facilitate construction from reusable components.

A topic suggested by Rick Spiewak of MITRE would be using software "building codes." These building codes might be developed to show the specific nonfunctional requirements for applications that need Food and Drug Administration (FDA) or Federal Aviation Authority (FAA) certification or may come under the Sarbanes–Oxley law.

Software building codes would show the specific criteria needed for various sizes and types of software applications, with the most stringent building codes being for large systems above 10,000 function points, which may involve valuable financial assets or have an impact on human life and safety; that is, air-traffic control, major banking systems, major weapons systems, medical device software, and so on.

The immediate next wave after Agile has run its course will probably be "pattern-based development" in which software applications are constructed using patterns of reusable architectures, reusable designs, reusable code segments, and reusable test cases.

Currently in 2016, not enough is known about software feature sets and generic software architectures and designs to do this, but probably the SEMAT essence can assist in this important stage of discovery, which might raise software reuse from today's averages of perhaps 15% up to over 95% within 10–15 years.

Large-scale reuse of software architectures, designs, code segments, and test materials can raise software productivity from 2016 norms of less than 8.00 function points per staff month up to more than 100.00 function points per staff month, which is where it should be to match the economics of other kinds of technical product construction such as automobiles and electronic devices. Today, software is a drag on the economical construction of complex physical systems due to poor software quality control and far too frequent software cost overruns and schedule delays.

When software reuse becomes common, the next stage after pattern-based development might possibly be robotic development. Robotic tools have transformed the manufacturing of complex physical products such as automobiles and have also improved medical practice via robotic surgery.

There is no intrinsic reason why robotic software development might not become possible in perhaps 10–15 years. When robotic software development is feasible, it will minimize today's common human errors associated with custom designs and manual coding. The book will discuss the technology stacks needed for the successful robotic development of software applications.

This new book is not a tutorial on methodologies, but rather a critique and quantified comparison of 60 of them. Metrics such as work hours per function point and defect removal efficiency (DRE) are used consistently on all 60 methodologies. A standard size of 1000 IFPUG function points is used in order for the comparisons of methodologies to have the same standard base size.

From doing the research for this book, quite a few methodologies have been proved to be successful, but their success is relative and specialized. For example, Agile is successful on smaller projects with less than 100 users. DevOps is successful on projects with extensive runs in data centers and high transaction rates. Mashups are successful for prototypes and some web applications. The capability maturity model integrated and iterative development are successful for defense projects. TSP and RUP are successful for large applications above 10,000 function points and those that need FDA or FAA certification, or those that must adhere to Sarbanes–Oxley regulations. No methodology is equally successful on all sizes and types of software.

The measurement practices and common metrics used by the software industry are so bad that hardly anybody knows the actual productivity and quality results of any methodology. This book will provide quantitative data on the productivity and quality results of all 60 methodologies using function points and DRE.

Lines of code penalize modern high-level languages and ignore requirements and design. Cost per defect penalizes quality and makes buggies applications look better than high-quality applications. Neither of these metrics is suitable for large-scale economic analysis or for comparisons of the 60 methodologies in this book.

Neither story points nor use-case points have ISO standards or even standard counting rules. Story points vary by over 400% from group to group and use-case points by over 200%. Software needs accurate metrics and not vague approximations. By contrast, function points do have ISO standards and also formal certification exams. Therefore, function points in 2016 are the only accurate metric for either economic comparisons or accurate quality comparisons.

The new technical debt metric does not have an ISO standard and varies widely from company to company. As a courtesy to readers, the book will show approximate technical debt using a fixed assumption that all 60 methodologies compared are 1000 function points in size and have 1000 users after applications are released.

This book uses the metrics of IFPUG function points, work hours per function point, function points per month, cost per function point, defect potentials per function point, DRE, delivered defects, delivered defects per function point, security flaws per function point, cost of quality (COQ), and total cost of ownership (TCO). COQ and TCO per function point are also used.

To minimize variations caused by the presence of over 3000 programming languages, this book assumes Java as the programming language for all 60 methodologies. In real life no doubt other languages such as C, Objective C, Ruby, Python, MySQL, and so on would be used with many methodologies but for consistency and ease of statistical analysis Java is the only language used in the 60 methodology comparisons.

The new software non-functional assessment process metric for non-functional requirements lacks sufficient data on COQ, technical debt, and TCO and is not used in the book. This metric is new and as of 2016 does not yet have a great deal of quantitative data available and especially so for software quality and TCO.

None of my clients used the software non-functional assessment process metric in 2016, so our benchmark data collections, which are the basis of the comparisons in the book, have zero SNAP data in 2016. It would have been far too expensive to attempt to retrofit SNAP into projects already benchmarked without SNAP.

Other function point methods such as Common Software Measurement International Consortium, the Netherlands Software Measurement Association, and the Finnish Software Measurement Association are not used in the book but if they had been, the results would have been similar and within 15% of the IFPUG function point results. All of the functional metrics use the ISO/IEC 2096 standard set in 2009.

I hope that this quantitative comparison of methodologies will be of interest to readers and perhaps useful in selecting the most suitable methods for large and complex software applications, which may be outside the effective scope of Agile.

Acknowledgments

Quantitative Analysis and Comparison of 60 Software Development Methodologies

October 21, 2016 Version 1.0

Capers Jones, VP and CTO, Namcook Analytics LLC
E-mail: Capers.Jones3@gmail.com Web: www.Namcook.com

As always, many thanks to my wife, Eileen Jones, for making this book possible. Thanks for her patience when I get involved in writing and disappear for several hours. Also thanks for her patience on holidays and vacations when I take my portable computer and write early in the morning.

Additional thanks to my neighbor and business partner, Ted Maroney, who handles contracts and the business side of Namcook Analytics LLC, which frees up my time for books and technical work. Thanks also to Aruna Sankaranarayanan for her excellent work with our Software Risk Master (SRM) estimation tool and our website. Thanks also to Larry Zevon for fine work on our blog and to Bob Heffner for marketing plans. Thanks also to Gary Gack and Jitendra Subramanyam for their work with us at Namcook and in contributing materials to this book.

Thanks also to many researchers, including Dr. Alain Abran, Mauricio Aguiar, Dr. Victor Basili, Dr. Barry Boehm, Dr. Fred Brooks, Manfred Bundschuh, Tom DeMarco, Dr. Reiner Dumke, Christof Ebert, Gary Gack, Dave Garmus, Tom Gilb, Scott Goldfarb, Mike Harris, David Herron, Peter Hill, Dr. Steven Kan, Dr. Leon Kappelman, Dr. Tom McCabe, Dr. Howard Rubin, Dr. Akira Sakakibara, Manfred Seufort, Paul Strassman, Dr. Gerald Weinberg, Cornelius Wille, the late Ed Yourdon, and the late Dr. Harlan Mills for their own solid research and for the excellence and clarity with which they communicate ideas about software. The software industry is fortunate to have researchers and authors such as these.

Thanks to my former colleagues at Software Productivity Research (SPR) for their hard work on our three commercial estimating tools (SPQR/20 in 1984; CHECKPOINT in 1987; KnowledgePlan in 1990): Doug Brindley, Chas Douglis, Lynn Caramanica, Carol Chiungos, Jane Greene, Rich Ward, Wayne Hadlock, Debbie Chapman, Mike Cunnane, David Herron, Ed Begley, Chuck Berlin, Barbara Bloom, Julie Bonaiuto, William Bowen, Michael Bragen, Doug Brindley, Kristin Brooks, Tom Cagley, Sudip Charkraboty, Craig Chamberlin, Michael Cunnane, Charlie Duczakowski, Gail Flaherty, Richard Gazoorian, James Glorie, Scott Goldfarb, David Gustafson, Bill Harmon, Shane Hartman, Bob Haven, Steve Hone, Jan Huffman, Peter Katsoulas, Richard Kauffold, Scott Moody, John Mulcahy, Phyllis Nissen, Jacob Okyne, Donna O'Donnel, Mark Pinis, Tom Riesmeyer, Janet Russac, Cres Smith, John Smith, Judy Sommers, Bill Walsh, and John Zimmerman. Thanks also to Ajit Maira and Dick Spann for their service on SPR's board of directors.

Appreciation is also due to various corporate executives who supported the technical side of measurement and metrics and methodology research by providing time and funding. From IBM, the late Ted Climis and the late Jim Frame both supported the author's measurement work and in fact commissioned several studies of productivity and quality inside IBM, as well as funding IBM's first parametric estimation tool in 1973. Rand Araskog and Dr. Charles Herzfeld at ITT also provided funds for metrics and methodology studies, as did Jim Frame, who became the first ITT VP of software.

Appreciation is also due to the officers and employees of the International Function Point Users Group (IFPUG). This organization started almost 30 years ago in 1986, and has grown to become the largest software measurement association in the history of software. When the affiliates in other countries are included, the community of function point users is the largest measurement association in the world.

There are other function point associations such as COSMIC, FISMA, and NESMA but all 16 of my software books have used IFPUG function points. This is in part due to the fact that Al Albrecht and I worked together at IBM and later at Software Productivity Research.

Author

Capers Jones is currently vice president and chief technology officer of Namcook Analytics LLC (www.Namcook.com). Namcook Analytic LLC designs leading-edge risk, cost, and quality estimation and measurement tools. Software Risk Master (SRM)™ is the company's advanced estimation tool with a patent-pending early sizing feature that allows sizing before requirements via pattern matching. Namcook Analytics also collects software benchmark data and engages in longer-range software process improvement, quality, and risk-assessment studies. These Namcook studies are global and involve major corporations and some government agencies in many countries in Europe, Asia, and South America.

Introduction

Executive summary: *This introduction discusses the problems of software measurement and software metrics that have blinded software engineers and software managers and concealed progress for over 50 years. In order to understand the advantages and disadvantages of software methodologies, it is necessary to measure both productivity and quality with accuracy. The traditional "lines of code" (LOC) metric is not accurate and indeed distorts reality. LOC metrics penalize high-level languages and can't measure requirements, design, and non-code activities. This book uses function point metrics, which are accurate, and this introduction provides examples of why function points are better than LOC metrics for economic analysis.*

Software engineering is unique in having created more variations of tools, methods, and languages than any other engineering field in human history. For sociological reasons, software engineering in 2016 has about 60 software development methodologies; 50 static analysis tools; 40 software design methods; 37 benchmark organizations; 25 size metrics; 20 kinds of project management tools; 22 kinds of testing; and dozens of other tool variations. Software also has at least 3000 programming languages, although less than 100 are in frequent use. New programming languages are announced about every 2 weeks. New tools tend to come out at rates of more than one a month and new methodologies arrive about every 10 months. The existence of all these variations is proof that none are fully adequate for all kinds of software, or else they would have become standards and eliminated all the others.

The large number of software development methodologies may surprise some readers, but software has more variations of things than any other industry in history for apparently sociological reasons since there are clearly no good technical reasons for 3000 programming languages and more than 60 software development methodologies.

This book does not deal with the thousands of programming languages and the scores of design methods. It concentrates instead on 60 named software development methodologies that were released between about 1966 and 2016; a 50-year period.

In other words, 60 software methodologies have been released over the past 50 calendar years. That is an average of about 1 new methodology every 10 months.

Turning the math around, the software industry is creating new development methodologies at a rate of about 1.2 methodologies per calendar year.

This book is primarily a quantitative comparison of the productivity rates and quality results of these 60 methodologies. There are also at least 30 hybrid methodologies such as Agile+waterfall. A few of the more common hybrids are included.

To explain the reason for all of these variations, it must be pointed out that software has the worst and most incompetent measurement practices and the worst metrics of any industry in human history. (A recent prior book by the author, *Selecting Software Measurements and Metrics*, CRC Press 2016, describes these metric problems in detail.)

Suffice it to say that lines of code (LOC) makes requirements and design invisible and also penalizes modern high-level programming languages. The "cost per defect" measure penalizes quality and makes buggy software look cheaper than modern high-quality software. Both LOC and cost per defect distort reality. Story points and use-case points have no International Standards Organization (ISO) standards and vary by about 400% from company to company and project to project. Use-case points have no ISO standards and also vary widely.

Even with function point metrics, there are a number of variations. There are 4 major variations and 16 minor variations of function point metrics in 2016. The four major variations are: (1) the Common Software Measurement International Consortium (COSMIC) function points, (2) the Finnish Software Measurement Association (FISMA) function points, (3) International Function Point Users Group (IFPUG) function points, and (4) the Netherlands Software Measurement Association (NESMA) function points. All four follow ISO/International Electrotechnical Commission (IEC) standards for functional software sizing and also have formal user groups and certification examinations for accuracy in counting.

Among the minor variations without formal user groups or ISO standards are automated function points (with an Object Management Group [OMG] standard), function points light, full function points, engineering function points, feature points, weighted micro function points, unadjusted function points, and Software Risk Master (SRM) function points. The new software non-functional assessment process (SNAP) metric is also a minor variation in 2016 since, as a new metric still under development, it has little data available.

Table 0.1 shows the approximate sizes in terms of 25 of the software sizing metrics, with the basic size of the example being 1000 IFPUG function points.

To complete the set of information about the 25 current software size metrics, Table 0.2 shows 5 additional size metrics used in 2016.

IFPUG function points are the oldest form of function point metrics and have existed since about 1978 when IBM placed function points in the public domain. There are primarily sociological reasons for the other function point variations. As a rule, some software metric researchers felt that IFPUG function points did not provide a large enough size for some kinds of software, such as systems software, so they developed variations that gave a larger value for their favorite software type.

Table 0.1 Function Point Variations 2016

	Function Point Method	*Size*
1	Automated function points	1,015
2	Backfired function points	990
3	COSMIC function points	1,086
4	Engineering function points	1,135
5	Feature points	1,000
6	FISMA function points	1,120
7	Full function points	1,170
8	IFPUG function points	1,000
9	NESMA function points	1,060
10	SNAP non-functional size points	800
11	SRM function points	1,000
12	Unadjusted function points	890
13	Weighted micro function points	1,012
14	Function points light	987
15	IntegraNova function points	1,012
16	SSSQI function points	1,100
17	Fast function points	1,023
18	Simple function points	970
19	Web-object points	950
20	Mark II function points	1,105
	Average	1,021

Some of the function point variations make claims of "improved accuracy" but this claim is hypothetical and cannot be verified. There is no cesium atom for software size that always stays constant. Even when certified counting personnel count the same application using the same counting rules there can be variations of + or – 15% based on several controlled studies.

Here is a small case study of why a specific function point variant metric was developed. The "feature point metric" was created in 1984 to solve a sociological

Table 0.2 Size in Other Software Metrics

	Size Metric	*Size*
21	Story points	333
22	Use-case points	200
23	RICE objects	4,429
24	Lines of code: Logical Java	53,333
25	Lines of code: Physical Java	215,555

problem. It was also co-developed by A.J. Albrecht the inventor of function points and by the author of this book.

In 1984, telecommunication software engineers in a client company asserted incorrectly that "function points only worked on IT projects and not on telecom projects."

To solve this sociological issue, Al Albrecht and I developed a minor variation called "feature points," which used the five basic function point size elements (inputs, outputs, inquiries, interfaces, and logical files) and added a sixth element: algorithms. Feature points and IFPUG function points were calibrated to return essentially the same results.

The telecom community had long used the word "features" as part of their standard software vocabulary, so "feature points" were sociologically acceptable to that community even though feature points and function points resulted in the same numeric totals and, except for algorithms, had the same counting rules.

A few years later, IFPUG added counting rules for systems software so the feature point metric was withdrawn from service, although even in 2016 it pops up occasionally from the systems software domain.

Two of the function point variations, automated function points by CAST Software and SRM function points by Namcook Analytics, were created to solve the technical problems of faster counting speeds. Certified manual function point counters can perform function point sizing at an average rate of about 500 function points per day. For this reason, function points are seldom used on large applications above 10,000 function points.

The automated function point method by CAST Software uses a tool to examine legacy application code and can produce function point size values at rates of perhaps 50,000 function points per day. However, this tool only works on existing legacy software. Automated function points cannot be used prior to having code available.

The SRM function point method uses pattern matching and essentially reports the sizes of existing applications that share the same taxonomy pattern with a new application being sized. The SRM variation is the fastest known as of 2016 and can size any application in about 1.8 minutes. This includes even huge applications such as enterprise resource planning (ERP) packages in the 250,000 function point size range.

The SRM method can also be used 30–180 days earlier than other size methods. So long as enough information exists to show the taxonomy pattern of an application, it can be sized in less than 2 minutes using the author's SRM tool.

To ensure accurate comparisons of methodologies, this book uses a standard size of 1000 function points using the function point metric defined by the IFPUG counting rules version 4.3. This metric has an ISO standard and a certification examination for personnel who need to count function points. The combination of the ISO standard and certification ensures that function point sizes are fairly accurate. Function points certainly don't vary by over 400% from company to company like LOC or story point metrics.

Productivity in this book is measured using both work hours per function point and its reciprocal, function points per staff month. The metric of "function points per staff month" is the most common productivity metric in the world in 2016 with over 100,000 global benchmarks. But it has a problem. The metric of "function points per staff month" varies widely based on local work hours per month!

The results for countries such as India and China where work hours per month can top 190 will be much better than for countries such as the Netherlands or Germany where work hours per month are below 120. The United States tends to work about 132 hours per month on average, but there are wide ranges. For example, during the writing of this book, Amazon cut work hours per week from 40 to 30 hours, resulting in about 120 hours per month. Technology start-ups may work over 200 hours per month; while some government software groups may work less than 125 hours per month. Now that Amazon has reduced employee work weeks from 40 to 30 hours, function points per month in Amazon will go down, although work hours per function point will not change as a result.

The metric of "work hours per function point" is valid in all countries and does not vary due to local working conditions. This metric is like a golf score: lower is better. Work hours per function point will vary based on team experience, programming languages, application sizes, and other factors. These are true technical reasons and not derived just from the quantity of local hours worked each month including unpaid overtime.

If you measure a full software project from requirements to delivery and you include specialists and management as well as programmers and testers, then work hours per month will be in the range of 6.00–22.00 based on methodology, application size, team skills, and complexity. To simplify statistical analysis, this book assumes 1000 function points and average team skills for all 60 methodology examples.

Variations in work hours per function point derive from using better methodologies and from having expert teams as opposed to novice or average teams. A major variation is due to the volume of reusable materials used in applications, which today average about 15% code reuse. Work hours per function point also vary by programming language. But this book uses Java for all 60 examples so it does not discuss the impact of the better high-level languages such as Objective C or the older lower-level languages such as Fortran. Using the same language with all 60 methodologies is merely a convenience to make statistical analysis straightforward and not muddy the issue with variables that are outside the scope of the methodologies themselves.

(The new SNAP metric for non-functional requirements is not used in this book due to it being a recent metric that lacks data on quality and productivity. None of the author's clients use the SNAP metric so none of the author's benchmarks include it as of 2016. This may change in future years.)

Readers may be interested in how many projects have been measured and added to various benchmark collections, some of which are commercially available.

Table 0.3 gives the total number of external benchmarks gathered by some 37 benchmark companies and associations through late 2016 using 25 software size metrics. The data in Table 0.3 is extrapolated from the client projects measured by the author's clients and colleagues.

Software quality in this book uses a combination of metrics developed by IBM around 1973. The combination includes three main metrics: (1) defect potentials measured with function points; (2) defect removal efficiency (DRE); and (3) delivered defects per function point. Additional metrics are used to show high-severity defects and security flaws still present when software is delivered to clients.

The reason why function points are used for quality data and defect potentials is because defects come from many sources and not just from source code. Table 0.1 shows the approximate average software defect potential for the United States in 2016.

Defect potentials and DRE metrics are widely used by technology companies as of 2016. These metrics are frequently used in software benchmarks. The approximate 2016 average for software defect potentials is shown in the following list:

Average Software Defect Potentials around 2016 for the United States

- Requirements: 0.70 defects per function point
- Architecture: 0.10 defects per function point
- Design: 0.95 defects per function point
- Code: 1.15 defects per function point
- Security code flaws: 0.25 defects per function point
- Documents: 0.45 defects per function point
- Bad fixes: 0.65 defects per function point
- Totals: 4.25 defects per function point

Note that the phrase *bad fixes* refers to new bugs accidentally introduced by bug repairs for older bugs. The current U.S. average for bad-fix injections is about 7%; that is, 7% of all bug repairs contain new bugs. For modules that are high in cyclomatic complexity and for "error-prone modules" bad-fix injections can top 75%.

Table 0.3 Software Benchmarks 2016

	Metric	*Benchmarks*	*Percentage*
1	IFPUG function points	80,000	35.54
2	Backfired function points	65,000	28.88
3	Physical code	35,000	15.55
4	Logical code	15,000	6.66
5	COSMIC function points	9,600	4.26
6	NESMA function points	7,000	3.11
7	FISMA function points	5,000	2.22
8	Mark II function points	3,000	1.33
9	Story points	1,000	0.44
10	Automated function points	1,000	0.44
11	SRM function points	700	0.31
12	RICE objects	550	0.24
13	Use-case points	450	0.20
14	Function points light	375	0.17
15	Unadjusted function points	350	0.16
16	Feature points	200	0.09
17	Fast function points	160	0.07
18	Full function points	140	0.06
19	IntegraNova function points	125	0.06
20	Engineering function points	105	0.05
21	Simple function points	95	0.04
22	SNAP non-functional size points	75	0.03
23	Weighted micro function points	65	0.03
24	Web-object points	60	0.03
25	SSSQI function points	50	0.02
	Total	225,100	100.00

Defect potentials are of necessity measured using function point metrics. The older LOC metric cannot show requirements, architecture, and design defects or any other defects outside the code itself.

DRE is also a powerful and useful metric. The current U.S. average for DRE in 2016 is roughly 92.5%. Best in class DRE results from top teams, static analysis, inspections, and mathematical test case design and certified test personnel can top 99.65%. Inexperienced teams and poor quality control can result in DRE values below 80%.

DRE is measured by keeping track of all bugs found internally during development, and comparing these with customer-reported bugs during the first 90 days of usage. If internal bugs found during development total 95 and customers report 5 bugs, DRE is 95%.

A key reason for the huge numbers of competing variations of languages, methodologies, tools, and so on is because software engineering is also unique in having the worst and most incompetent measurements of any engineering field in human history.

No other engineering field has ever used a metric such as LOC that ignores most of the work that must be done, penalizes high-level programming languages, and ignores fixed costs. (LOC was developed when assembly was the only programming language and coding was 90% of the work; the modern LOC problems are because there are now over 3000 languages of various levels and coding is less than 30% of the work.)

It is likely that all of these variations in languages, tools, and methodologies are merely because hardly anyone in software engineering knows how to measure either productivity or quality or how to apply standard economics to software.

Both LOC and cost per defect violate standard economics and both distort reality, making economic and quality progress invisible. A recent earlier book by the author, *Selecting Software Measurements and Metrics*, CRC Press 2016, summarizes the major software metrics and measurement problems.

For over 50 years, the software industry has blundered and bungled with poor metrics and leaky measurements that omit over half of the actual software work performed on major software projects. The 12 chronic and acute problems of software measurements are

1. Continued use for over 50 years of metrics that violate standard economic assumptions such as cost per defect and LOC. Cost per defect penalizes quality. LOC reverses real economic productivity and makes assembly language look better than Ruby or Objective C.
2. Totally ignoring the impact of fixed costs on software development when measuring both quality costs and overall productivity. Ignoring fixed costs is the main source of error for both LOC and cost per defect. All other industries have understood the impact of fixed costs for over 200 years.

3. Chronic gaps and errors in measuring software productivity by leaving out topics such as unpaid overtime, management effort, and the work of specialists such as business analysts, technical writers, and quality assurance. The "design, code, and unit test" (DCUT) metric is one of the most foolish and irresponsible metric concepts in human history. Software need full activity-based cost data and not just partial data such as DCUT, which barely includes 30% of true software development. Phase-based costs and single-point productivity values are also amateurish and irresponsible.
4. Chronic gaps in measuring quality by omitting to measure early defects found via desk checking, static analysis, walk-throughs, unit testing, phase reviews, and other pre-test defect removal activities. Measured quality data barely includes 25% of actual bugs.
5. Concentrating only on code defects and ignoring the more effective topic of "defect potentials," which include requirements defects, architecture defects, design defects, code defects, document defects, and bad-fix defects. The defect potential metric was developed by IBM around 1973 and was first used to validate the effectiveness of software inspections. The current U.S. average for defect potentials is about 4.25 defects per function point. Every company and government group in the world should know their defect potentials for every major project.
6. Persistent failure to use proven quality metrics such as DRE, which was developed by IBM around 1973 and has more than 40 years of success in high-technology companies. This metric shows the percentage of bugs or defect removal prior to the release of software to customers and the current U.S. average is about 92.5% DRE. This metric should be used by all software companies and government groups everywhere in the world. This metric is fairly inexpensive and easy to use.
7. Continued use of totally worthless measures such as single-point productivity values or phase-based productivity values rather than adopting activity-based costs like all other industries. Neither phase data nor single-point productivity data can be validated or used for serious economic studies of software costs and productivity.
8. Developing a series of quirky and non-standard metrics that vary by over 400% from company to company and project to project such as story points and use-case points.
9. Releasing metrics and methodologies with zero validation prior to release and essentially no empirical data of their success (or failure). Technical debt and pair programming are two recent examples of topics that should have been validated prior to release to the general public. Story points and use-case points are two older examples.
10. Failing to adopt the proven and effective function point metrics that have been shown to match standard economic assumptions. The function point metrics supported by the COSMIC, FISMA, IFPUG, and NESMA function

point associations all can measure software economic costs and quality in terms of defect potentials. Function points were put into the public domain by IBM in 1978 but even after more than 40 years of availability the overall usage of function point metrics is less than 10% of all companies and less than 5% of all software projects. A few countries such as Brazil, South Korea, Malaysia, and Japan now mandate function points for government software contracts, but these are exceptions that are more far-sighted than most countries. The U.S. government and the Department of Defense (DoD) still use obsolete and inaccurate metrics such as LOC.

11. Universities are essentially worthless for software metrics education. Most do not teach or even know about the metrics problems discussed here. Far too many continue to teach students using inaccurate metrics such as "lines of code" and "cost per defect." A few universities talk about a "debate" between LOC and function points, but there is no debate because there are no facts supporting the LOC side. That would be like a debate among medical doctors as to whether to use modern antibiotics or arsenic for treating infections.
12. Software quality and tool companies (static analysis, automated testing, project management, etc.) are essentially worthless for software metrics. Most make vast claims for improved productivity and quality but provide zero supporting data in any metric, much less using accurate metrics such as defect potentials using function points and DRE. Almost the only tool category that does a decent job with metrics is the parametric estimation industry, all of which now support function point metrics. Only a few estimating tools such as SRM also support defect potential estimates and DRE estimates.

For this book, function points are accurate enough for reliable economic studies and quality studies. But they need to be used in conjunction with "activity-based costs" and not just with total projects or with phases, which are unreliable and lack standard definitions. (The "testing phase" can encompass as few as 1 kind of testing or as many as 18 kinds of testing. This is why activity-based cost analysis is more accurate than phase-based cost analysis. Nobody knows what a "phase" actually consists of.)

The overall results in this book assume activity-based costs using the author's standard list of 40 software development activities, as shown in Table 0.4. Table 0.4 assumes an application of 1,000 function points coded in Java and monthly costs of $10,000 for all activities.

Although individual activity production rates can get up into the thousands of function points per month, the aggregate results of all 40 activities in the example is 6.02 function points per staff month or 21.91 work hours per function point. This is not uncommon when a full project is measured using activity-based costs.

Of course, partial measurements that omit the work of specialists such as quality assurance and technical writers would be higher. One of the fundamental problems of software measurement is that of "leakage" or leaving out too many activities.

Table 0.4 Example of Activity-Based Costs for 40 Software Activities

	Development Activities	*Staff Funct. Pt. Assignment* *Scope* *ASCOPE*	*Monthly Funct. Pt. Production* *Rate* *PRATE*	*Work Hours per Funct. Pt.*	*Burdened Cost per Funct. Pt. ($)*
1	Business analysis	5,000	25,000.00	0.01	0.40
2	Risk analysis/sizing	20,000	75,000.00	0.00	0.13
3	Risk solution planning	25,000	50,000.00	0.00	0.20
4	Requirements	2,000	450.00	0.29	22.22
5	Requirement—inspection	2,000	550.00	0.24	18.18
6	Prototyping	10,000	350.00	0.38	28.57
7	Architecture	10,000	2,500.00	0.05	4.00
8	Architecture—inspection	10,000	3,500.00	0.04	2.86
9	Project plans/estimates	10,000	3,500.00	0.04	2.86
10	Initial design	1,000	200.00	0.66	50.00
11	Detail design	750	150.00	0.88	66.67
12	Design inspections	1,000	250.00	0.53	40.00
13	Coding	150	20.00	6.60	500.00
14	Code inspections	150	40.00	3.30	250.00
15	Reuse acquisition	20,000	75,000.00	0.00	0.13
16	Static analysis	150	15,000.00	0.01	0.67
17	COTS package purchase	10,000	25,000.00	0.01	0.40
18	Open-source acquisition	50,000	50,000.00	0.00	0.20
19	Code security audit	33,000	2,000.00	0.07	5.00
20	Ind. verif. & valid.	10,000	10,000.00	0.01	1.00
21	Configuration control	15,000	5,000.00	0.03	2.00
22	Integration	10,000	6,000.00	0.02	1.67
23	User documentation	10,000	500.00	0.26	20.00
24	Unit testing	150	125.00	1.06	80.00
25	Function testing	250	140.00	0.94	71.43
26	Regression testing	350	90.00	1.47	111.11
27	Integration testing	450	125.00	1.06	80.00

(Continued)

Table 0.4 (Continued) Example of Activity-Based Costs for 40 Software Activities

	Development Activities	Staff Funct. Pt. Assignment Scope ASCOPE	Monthly Funct. Pt. Production Rate PRATE	Work Hours per Funct. Pt.	Burdened Cost per Funct. Pt. ($)
28	Performance testing	1,200	500.00	0.26	20.00
29	Security testing	1,250	350.00	0.38	28.57
30	Usability testing	1,500	600.00	0.22	16.67
31	System testing	300	175.00	0.75	57.14
32	Cloud testing	3,000	2,400.00	0.06	4.17
33	Field (Beta) testing	600	4,000.00	0.03	2.50
34	Acceptance testing	600	5,000.00	0.03	2.00
35	Independent testing	5,000	6,000.00	0.02	1.67
36	Quality assurance	10,000	750.00	0.18	13.33
37	Installation/training	10,000	4,000.00	0.03	2.50
38	Project measurement	25,000	9,500.00	0.01	1.05
39	Project office	10,000	550.00	0.24	18.18
40	Project management	2,500	75.00	1.76	133.33
	Cumulative results	332	6.02	21.91	1,660.08

Table 0.4 introduces two terms that may be unfamiliar to readers unless they have read other books by the author such as *Applied Software Measurement* and *Estimating Software Costs*, which used the same terms. In fact, the author has been using, measuring, and publishing *assignment scope* and *production rate* data since 1984 and it has been used in all of the author's parametric estimation tools as well.

The first term is *assignment scope* (ASCOPE) or the quantity of function points assigned to one person. The assignment scopes vary by occupation. For technical writers, a typical assignment scope might be 1000 function points. For coders and testers, typical assignment scopes might be 200 function points.

The second term is *production rate* (PRATE) or the number of function points that one person can complete in a calendar month. In this book, PRATE is based on the assumption of 132 work hours per calendar month. Needless to say, PRATE will vary between countries such as China and India that work over 190 hours per month and countries such as the Netherlands and Germany that work less than 125 hours per month. Here, too, production rates vary by occupation.

In summation, this book avoids using metrics that distort reality such as LOC, cost per defect, story points, use-case points, and other kinds of traditional software metrics that have never been validated, lack ISO standards, lack certification exams, lack formal counting rules, and vary by hundreds of percentage points from company to company and project to project.

It also avoids using incomplete "phase" metrics, which are impossible to validate and usually leak away almost 50% of true software effort.

Even worse are single-point project metrics with no inner structure at all. A productivity report such as "project X had a productivity rate of 8.5 function points per month" is essentially worthless for careful economic analysis because the lack of inner structure makes it impossible to validate whether the data is complete or leaked away over 60% of total effort! In this book, at least, summary project results can be assumed to include all of the 40 activities shown in Table 0.2.

The purpose of the book is to show readers accurate productivity and quality comparisons of 60 methodologies results using valid function point metrics and standard economic practices such as activity-based cost analysis.

The book's quality results use the IBM approach of combining software defect potentials with DRE. Ambiguous and subjective quality definitions such as "fitness for use" and "conformance to requirements" have no place in the economic analysis of COQ.

It is necessary to know tangible facts about quality such as actual numbers of bugs that occur and actual DRE and actual costs of defect prevention, pre-test defect removal, and all forms of testing plus post-release defect repairs.

Poor measurement practices and bad metrics that distort reality have slowed software progress for over 50 years. Poor measures and bad metrics are a professional embarrassment to the software industry, which is now among the largest and wealthiest in human history.

In 2016, it is now technically possible to measure productivity and quality with high precision, and it is time that software stepped up to the job and abandoned the bad metrics and incomplete measures that have concealed true progress since software began.

Chapter 1

Overview of 60 Software Development Methodologies

Executive summary: *Software has developed 60 different methodologies due in part to sociological reasons and in part to bad metrics and measurement practices which prevent accurate measurements of methodology productivity and quality.*

Sociological factors are the primary reason why software has over 25 size metrics, 60 methodologies, and over 3000 programming languages. The existence of so many variations is proof that none are fully adequate for all sizes and types of software. Truly effective general-purpose methodologies or programming languages would become standards practices, and none are good enough for this to happen as of 2017.

No matter what software development methodology is selected, all software projects have certain common features and common tasks that must be performed. The following section lists 20 essential activities that are part of all 60 methodologies discussed in this book.

Essential Activities for All 60 Software Development Methodologies

1. Cost and schedule estimates must be created.
2. Quality and risk estimates must be created.

3. User requirements must be studied and documented.
4. Government mandates such as Federal Aviation Administration (FAA) or Food and Drug Administration (FDA) approval must be planned.
5. For large applications, a formal architecture may be needed.
6. Software designs must be created for features and interfaces.
7. Software code must be developed or reusable code acquired.
8. Software code must be integrated and controlled.
9. Pre-test defect removal such as inspections and static analysis should be performed.
10. Test cases must be designed and developed to test the application.
11. Testing must be performed either manually or with automated testing tools.
12. Bugs must be recorded and repaired.
13. Configuration control for all changes must be recorded.
14. User documents and educational materials must be prepared.
15. Marketing materials must be prepared for commercial software.
16. Progress and cost data needs to be collected and analyzed.
17. Requirements changes and change control must take place.
18. The application must be put together for delivery to clients.
19. Software quality assurance (SQA) needs to review key deliverables.
20. Executives and project management must approve delivery.

These 20 generic development activities are part of all 60 software development methodologies. The 60 methodologies vary in how they envision these activities and the assumptions they make about them:

- Waterfall development assumes that almost all requirements can be defined at the start.
- Waterfall development assumes few changes in requirements and design.
- Iterative and Agile assume that requirements grow throughout development.
- Iterative and Agile assume frequent changes in designs and code during development.
- Team software process (TSP) assumes that defects are common and should be planned for throughout.
- TSP assumes that every major deliverable should be inspected.
- Pattern-based development assumes that modern software has similar features to legacy software and can be described via a standard pattern of design.
- DevOps assumes that the installation and execution of software with high transaction rates are as important as development and need to be planned together.
- Reuse-based development assumes that standard reusable components are possible and can even top 50% of the total volume of materials in modern software projects.
- Mashup development assumes that little or no unique design or coding is needed and that at least some useful applications can be constructed entirely

from segments of existing application. Mashups are used mainly for modern web and smartphone software.

- Robotic development assumes that with standard reusable components, robots could do a better job of building software than humans. Robots can't handle novel or unique requirements, but they can construct generic software applications if a blueprint and a parts inventory are available.

Since methodology development started in the 1960s, when new development projects were the norm, it always assumes "new development" as the basis of the methodology. In today's world of 2017, most software projects are not new but are either replacements for legacy software, enhancements to legacy software, or changes to commercial off-the-shelf (COTS) packages.

It is interesting to compare the mix of software projects circa 2016 with the mix of software projects circa 1966, or 50 years ago. Table 1.1 shows a modern mix of 100 software projects in a Fortune 500 company (one of the author's clients). Table 1.1 is sorted by work hours.

Table 1.1 Distribution of 100 Software Projects in a Fortune 500 Company circa 2016

	Type of Project	*Number*	*Function Points*	*Work Hours*	*Work Hours per Function Point*
1	Legacy enhancements	22	8,030	109,208	13.60
2	New development	7	7,350	84,525	11.50
3	ERP modifications	12	4,260	55,167	12.95
4	Legacy renovations	4	2,300	35,995	15.65
5	COTS modifications	16	2,160	31,644	14.65
6	Open-source modifications	3	285	2,950	10.35
7	Legacy repairs	30	270	2,025	7.50
8	Cyber-attack recovery	6	105	1,171	11.15
	Total	100	24,655	322,685	
	Average		247	3,227	13.09

In Table 1.1, there are only seven new development projects, and these are only in second place in terms of overall work hours. In 2016, legacy enhancements use more software work hours than new development projects.

Note also that cyber-attack recovery is a rapidly growing activity, which is starting to use significant resources. Cyber-attacks are also starting to cause major losses due to the theft of proprietary information or even the theft of funds and securities.

None of the 60 methodologies discussed in this book is very strong for the maintenance and enhancement of legacy software or for modifying COTS and enterprise resource planning (ERP) packages. Worse, none of the methodologies, with the possible exception of TSP, even considers cyber-attack threats as a topic to be addressed during software development.

By contrast, Table 1.2 shows the mix of 100 software projects in the same company circa 1966, or 50 years ago.

As can be seen from Table 1.2, 50 years ago, new development projects were the major activity of Fortune 500 companies, as they applied computers and software to business sectors that had never before been automated.

Table 1.2 Distribution of 100 Software Projects in a Fortune 500 Company circa 1966

	Type of Project	*Number*	*Function Points*	*Work Hours*	*Work Hours per Function Point*
1	New development	80	37,200	552,420	14.85
2	Legacy enhancements	5	625	8,406	13.45
3	Legacy repairs	15	300	2,685	8.95
4	ERP modifications	0	—	—	—
5	Legacy renovations	0	—	—	—
6	COTS modifications	0	—	—	—
7	Open-source modifications	0	—	—	—
8	Cyber-attack recovery	0	—	—	—
	Total	100	38,125	563,511	
	Average		381	5,635	14.78

Software development projects were larger than they are today, and low-level languages meant lots of work hours and also lots of bugs.

Modern commercial software packages, such as ERP packages, including SAP and Oracle, and financial packages, such as those from Computer Associates, did not even exist. Project management tools such as the Automated Project Office (APO) did not exist. Parametric estimating tools such as SEER, Software Lifecycle Management (SLIM), and Software Risk Master (SRM) did not exist. Very few COTS packages were available, so companies needed to develop almost all of their own applications, except for a few, such as sorts and currency conversion routines, which are among the oldest reusable features in the software industry.

The majority of all software applications were fairly new in 1966, so maintenance and enhancements had not become the major tasks of software organizations, as they are today in 2016. Cyber-attacks did not exist in 1966, or if they did exist, they had not yet become major threats to Fortune 500 companies, as they are in 2016.

Today in 2016, almost every known business process has been fully automated for years, so the bulk of modern software work is no longer centered in new software development but rather, in modifications to existing software, either legacy applications or external applications such as COTS and ERP packages and even open-source packages.

Unfortunately, most of the 60 methodologies in this book are weak on maintenance and enhancements and don't consider COTS modifications at all. Nor are they strong on cyber-attack deterrence or post-attack repairs.

This is why the correct starting place of a modern software methodology should not be new requirements but rather, extracting the business rules and knowledge patterns from similar applications that are already deployed and in daily use, including their modification history and their susceptibility to cyber-attacks.

One of the most important aspects of software methodologies is whether they are "quality strong" or "quality weak." Because the #1 cost driver of the entire software industry is the cost of finding and fixing bugs, quality-strong methodologies will be much more successful than quality-weak methodologies for large systems.

A "quality-strong" methodology will lower software defect potentials and raise software defect removal efficiency above average. Table 1.3 shows a side-by-side comparison of the factors that lead to being quality strong or quality weak. As can be seen from Table 1.3, quality-strong methods are much better in defect prevention and pre-test defect removal than quality-weak methods. Table 1.4 shows some of the quantitative results of quality-strong and quality-weak methodologies. In every important quantitative result, such as defect potentials, defect removal efficiency (DRE), cost of quality, and technical debt, the quality-strong methodologies are much more cost-effective than the quality-weak methodologies.

Table 1.3 Quality-Strong and Quality-Weak Technology Stacks

Software Quality Technology Stacks	*Quality Strong*	*Quality Weak*
Quality Measures		
Defect potentials in function points	Yes	No
DRE	Yes	No
Delivered defects: All severity levels	Yes	Yes
Delivered defects: High-severity defects	Yes	Yes
Delivered defects: Security flaws	Yes	No
Defect removal ($ per function point)	Yes	No
Cost of quality (COQ) per function point	Yes	No
Technical debt per function point	Yes	No
Defect density: Thousands of lines of code (KLOC)	No	Yes
Cost per defect	No	Yes
Defect Prevention		
Joint application design (JAD)	Yes	No
Quality function deployment (QFD)	Yes	No
Requirements models	Yes	No
Early risk estimates	Yes	No
Early quality estimates	Yes	No
Defect potential estimates: Function points	Yes	No
DRE estimates	Yes	No
SEMAT essence	Yes	No
Pre-Test Defect Removal		
Quality inspections	Yes	No
Security inspections	Yes	No
Static analysis of all code	Yes	No
Text static analysis (requirements, design)	Yes	No
Fog index of requirements, design	Yes	No

(*Continued*)

Table 1.3 (Continued) Quality-Strong and Quality-Weak Technology Stacks

Software Quality Technology Stacks	*Quality Strong*	*Quality Weak*
Desk check	Yes	Yes
Automated correctness proofs	Yes	No
Pair programming	No	Yes
Race condition analysis	Yes	No
Refactoring	Yes	No
Test Technologies		
Certified testers	Yes	No
Design of experiments test case design	Yes	No
Cause–effect graph test case design	Yes	No
Test coverage analysis tools	Yes	No
Cyclomatic complexity tools	Yes	No
Automated test tools	Yes	No
Reusable test cases	Yes	No
Test library control tools	Yes	No
Test Stages		
Unit test	Yes	Yes
Function test	Yes	Yes
Regression test	Yes	Yes
Performance test	Yes	Yes
Security test	Yes	No
Nationalization test	Yes	No
Usability test	Yes	No
Stress and limits test	Yes	No
Component test	Yes	No
System test	Yes	Yes
Beta test	Yes	No

(*Continued*)

Table 1.3 (Continued) Quality-Strong and Quality-Weak Technology Stacks

Software Quality Technology Stacks	*Quality Strong*	*Quality Weak*
Acceptance test	Yes	Yes
Post-Release Quality		
Complete defect tracking	Yes	No
Root-cause analysis: High-severity bugs	Yes	No
Root-cause analysis: Security flaws	Yes	No
DRE analysis	Yes	No
Delivered defects per function point	Yes	No
Delivered defects per KLOC	Yes	No
Defect origin analysis	Yes	No
Defect severity analysis	Yes	No
Defect consequence analysis	Yes	No
Security flaw analysis	Yes	No
Ethical hackers	Yes	No
Bad test case analysis	Yes	No
Bad-fix analysis	Yes	No
Error-prone module (EPM) analysis	Yes	No
COQ analysis	Yes	No
Technical debt analysis	Yes	No

Table 1.5 summarizes a sample of 20 quality-strong methodologies with high DRE levels and low levels of delivered defects per function point. It can be seen that the methods that use substantial volumes of certified reusable components are at the very top of the list in terms of software quality. Pattern-based development is also a quality-strong methodology, as is the Japanese method Kaizen. The methods that use Software Engineering Method and Theory (SEMAT) are also good in terms of software quality.

By contrast, Table 1.6 summarizes 10 methodologies with the lowest values for DRE and the highest values for delivered defects per function point. It can be seen

Table 1.4 Differences between Quality-Strong and Quality-Weak Technology Stacks

	Quality Strong	*Quality Weak*
Requirements defects per function point	0.35	0.70
Architecture defects per function point	0.05	0.20
Design defects per function point	0.60	1.00
Code defects per function point	0.80	1.50
Document defects per function point	0.20	0.30
Bad-fix defects per function point	0.30	0.80
Total defect potential per function point	2.50	4.50
Pre-test DRE (%)	>90.00	<35.00
Test DRE (%)	>90.00	<80.00
Total DRE (%)	>99.50	<85.00
Delivered defects per function point	0.0125	0.675
High-severity delivered defect (%)	<7.50	>18.00
Security flaw delivered defect (%)	<0.01	>1.5
Bad-fix injection (%)	<1.00	>9.00
Bad test cases in test library (%)	<1.00	>15.00
EPM (%)	<0.01	>5.00
Reliability (MTTF) (days)	>125	<2.00
Reliability (MTBF) (days)	>100	<1.00
Stabilization (months to reach zero defects)	<1.5	>18.0
Pre-test defect removal ($ per function point)	150.00	25.00
Test defect removal ($ per function point)	250.00	650.00
Post-release defect removal ($ per function point)	100.00	825.00
COQ ($ per function point)	500.00	1500.00
Technical debt ($ per function point)	75.00	550.00

(Continued)

Table 1.4 (Continued) Differences between Quality-Strong and Quality-Weak Technology Stacks

	Quality Strong	*Quality Weak*
Test coverage (risks) (%)	>97.00	<70.00
Test coverage (code and branches) (%)	>99.00	<70.00
Average cyclomatic complexity	<10.00	>20.00
Customer satisfaction	High	Low
Risk of project cancellation (%)	<5.00	>35.00
Risk of deferred features due to poor quality (%)	<6.5	>65.00
Risk of schedule delays (%)	<10.00	>70.00
Risk of cost overruns (%)	<10.00	>50.00
Risk of litigation for poor quality (%)	<1.00	>15.00
Risk of successful cyber-attacks (%)	<2.50	>35.00

Note: MTBF: mean time between failures; MTTF: mean time to failure.

Table 1.5 Twenty Quality-Strong Methodologies 2016

	Methodologies	*Defect Removal Efficiency 2016 (%)*	*Delivered Defects per Function Point 2016*
1	Robotic development with 99% standard parts	99.65	0.002
2	Reuse oriented (85% reusable materials)	99.75	0.002
3	Pattern-based development	99.50	0.009
4	Virtual reality global development	99.20	0.016
5	T-VEC development	98.50	0.023
6	IntegraNova development	98.50	0.024

(*Continued*)

Table 1.5 (Continued) Twenty Quality-Strong Methodologies 2016

	Methodologies		*Defect Removal Efficiency 2016 (%)*	*Delivered Defects per Function Point 2016*
7	Animated, 3-D, full color design development		99.75	0.025
8	Kaizen development		98.50	0.026
9	Container development (65% reuse)		97.00	0.030
10	Model-driven development		98.50	0.032
11	Clean room development		98.50	0.035
12	TSP		98.50	0.035
13	Feature-driven development (FDD)		98.50	0.037
14	Personal software process (PSP)		98.50	0.038
15	Specifications by example		98.38	0.038
16	CMMI Level 3 development		96.00	0.040
17	Micro service development		95.50	0.061
18	Evolutionary development (EVO)		97.80	0.064
19	Rational Unified Process (RUP) from IBM		98.00	0.068
20	Prototypes:	Disposable	97.00	0.072
		Average	98.28	0.034

that there is about a 10% differential between the 10 quality-strong methodologies and the 10 quality-weak methodologies in DRE. Since DRE values below 90% are hazardous and lead to expensive post-release bug repairs, low DRE values should be avoided for all software projects.

Fortunately, a synergistic combination of inspections, static analysis, and formal testing by certified test personnel can result in DRE values above 99.50%. High DRE values also come with shorter schedules and lower development costs than low DRE values, so this is a win-win-win situation.

Table 1.6 Ten Quality-Weak Methodologies 2016

	Methodologies	*Defect Removal Efficiency 2016 (%)*	*Delivered Defects per Function Point 2016*
1	Agile+Scrum	92.50	0.236
2	Rapid application development (RAD)	92.50	0.240
3	Reverse engineering	92.00	0.264
4	V-Model development	92.50	0.270
5	Reengineering	91.50	0.289
6	Cowboy development	94.00	0.300
7	ERP modification development	90.49	0.352
8	Waterfall development	88.00	0.450
9	COTS modifications	85.00	0.860
10	Anti-patterns	76.00	1.776
	Average	89.45	0.504

Other popular methodologies, such as DevOps, are mid-range in quality and near U.S. averages in DRE values. The current overall U.S. average for DRE is about 92.50%, which is close to the Agile average values, also at 92.50%, while DevOps is about 94% in DRE. The older waterfall method is only about 88.00% in DRE unless it is used on defense software projects at or above capability maturity model integrated (CMMI) Level 3. In this case, waterfall DRE values usually top 95.00%.

The really low DRE values are for the last three methods: waterfall, COTS modifications, and anti-patterns.

Chapter 2

Life Cycles and Life Expectancies of Software Development Methodologies

Software development methodologies are created to solve problems, are put into production, and after a period of usage, will either expand into the mainstream or drop out of service. A few, such as Agile development, will expand rapidly and become successful cults. Others, such as computer-aided software engineering (CASE), will disappear after a few years due to disappointing results. Still others will continue on with flat use for many years, while others perhaps will have declining use, such as waterfall. As this methodology use cycle continues, new methodologies will keep appearing at approximately 7-month intervals, as they have for over 25 years.

Four interesting questions about software development methodologies are

1. Why does software have more than 60 different development methodologies?
2. Why are new methodologies being developed about every 7 months?
3. Why do some methodologies become cults and grow rapidly?
4. How long do methodologies remain in active service?

The most probable answer to Questions 1 and 2 is that software has had the worst sets of measures and metrics of any industry in human history.

Software develops lots of variations of programming languages, tools, and methodologies because hardly anybody knows how to measure software productivity and quality, so there is no easy way to make rational selections.

Methodologies, programming languages, test tools, static analysis tools, and other areas with lots of variations have only subjective claims of better quality and productivity; there is scarcely any reliable empirical data available.

The metric of *cost per defect* penalizes quality and is worthless for showing the economic value of quality or for comparing one methodology with another. Cost per defect actually penalizes quality and is cheapest for the buggiest software.

The metric of *lines of code* (LOC) penalizes modern high-level languages and is also worthless for comparing one methodology with another, unless the comparisons happen to use the same programming languages. LOC also makes requirements and design invisible, even though they may cost more than the code itself.

The metrics of *story points* and *use-case points* are not standardized and have no certification examinations for counting consistency. They have been observed to vary by over 400% from company to company for similar projects.

For example, you might compare two methodologies with LOC if they both used the Java programming language, but if one used Java and the other used Objective C, then the results would be wrong, and the Objective C language would be penalized.

This book uses *function point metrics* as defined by the International Function Point Users Group (IFPUG), version 4.3. These metrics are capable of making valid comparisons between methodologies, but there are about 20,000,000 people working in software on a global basis, while all function point methods combined probably have fewer than 100,000 users on a global basis in 2016. As already demonstrated in Chapter 1 of this book, function points can be used to compare methodologies, tools, programming languages, and so on.

The new Software Non-functional Assessment Processs (SNAP) metric for non-functional requirements might be used in the future, but as of 2016, there is not enough empirical data on SNAP and various methodologies for it to be included in this book. Also, the actual math of using SNAP for methodology comparisons was ambiguous as of 2014.

Poor measures and bad metrics also play a part in Question 3, or why new methods keep popping up about every 7 months. The simple answer is that since none of the older methodologies have reliable quantitative data available, there is no actual proof that any of them work well. Therefore, new methodologies or variations of existing methodologies are likely to keep appearing at frequent intervals. For example, there are now about six variations of the Agile development methodology, none of which has quantified data available (except for the data shown in this book).

A more difficult question is why some methodologies become popular cults while others, which may be just as effective, have low use. Several methods come to mind that spread rapidly and then faded away or at least stopped gaining new users. Two of these are CASE and rapid application development (RAD). Other methods have had lasting value and are still useful today in 2016. Still other methods are popular in 2016, but the jury is still out as to whether their value will be permanent or transient. The following section gives a few examples of each category.

Methodologies with Permanent, Transient, and Uncertain Value circa 2016

Permanent value (more than 10 years of successful use):

1. Structured development
2. Object-oriented development
3. Reuse-based development
4. Rational unified process (RUP)
5. Personal software process (PSP)
6. Team software process (TSP)
7. Capability maturity model integrated (CMMI) levels >3
8. Agile/Scrum software development
9. Disposable prototypes
10. Open-source development

Transient Value (less than 5 years of use before use declined)

1. CASE
2. RAD
3. Clean-room software engineering
4. Specifications by example
5. Agile unified process (AUP)
6. Waterfall development

Popular but Uncertain Value due to Lack of Data as of 2016

1. Mashup development
2. Micro service development
3. Microsoft Sharepoint
4. DevOps
5. Product Line Engineering
6. Feature-driven development (FDD)
7. Software Engineering Method and Theory (SEMAT)
8. Git development
9. Lean development

Experimental Methods not yet Widely Used or Measured but Promising

1. Robotic development from standard parts
2. Animated, 3-D, full color design methods
3. 24-hour global development in multiple time zones
4. Disciplined Agile Delivery (DAD)
5. Pattern-based development

Methods with No Tangible Value Based on Accurate Measures

1. Cowboy development
2. Evolutionary prototypes
3. Pair programming
4. Manual proofs of correctness
5. Anti-patterns

The life expectancy of a typical software development methodology from release until use starts to decline or reaches a steady state seems to be in the 5–7 calendar year range. By coincidence, this is about the same interval as men's clothing styles. (Women's clothing styles seem to change more frequently, at 2–3-year intervals.) The life cycle of a software development methodology is approximately as shown in Table 2.1.

In a perfect world, there would be eight additional stages for software methodology development that do not actually happen with any of the current methodologies such as waterfall, Agile, DevOps, or Microservices.

Missing Stages for Software Methodology Development

3A: Measure the productivity of the methodology with function points.
3B: Measure the quality of the methodology with defect removal efficiency (DRE).
5A: Identify possible harmful side effects and negative results of the methodology.
5B: Identify possible security vulnerabilities in critical software.
5B: Identify optimal sizes and types of methodology use.
5C: Identify hazardous sizes and types of methodology use.
10A: Publish statistical ranges of methodology productivity and quality results.
10B: Publish warnings of harmful side effects or unsuitable applications.

It can be seen that the 10 actual stages of software methodology development and deployment tend to be somewhat informal and have not yet been accompanied by the 8 measurement stages of methodology productivity rates and quality results and the discovery of possible harmful side effects and security flaws.

Table 2.1 Ten Stages of a Typical Software Methodology Life Cycle

Stage 1:	Identification of a problem not solved by existing methodologies.	
	1–3 months	1–20 people
Stage 2:	Formulating a new or modified methodology to solve the problem.	
	3–6 months	1–20 people
Stage 3:	Trying the new methodology on one or more projects.	
	1–4 months	5–30 people
Stage 4:	Announcing the new methodology to other organizations.	
	1–3 months	1–5 people
Stage 5:	Other organizations use the new methodology on their projects.	
	6–12 months	25–250 people
Stage 6:	Publication of results from initial adopters.	
	3–6 months	10–30 people
Stage 7:	Additional adoption of methodology based on early adopters.	
	6–18 months	500–50,000 people
Stage 8:	New methodology found to have problems of its own.	
	4–6 months	10–50 people
Stage 9:	New methodology modified to solve observed problems if possible.	
	4–6 months	5–20 people
Stage 10:	Methodology use either expands or contracts based on user perceptions.	
	12–36 months	–250 to +50,000 people
Total methodology life cycle: 36–100 months (3–8 years)		

If software methodologies were developed in the same way as medical products, careful measures would occur, and possible harmful side effects would be identified and published. For example, no known methodology is 100% effective for all sizes, types, and classes of software. This knowledge should be a standard part of software methodology development, but in fact, it never occurs. Software methodologies are developed and released with no warnings at all. This explains why almost every methodology has multiple variations and why some methodologies don't last very long.

For example, there are at least six variations of Agile in 2016, including the original Agile/Scrum, Agile light, DAD, AUP, several hybrids such as Agile+waterfall and Agile+CMMI, and so forth.

Agile is the most popular methodology in 2016, but its limits and side effects should have been evaluated during development and published, as are the risks of a new medical product or a new drug.

Obviously, Agile was created for fairly small projects with a pool of actual users readily available. This means that Agile is not optimal for such things as cochlear implant devices, where actual users have no part in the requirements, which are determined by medical doctors and electrical engineers and are also governed by Food and Drug Administration (FDA) mandates for efficacy, reliability, and side effects testing prior to release.

The Agile replacement of large written specifications with embedded users is very helpful for small information technology projects but is not really suitable for software with government certification by the FDA, the Federal Aviation Administration (FAA), the Federal Communications Commission (FCC), the Securities Exchange Commission (SEC), or Sarbanes–Oxley, where a large number of written reports are included in the certification process and must be created as part of certification.

For software methodologies, there is no organization like the FDA that requires proof of efficacy and documentation of possible side effects prior to releasing the methodology for general use.

As long as software methodologies are built informally and do not have accurate measures of either productivity in function points or quality in terms of DRE, the industry will probably continue to churn out new and modified software development methodologies several times a year forever. Probably 80% of these new and modified methodologies will be of some value for some projects, but the other 20% may have negative value for some projects, such as pair programming, which is expensive and less effective than using static analysis or inspections.

Because software is the key operating tool of every industry, including governments and military operations, it is probably time to consider the creation of a Software Safety Administration, along the lines of the FDA, which would set standards for reliability, security, and the identification of software hazards and possible side effects prior to release.

The Sarbanes–Oxley legislation does this for financial software applications, but the concept might be expanded to other forms, such as medical equipment software and software used to process sensitive government data and probably medical data.

As of 2016, new software development methodologies come out about every 7 months and make vast claims to being panaceas. None of these new methodologies will have cautions, warnings, or information about possible side effects or applications for which the methodology may not be suitable.

Hopefully, the warnings and cautions about the 60 methodologies shown later in this book will provide a starting point for moving software along the path already followed by medicine in terms of validating new methodologies before releasing them to the world.

Since it is easier to see the patterns of software methodology development if they are listed sequentially, all 18 of the methodology deployment steps are shown here in sequential order.

Proposed 18 Stages of Software Methodology Development

Stage 1: Identifying problems not solved by existing methodologies
Stage 2: Formulating new or modified methodologies to solve the problems
Stage 3: Trying the new methodologies on one or more projects
Stage 4: Measuring the productivity of the methodology in function points
Stage 5: Measuring the quality of the methodology with DRE
Stage 6: Announcing the new methodology to other organizations
Stage 7: Other organizations using the new methodology on selected projects
Stage 8: Identifying possible harmful side effects and negative results of the methodology
Stage 9: Identifying possible security weaknesses of the new methodology
Stage 10: Identifying optimal sizes and types of applications where the methodology is effective
Stage 11: Identifying sizes and types of software where the methodology is ineffective
Stage 12: Publishing results from initial adopters
Stage 13: Additional users trying the methodology based on favorable results
Stage 14: Finding that the new methodology has problems of its own
Stage 15: Modifying the new methodology to solve observed problems if possible
Stage 16: Seeing whether the new methodology either expands or contracts based on accumulated results
Stage 17: Publishing statistical ranges of methodology productivity and quality
Stage 18: Publishing warnings and cautions of harmful side effects or unsuitable applications

In conclusion, as of 2016, the software industry has some 60 named methodologies and perhaps as many as 30 hybrids, such as Agile+waterfall. New software methodologies come out above every 10 months. About a dozen of these 60 methodologies, such as Agile and DevOps, are popular and are used for thousands of applications. About a dozen, such as clean-room and CASE, are now obsolete and no longer used. The rest of the methodologies are used from time to time, based mainly on inertia and prior use patterns.Unfortunately, as of 2016, software development methodology selection is not usually a technical decision based on actual data, but is more akin to joining a religious cult. Methodologies all seem to claim that they are panaceas, and none publishes information about quantified results for quality and productivity, or about harmful side effects or out-of-range applications where the methodologies are ineffective.

Chapter 3

A Method of Scoring Software Methodologies

Executive summary: *This chapter discusses and illustrates a quantitative method for exploring the strengths and weaknesses of various software methodologies and assigning overall numeric scores. The scoring method has two parts: one part examines the types of software that the methodology might be used for such as embedded or web software; the second part examines the development activities such as requirements, design, and coding where the methodology is strong or weak.*

This book is the first attempt at a large-scale quantitative comparison of 60 software development methodologies. Obviously, productivity and quality results are important, but other factors are important too.

The method of scoring and evaluating software methodologies in this book uses an identical set of factors for all 60 methods. This method was developed more than 5 years ago and has been used on all the methods in this book, and also on proprietary methods developed by clients for their own internal use but not released to the public.

The main purpose of the evaluation is to establish guidelines for readers about which methods are best suited (or worst suited) to specific types of software and also to specific size ranges spanning from 100 to more than 100,000 function points in size. We also show best and worst methodologies by software types such as embedded software, information systems, web applications, and defense software.

Below 100 function points, methodologies are almost irrelevant, and any of them will be sufficient. Above 10,000 function points, methodologies are very important and need to be selected carefully with special emphasis on quality-strong and security-strong methodologies, which alone are safe to use for large software systems.

From examining a total of 60 software development methodologies, it soon became obvious that many different topics needed to be considered. For example, software projects that used *information engineering* (IE) methodology all had large databases or repositories, and hence, all included database designers and carried out extensive work on data modeling. User requirements were often quite specific about data topics.

By contrast, small embedded projects using *personal software process/team software process* (PSP/TSP), such as cochlear implant medical devices, have hardly any outside data other than sensor-based inputs. There were no database analysts and no business analysts on these projects. There were no actual user requirements, since the users are medical patients and not really qualified to specify requirements. Instead of "user" requirements by actual users, there were requirements specified by medical doctors and electrical engineers. There may also be significant non-functional requirements because of Food and Drug Administration (FDA) mandates that affect medical device safety and reliability. Most requirements for medical devices would probably be unknown to the majority of actual users.

It is not realistic for one kind of software development methodology to encompass both large database methodologies and small embedded software projects. It may not make sense to use the same methodology for two such unlike software types.

But the question is: how can users of software development methodologies know which specific methodology might be optimal for the kinds and sizes of projects they need to develop, since none of the methodologies come with warnings or cautions about suitable or unsuitable application sizes and types? All methodologies claim to be universal panaceas, but none of them are really suited for all sizes and types of software projects.

Another important factor is security and methodology resistance to cyber-attacks. Large database projects using IE are often subject to cyber-attacks after deployment and hence, need rigorous security analysis during development.

Small embedded projects such as cochlear implants are not quite immune to cyber-attacks but certainly are at very low risk compared with large database projects. As far as can be determined, no small medical embedded projects circa 2016 deal with topics such as deploying ethical hackers and carrying out penetration testing, which might be part of "big data" projects using IE.

Yet another important topic is that of "hybrid" methodologies based on the fusion of at least two "pure" methods such as Agile and waterfall. Hybrids often have better results than the "pure" methods. Some of the many forms of hybrid methodology that have been observed among Namcook clients include the following.

Common Forms of Hybrid Methodologies in 2016

1. Agile+CMMI 3
2. Agile+Microsoft solutions
3. Agile+reuse
4. Agile+Spiral
5. Agile+TSP
6. Agile+waterfall
7. Agile+DevOps+TSP
8. Agile+container development
9. CMMI3+TSP
10. CMMI3+RUP
11. CMMI3+Spiral
12. CMMI3+XP
13. Continuous development+reuse
14. DevOps+Agile
15. DevOps+Scrum
16. DevOps+reuse
17. DevOps+Microsoft SharePoint
18. Microsoft SharePoint+Agile
19. Microsoft solutions+reuse
20. Microsoft solutions+feature driven
21. RUP+reuse
22. RUP+waterfall
23. RUP+TSP+patterns
24. SEMAT+Agile
25. SEMAT+TSP
26. TSP+Kaizen
27. TSP+models
28. TSP+patterns
29. TSP+PSP
30. TSP+reuse

Neither the general software engineering literature nor the methodology literature has focused on hybrids, but to date, most Namcook benchmarks that include them have very favorable results from many hybrids.

The topic of hybrid methodologies is one that needs a great deal more study, since it has proved to be quite successful. CAST software has also studied hybrid methods and also reports successful outcomes compared with "pure" methodologies.

The Agile+waterfall form of hybrid will be discussed in this book, but if all the possible hybrids were included, the number of methodologies in the book would approach about 125 instead of the 60 selected by the author. Perhaps at some point, a separate book on hybrid software development methods would be valuable for the software engineering literature.

Since new methodologies come out more than once a year, there are some 60 named software methodologies at the end of 2016. No doubt, new methodologies will continue to appear in the future as often as they have in the past. This plethora of software development methodologies is actually a proof that the "one size fits all" concept cannot work for software due to the diversity of application sizes and types.

The Scoring Technique for Evaluating Methodologies

The scoring technique looks at two discrete sets of factors for each of the 60 methodologies. First, we rank each method on how well or poorly it fits the needs of 10 different kinds and sizes of software applications.

Project Size and Nature Evaluation

1. Small projects <100 function points
2. Medium projects
3. Large projects >10,000 function points
4. New projects
5. Enhancement projects
6. Maintenance projects
7. IT projects
8. Systems/embedded projects
9. Web/cloud projects
10. High-reliability, high-security projects

Then, we look inside each of the 60 methodologies and use an abstract model to evaluate how well or poorly each methodology handles critical software development activities.

Software Activity Strength Evaluation

1. Pattern matching
2. Reusable components
3. Planning, estimating, measurement
4. Requirements—functional/non-functional
5. Requirements changes
6. Architecture
7. Models
8. Design

9. Coding
10. Change control
11. Integration
12. Defect prevention
13. Pre-test defect removal
14. Test case design/generation
15. Testing defect removal

While most of the terms are common and well-understood software activities, several should be defined.

The term *pattern matching* asks whether the methodology starts by examining the relevant patterns from similar projects that have already been constructed. Too many projects jump off into the dark and start looking at new requirements, when in fact, over 90% of "new" applications are merely small variations to existing legacy applications. The proper starting point for any methodology is examination of the results of similar projects, ideally using a formal taxonomy to identify similar historical projects.

Custom designs and manual coding are intrinsically expensive, slow, and error prone no matter what methodology is used. The only cost-effective method that can also achieve short schedules and high quality is constructing software from standard certified reusable components. This approach should be part of every methodology, but in fact, is absent from most. We evaluate every methodology on whether or not it seeks to use standard reusable components. This is why the *reusable components* topic is more important than almost any other.

Incidentally, seeking out and using reusable components is absent from the International Organization for Standardization (ISO) and Object Management Group (OMG) meta-models for methodologies, although it should be included in both. Meta-models seem to assume custom design and manual coding rather than construction from standard reusable components.

The Namcook Analytics evaluation approach uses a scale that ranges from +10 to −10. Positive numbers indicate that a method provides tangible benefits for the topic in question; negative numbers imply harm.

To deal with this situation, a scoring method has been developed that allows disparate topics to be ranked using a common scale. Methods, practices, and results are scored using a scale that runs from +10 to −10 using the criteria shown in Table 3.1.

Both the approximate impact on productivity and the approximate impact on quality are included.

One important topic needs to be understood. Quality needs to be improved faster and to a higher level than productivity in order for productivity to improve at all. The reason for this is that finding and fixing bugs is overall the most expensive activity in software development. Quality leads, and productivity follows. Attempts to improve productivity without improving quality first are not effective.

Table 3.1 Scoring Ranges for Software Methodologies and Practices

Score	*Productivity Improvement (%)*	*Quality Improvement (%)*
10	25	35
9	20	30
8	17	25
7	15	20
6	12	17
5	10	15
4	7	10
3	3	5
2	1	2
1	0	0
0	0	0
−1	0	0
−2	−1	−2
−3	−3	−5
−4	−7	−10
−5	−10	−15
−6	−12	−17
−7	−15	−20
−8	−17	−25
−9	−20	−30
−10	−25	−35

While this scoring method is not perfect and is somewhat subjective, it has proved to be useful in evaluating software methodologies, technical and management tools, programming languages, and even organization structures.

(Although the scoring system is discussed here in a software context, the same scoring logic has been applied to state and Federal government issues. For example, offshore wind farms are extremely expensive and unreliable and hence, score −5 as a power generation method. Geothermal scores +10 and is currently the most beneficial power generation method.)

Using the Namcook methodology evaluation approach, the maximum positive score would be +250, and the maximum negative score would be -250. A high positive score does not mean that a specific method is necessarily better than one with a lower score, but it does imply that high-scoring methods are the most versatile and have the widest range of applicability. However, in this book, the range is between +237 and -115.

The highest positive score in the Namcook evaluation is +237 for robotic construction using 99% reusable components. The lowest negative score is -115 for anti-patterns.

An example of the full Namcook scoring approach is shown in Table 3.2 for one of the most important emerging development methodologies: pattern-based development.

As it happens, collections of harmful techniques called *anti-patterns* have the lowest overall score and should be avoided at all costs. Table 3.3 shows the lowest score for any of the current methodologies in this book.

Table 3.4 shows side-by-side results of these two very different methodologies. It is a sign of the poor measurements, bad metrics, and sloppy research techniques of the software industry that the extremely harmful anti-pattern development approach is more widely deployed than the highly effective pattern-based development. This is because many companies do not evaluate development methodologies or even know how to evaluate them, and therefore, tend to make the same mistakes over and over again on project after project.

This problem is due in part to the chronic use of bad metrics, such as cost per defect and lines of code (LOC), which distort reality and reverse real economic facts so that bad methods and low-level languages seem to be better than good methods and high-level languages.

As already noted, the vast majority of software-producing companies and also universities do not have a clue as to how either productivity or quality should be measured. Software has the worst and most inept set of measures and metrics of any industry in human history.

Usage statistics do not correlate strongly with methodology effectiveness, as shown by the 11,000 projects using harmful anti-patterns as opposed to only 2300 projects using the very beneficial pattern-based methodology.

In software, there is close to zero correlation between the value of a methodology and the number of projects that actually use a methodology. In fact, the usage of harmful methodologies outnumbers the usage of effective methodologies due to cognitive dissonance and total lack of understanding of effective measures and metrics for productivity and quality.

Note: From this point on, some of the tables use subsets of 30 software methodologies from the full total of 60 software methodologies discussed overall. This is primarily to save space and condense the tables to fit into a single page. A table with 60 methodologies of necessity spans several pages. Also, some of the less relevant methodologies with little or no benefit or comparatively few users don't really need

Table 3.2 Pattern-Based Development

Project Size and Nature Scores		**Patterns**
1	Small projects <100 FP	10
2	Medium projects	10
3	New projects	10
4	IT projects	10
5	Systems/embedded projects	10
6	Web/cloud projects	10
7	High-reliability, high-security projects	10
8	Large projects >10,000 FP	9
9	Enhancement projects	6
10	Maintenance projects	3
	Subtotal	88
Project Activity Strength Impacts		**Patterns**
1	Pattern matching	10
2	Reusable components	10
3	Planning, estimating, measurement	10
4	Requirements	10
5	Architecture	10
6	Models	10
7	Design	10
8	Coding	10
9	Integration	10
10	Defect prevention	10
11	Test case design/generation	10
12	Testing defect removal	10
13	Change control	8
14	Requirements changes	7

(*Continued*)

Table 3.2 (Continued) Pattern-Based Development

15	Pre-test defect removal	7
	Subtotal	142
	Total	230
Regional Usage (Projects in 2016)		**Patterns**
1	North America	1000
2	Pacific	500
3	Europe	400
4	South America/Central America	300
5	Africa/Middle East	100
	Total	2300
Defect Potentials per Function Point		**Patterns**
	(1000 function points; Java language)	
1	Requirements	0.10
2	Architecture	0.15
3	Design	0.20
4	Code	0.80
5	Documents	0.30
6	Bad fixes	0.15
	Total	1.70
	Total Defect Potential	1700
Defect Removal Efficiency (DRE)		
1	Pre-test defect removal (%)	95.00
2	Testing defect removal (%)	95.00
	Total DRE (%)	99.75
	Delivered defects per function point	0.004
	Delivered defects	4
	High-severity defects	0

(Continued)

le 3.2 (Continued) Pattern-Based Development

	Security defects	0
	Productivity for 1000 function points	Patterns
	Work hours per function point 2016	9.50
	Function points per month 2016	13.80
	Schedule in calendar months	10.00

Pattern-Based Scores
Best = +10
Worst = −10

to be shown. Further, some of the data is in rapid evolution, so the tables will no doubt change on an annual basis.

While Agile is the largest recent pure development method, the older waterfall method still has many users circa 2016. The older semi-methodology of "disposable prototypes" is actually also very popular and fairly effective. However, in 2016, repairs to aging legacy applications outnumber new development. The global #1 method is Git development, probably used by over half of the programmers in the United States for version control, and by almost half of the programmers in Europe and Asia (Table 3.5).

The usage data comes from extrapolation from Namcook clients. For example, the author has had 12 insurance companies, 9 banks, 8 telecom manufacturing companies, and 6 telecom operating companies as current clients. It is reasonable to assume that the ratios of methodology use from the author's client lists will occur in all other banks, insurance companies, and telecom companies. This may not be a totally valid assumption, but it is probably acceptable, since no other source of this information exists.

The usefulness of the Namcook scoring method can easily be seen by merely listing the top 15 software development methodologies for three size ranges: small projects below 100 function points; medium projects; and large systems above 10,000 function points (Table 3.6).

For small projects below 1000 function points, methodology selection is fairly minor. There are some 47 methodologies out of the 60 methodologies examined that would be suitable for small projects. Therefore, the small-project list merely shows the ones that are slightly better than average, such as Agile. In fact, for small projects, team experience and programming languages have more impact than any methodology.

For mid-range projects between 1,000 and 10,000 function points, complexity goes up, and methodology selection becomes more important. Quality-strong methodologies such as team software process/personal software process (TSP/PSP) begin to be significant. However, reuse is still the top-ranked method for all sizes. However, while 99% reuse may be possible for small projects, it is unlikely to be possible above 1000 function points.

Table 3.3 Software Anti-Pattern Scores

Project Size and Nature Scores		**Anti-Patterns**
1	Large projects >10,000 FP	−10
2	Web/cloud projects	−10
3	High-reliability, high-security projects	−10
4	Medium projects	−3
5	Systems/embedded projects	−3
6	Small projects <100 FP	−1
7	New projects	−5
8	Enhancement projects	−5
9	Maintenance projects	−5
10	IT projects	−5
	Subtotal	−57
		Anti-Patterns
1	Change control	−10
2	Planning, estimating, measurement	−10
3	Requirements changes	−7
4	Coding	−5
5	Requirements	−1
6	Models	−1
7	Design	−1
8	Defect prevention	−1
9	Pre-test defect removal	−1
10	Testing defect removal	−1
11	Pattern matching	−1
12	Reusable components	−5
13	Architecture	−5
14	Integration	−1

(*Continued*)

Table 3.3 (Continued) Software Anti-Pattern Scores

15	Test case design/generation	−8
	Subtotal	−58
	Total	−115
Regional Usage (Projects in 2016)		**Anti-Patterns**
1	North America	3,000
2	Pacific	2.000
3	Europe	2,000
4	South America/Central America	1,000
5	Africa/Middle East	3,000
	Total	11,000
Defect Potentials per Function Point (1000 function points; Java language)		**Anti-Patterns**
1	Requirements	1.20
2	Architecture	0.80
3	Design	2.00
4	Code	1.80
5	Documents	0.60
6	Bad fixes	1.00
	Total	7.40
	Total Defect Potential	7,400
Defect Removal Efficiency (DRE)		**Anti-Patterns**
1	Pre-test defect removal (%)	40.00
2	Testing defect removal (%)	60.00
	Total DRE (%)	76.00
	Delivered defects per function point	1.776
	Delivered defects	1,776
	High-severity defects	275
	Security defects	25

(Continued)

Table 3.3 (Continued) Software Anti-Pattern Scores

Productivity for 1000 Function Points		Anti-Patterns
	Work hours per function point 2016	26.00
	Function points per month 2016	5.08
	Schedule in calendar months	28.00

Anti-Pattern Scores
Best = +10
Worst = −10

For larger sized projects above 10,000 function points, quality-strong methodologies become increasingly important. Except for constructing applications from 85% reusable materials, the lists are very different. For example, Agile does not even appear in the top 10 for mid-sized and large projects. On the other hand, TSP/PSP, which is one of the best for large systems, does not appear at all for small projects, due mainly to high overhead.

Table 3.8 shows clearly that "one size fits all" is not a good match for software development methodologies. Methodologies need to be matched to specific application size and type ranges to be effective and achieve strong and successful results.

Because large systems are very difficult to complete successfully, it is also important to know the worst methodologies, which should be avoided.

Ten Worst Methodologies for Large Systems

1. Reverse engineering
2. DevOps
3. Agile Scrum
4. Test-driven development (TDD)
5. Lean development
6. Prototypes: evolutionary
7. Pair programming
8. Extreme programming (XP)
9. Cowboy
10. Anti-patterns

The same kind of ranking can be done by application types, as shown in Table 3.7.

For IT projects, more than 50 of the 60 methodologies are suitable. This is because many methodologies, such as Agile, concentrate on IT projects and more or less ignore systems and embedded applications. Therefore, many methods besides the 10 shown in Table 3.9 would work.

For systems and embedded software, there are fewer effective methods, and these tend to be "quality-strong" methods such as TSP/PSP, patterns, and high

Table 3.4 Comparison of Patterns and Anti-Pattern Methodologies

Project Size and Nature Scores		Patterns	Anti-Patterns
1	Small projects <100 FP	10	−10
2	Medium projects	10	−10
3	New projects	10	−10
4	IT projects	10	−3
5	Systems/embedded projects	10	−3
6	Web/cloud projects	10	0
7	High-reliability, high-security projects	10	0
8	Large projects >10,000 FP	9	0
9	Enhancement projects	6	0
10	Maintenance projects	3	0
	Subtotal	88	−36
Project Activity Scores		**Patterns**	**Anti-Patterns**
1	Pattern matching	10	−10
2	Reusable components	10	−10
3	Planning, estimating, measurement	10	−7
4	Requirements	10	−5
5	Architecture	10	−1
6	Models	10	−1
7	Design	10	−1
8	Coding	10	−1
9	Integration	10	−1
10	Defect prevention	10	−1
11	Test case design/generation	10	0
12	Testing defect removal	10	0
13	Change control	8	0
14	Requirements changes	7	0

(*Continued*)

Table 3.4 (Continued) Comparison of Patterns and Anti-Pattern Methodologies

15	Pre-test defect removal	7	0
	Subtotal	142	–38
	Total	230	–74
Regional Usage (Projects in 2016)		**Patterns**	**Anti-Patterns**
1	North America	1,000	3,000
2	Pacific	500	2,000
3	Europe	400	2,000
4	South America/Central America	300	1,000
5	Africa/Middle East	100	3,000
	Total	2,300	11,000
Defect Potentials per Function Point (1000 function points; Java language)		**Patterns**	**Anti-Patterns**
1	Requirements	0.10	1.20
2	Architecture	0.15	0.80
3	Design	0.20	2.00
4	Code	0.80	1.80
5	Documents	0.30	0.60
6	Bad fixes	0.15	1.00
	Total	1.70	7.40
Defect Removal Efficiency (DRE)			
1	Pre-test defect removal (%)	95.00	40.00
2	Testing defect removal (%)	95.00	60.00
	Total DRE (%)	99.75	76.00
	Delivered defects per function point	0.016	1.776
	Productivity for 1000 function points		

(*Continued*)

Table 3.4 (Continued) Comparison of Patterns and Anti-Pattern Methodologies

	Work hours per function point 2016	9.50	24.00
	Schedule in calendar months	10.00	26.00

Data source: Namcook Analysis of 600 companies.

Scoring
Best = +10
Worst = −10

capability maturity model integrated (CMMI) levels. About 20 out of 60 methods are suitable for systems and embedded applications.

Web and cloud applications also have many potential effective development methods: about 50 out of the 60 we have evaluated are suitable for web applications.

When it comes to high-security high-value applications, some popular methods, such as Agile, are not particularly effective. Among the author's client set, Agile projects that required FDA or Federal Aviation Administration (FAA) certification took about a month longer than TSP/PSP projects of the same size to achieve the mandatory certification. The methodologies that emphasize quality are the top candidates for high-security applications. About 10 out of the 60 methodologies studied are suitable for high-security applications.

It is also important to know the worst methodologies to avoid when building secure systems. The five worst methodologies, four of which have negative scores, are listed here.

Five Worst Methodologies for Secure Systems

1. Waterfall
2. Continuous development
3. Open-source
4. Cowboy
5. Anti-patterns

In addition to evaluating the 60 methodologies, the author has also provided some quantified data on their usage and results. This data includes

1. Global usage of the methodologies
2. Defect potentials
3. DRE
4. Delivered defects
5. Work hours per function point for 1000 function points
6. Schedule calendar months for 1000 function points

Table 3.5 Global Usage of 30 Development Methodologies

	Methodologies	*Approximate Method Start Year*	*Global Method Usage 2016*
1	Git development	2005	2,200,000
2	Legacy repair development	1960	775,000
3	COTS Modifications	1969	490,000
4	Agile/Scrum	2001	435,000
5	Waterfall development	1960	385,000
6	Prototypes: disposable	1959	275,000
7	Container development (65% reuse)	2012	76,500
8	Microsoft solutions	1999	73,000
9	Structured development	1973	65,000
10	Mashup development	2006	63,000
11	Legacy renovation	1995	61,000
12	ERP modification development	1996	60,000
13	Object-oriented (OO) development	1985	57,000
14	RUP from IBM	1996	48,000
15	Legacy replacement development	1989	47,000
16	Lean development	2003	46,500
17	DevOps development	2010	45,990
18	Iterative development	1990	43,000
19	Reengineering	1999	42,000
20	Spiral development	1983	36,000
21	CMMI development	1985	35,000
22	Prototypes: evolutionary	1965	34,000
23	TDD	2005	30,000
24	Micro service development	2014	23,400
25	Kaizen development	1955	23,000

(Continued)

Table 3.5 (Continued) Global Usage of 30 Development Methodologies

	Methodologies	*Approximate Method Start Year*	*Global Method Usage 2016*
26	Model-driven development	2009	18,000
27	Evolutionary development (EVO)	1993	17,885
28	Anti-patterns	1955	17,000
28	Cowboy development	1955	16,863
29	Feature driven (FDD)	2007	16,500
30	IE	1980	15,000

Table 3.6 Best Methods for Small, Mid-Range, and Large Software Projects

	Small <1000 function points)	*Mid-Range*	*Large (>10,000 function points)*
1	Reuse-oriented (99%)	Reuse-oriented (85%)	Reuse-oriented (85%)
2	Agile+Scrum	Pattern-based	Animated, 3D, full color
3	SEMAT+Agile	Container (65% reuse)	Pattern-based
4	Container (65% reuse)	Micro service	Hybrid (CMMI3+TSP)
5	Micro service	Mashup	Prototypes: disposable
6	Model-driven	PSP	Global 24-h
7	PSP	FDD	TSP+PSP
8	FDD	Application generator	Virtual reality global
9	Service-oriented	DevOps	Model-driven
10	Application generator	DAD	Kaizen
11	OO	Agile+Scrum	DAD
12	Microsoft solutions	SEMAT+Agile	Product line engineering
13	DAD	TSP+PSP	FDD
14	Pattern-based	Model-driven	EVO
15	Open-source	Microsoft Sharepoint	Container development

Table 3.7 Top 10 Methods by Application Type

	IT	*Systems*	*Cloud*	*Secure*
1	85% reuse	85% reuse	85% reuse	85% reuse
2	24 hour global	24 hour global	Mashup	TSP/PSP
3	DAD	TSP/PSP	Microsoft	CMMI Level 3
4	Agile/Scrum	Model-based	Patterns	Patterns
5	DevOps	Pattern-based	Specs by Example	Models
6	Features	CMMI Level 3	Features	Spiral
7	Mashup	Iterative	Container Dev	Structured
8	Microsoft	Spiral	OO	Specs by Example
9	Patterns	Features (FDD)	Agile	T-VEC
10	RUP	OO	Crystal	RUP

Because quality is the most important factor in software development, Table 3.8 shows the 30 methodologies with the lowest values for delivered defects per function point.

Note that Agile/Scrum is not among the top 30 in terms of quality, although it is the most popular methodology in 2016. This shows that methodologies are not actually evaluated on their results, but rather, are adopted due to popularity. Since Agile is effective for small projects, and there are many times more small projects than large, it is easy to see why Agile is the most popular.

Because quality is so important, Table 3.9 shows the 10 largest numbers of delivered defects per function point.

Agile has reasonably good quality for small projects, but is not one of the "quality-strong" methodologies, even though it is twice as good as waterfall. Agile averages about 92.50% in DRE based on more than 400 projects. Note that Agile quality is better than waterfall or cowboy but is not equal to TSP, Rational Unified Process (RUP), and other quality-strong methodologies.

Because schedules are probably the #1 topic of interest to many C-level executives, Table 3.10 shows the schedule duration in calendar months for the 31 shortest development cycles.

It is obvious that custom designs and manual coding cannot possibly achieve the same schedule as reuse, which is why methods with substantial reuse, such as "85% reuse" and "mashups," take less than half the time of the other development methods.

Table 3.8 Methodologies with Lowest Delivered Defects

	Methodologies	*Delivered Defects per FP 2016*
1	Robotic development with 99% standard parts	0.002
2	Reuse-oriented (85% reusable materials)	0.002
3	Animated, 3D, full color design development	0.003
4	Pattern-based development	0.009
5	Virtual reality global development	0.016
6	T-VEC development	0.023
7	IntegraNova development	0.024
8	Kaizen development	0.026
9	Container development (65% reuse)	0.030
10	Model-driven development	0.032
11	Clean room development	0.035
12	TSP+PSP	0.035
13	FDD	0.037
14	PSP	0.038
15	Specifications by Example	0.038
16	CMMI Level 3 development	0.040
17	Micro service development	0.061
18	EVO	0.064
19	RUP from IBM	0.068
20	Prototypes: disposable	0.072
21	Open-source development	0.079
22	OO development	0.080
23	Global 24-h development	0.098
24	DAD	0.100
25	Product line engineering	0.107
26	Service-oriented modeling	0.108

(*Continued*)

Table 3.8 (Continued) Methodologies with Lowest Delivered Defects

	Methodologies	*Delivered Defects per FP 2016*
27	Mashup development	0.116
28	Prototypes: evolutionary	0.117
28	IE	0.119
29	Crystal development	0.119
30	XP	0.131

Table 3.9 Methodologies with Highest Delivered Defects

	Methodologies	*Delivered Defects per FP 2016*
1	Anti-patterns	1.776
2	COTS modifications	0.860
3	Waterfall development	0.450
4	ERP modification development	0.352
5	Cowboy development	0.300
6	Reengineering	0.289
7	V-Model development	0.270
8	Reverse engineering	0.264
9	RAD	0.240
10	Agile/Scrum	0.236

Because schedules are so important to C-level executives, Table 3.11 shows the 100 longest schedules in terms of calendar months.

Pair programming is the most sluggish of the popular development methods. Pair programming exists mainly because the software industry does not measure well, does not understand economic topics very well, and especially, does not understand effective quality control.

Productivity is an important but ambiguous topic. Namcook Analytics measured productivity in terms of "work hours per function point" for every activity between requirements and delivery, and for every occupation group, including, but not limited to, managers, architects, designers, business analysts, programmers, testers, quality assurance, technical writers, and often even more (a total of 126

Table 3.10 Methodologies with Shortest Schedules

	Methodologies	*Schedule Calendar Months 2016*
1	Robotic development with 99% standard parts	2.00
2	Reuse-oriented (85% reusable materials)	2.82
3	Mashup development	5.00
4	IntegraNova development	5.70
5	Service-oriented modeling	7.00
6	Global 24 h development	7.00
7	Animated, 3D, full color design development	9.46
8	Legacy repair development	10.00
9	Pattern-based development	10.00
10	Virtual reality global development	10.90
11	Container development (65% reuse)	11.20
12	Model-driven development	11.50
13	DAD	12.30
14	Open-source development	12.50
15	Micro service development	12.62
16	Continuous development	12.75
17	Prototypes: disposable	12.90
18	Kaizen development	13.20
19	Product line engineering	13.30
20	Dynamic system development method (DSDM)	13.50
21	PSP	13.50
22	Hybrid (Agile+waterfall)	13.60
23	Crystal development	13.65
24	Lean development	13.70
25	Agile/Scrum	13.80
26	Git development	13.80

(Continued)

Table 3.10 (Continued) Methodologies with Shortest Schedules

	Methodologies	*Schedule Calendar Months 2016*
27	FDD	13.80
28	RAD	13.90
29	Microsoft solutions	14.00
30	Legacy replacement development	14.05
31	Legacy data mining	14.10

Table 3.11 Methodologies with Longest Schedules

	Methodologies	*Schedule Calendar Months 2016*
1	Anti-patterns	28.00
2	Cowboy development	22.00
3	V-Model development	19.50
4	Reengineering	17.60
5	Merise development	17.30
6	Waterfall development	17.20
7	COTS modifications	17.10
8	Pair programming development	17.00
9	ERP modification development	16.50
10	Prince 2 development	16.50

occupation groups are employed by large software organizations such as Microsoft, IBM, Siemens, Nippon Telephone, and others).

Table 3.12 shows work hours per function point for the top 31 methodologies. This is the only metric that is universally accurate. LOC penalizes high-level languages and distorts reality. Function points per month are lowest for countries with long work months, such as India at 190 h/month and China at 186 h/month. Work hours per function point works in all countries and for all sizes and types of software.

Here, too, the methods with high volumes of reuse, mashups and 85% reuse, are far better than methodologies in which custom designs and manual coding are the

Table 3.12 Methodologies with Lowest Work Hours

	Methodologies	*Work Hours per FP 2016*
1	Robotic development with 99% standard parts	1.50
2	Reuse-oriented (85% reusable materials)	1.76
3	IntegraNova development	6.00
4	Mashup development	6.50
5	Pattern-based development	6.86
6	Animated, 3D, full color design development	7.50
7	Service-oriented modeling	7.81
8	Virtual reality global development	9.04
9	Container development (65% reuse)	10.31
10	Model-driven development	10.80
11	FDD	11.19
12	PSP	11.30
13	TSP+PSP	11.35
14	Agile/Scrum	11.40
15	Crystal development	11.50
16	Micro service development	11.81
17	DAD	11.84
18	DSDM	11.90
19	Kaizen development	12.20
20	Lean development	12.40
21	Open-source development	12.61
22	Hybrid (Agile+waterfall)	12.63
23	Microsoft solutions	12.85
24	Continuous development	13.00
25	Git development	13.19

(Continued)

Table 3.12 (Continued) Methodologies with Lowest Work Hours

	Methodologies	*Work Hours per FP 2016*
26	T-VEC development	13.20
27	EVO	13.24
28	RUP from IBM	13.30
29	Legacy replacement development	13.34
30	OO development	13.40
31	Legacy renovation	13.60

basic process. Reuse is better for quality, for costs, for schedules, and for economic productivity. But, unfortunately, reuse has a rather high up-front investment.

By contrast, Table 3.13 shows the 10 worst productivity rates in terms of work hours per function point.

It can be seen that there are big differences in schedule results between the best and the worst methodologies.

Table 3.13 Methodologies with Highest Work Hours

	Methodologies	*Work Hours per FP 2016*
1	Anti-patterns	26.00
2	Cowboy development	20.30
3	Pair programming development	18.51
4	Clean room development	18.49
5	V-Model development	18.49
6	CMMI development	17.51
7	Reengineering	17.50
8	ERP modification development	17.21
9	Prince 2 development	17.01
10	Waterfall development	16.86

Chapter 4

Detailed Evaluations of 60 Software Development Methodologies

Executive summary: *This chapter is the main focus of this book and provides information on 60 software methodologies in terms of strengths, weaknesses, and quantified results for software productivity and quality. To ensure consistency, all 60 examples are based on applications of 1000 function points in size, coded in the Java programming language.*

Each methodology is discussed with an executive overview, a text discussion, and then, numeric scores using the Namcook evaluation method. Schedules, productivity, and quality are also shown using a fixed size of 1000 IFPUG 4.3 function points and the Java programming language.

The new SNAP metric for non-functional requirements is not used in this book, since it is still in evolution and has very little measured empirical data available as of 2016.

Using the same application size and the same programming language makes it easier to judge the pros and cons of the methodologies and simplifies statistical analysis.

Monthly work hours of 132 h are assumed for all of the methodology examples, with 0 h for unpaid overtime. In real life, work hours can vary from less than 120 h/month in countries such as the Netherlands to more than 190 h/month in countries such as China. Unpaid overtime can vary from 0 to more than 20 h/month.

Varying numbers of work hours per month throws off the accuracy of the "function points per month" metric, so "work hours per function point" is the standard metric used throughout.

Unpaid overtime throws off the "cost per function point metric." Paid overtime is not shown either. In much of Europe and Australia, where many programmers and other workers are unionized, paid overtime is common. Most U.S. programmers and other software personnel are usually considered to be "exempt employees" and are not eligible for paid overtime.

An additional assumption used in the comparisons involves experience levels or skill sets: average team skills are assumed for development personnel, project managers, maintenance personnel, and also clients. In real life, experts can outperform novices by more than 5 to 1 in productivity and more than 7 to 1 in quality.

Unless the methodology deals specifically with CMMI, no use of CMMI is assumed. This is because the CMMI levels are primarily used for defense applications and seldom occur in the civilian sector.

Not all the data in the book was collected in calendar year 2016, especially not for some of the older development methodologies. The author's data collection of about 26,000 projects runs from 1984 to the middle of 2016, which is when the book is being written.

Chapter 5

Agile/Scrum Software Development

Executive summary: *Currently, Agile with Scrum is the #1 development method in the world. Agile is among the best software methodologies for small projects below 500 function points. The average size of Agile projects among the author's clients is about 275 function points. Agile is best below 100 internal users, below 10 development personnel, and below 500 function points. Agile effectiveness declines as applications and customer numbers grow larger. Agile is not effective for projects needing certification by government agencies such as the Food and Drug Administration (FDA), the Federal Aviation Administration (FAA), or the Sarbanes–Oxley legislation.*

Agile is partly an evolutionary method based on iterative development and partly a new approach based on the famous "Agile Manifesto." In the year 2001, some 17 well-known software experts met at the Snowbird resort in Utah to discuss software development problems and the potential of solving the problems. The result of this meeting was the Agile Manifesto, published in February 2001. The main principles of the Agile Manifesto are as follows:

- Individuals and interactions are better than formal processes and tools.
- Working software is better than comprehensive documentation.
- Customer collaboration is better than comprehensive contracts.
- Responding to change is better than following a rigid plan.

As of 2016, Agile is the world's most popular software development method. It has come to include a number of practices and techniques, although these are not always followed on every project, and some are used with other methodologies:

1. Embedded users who provide requirements
2. User stories for requirements analysis
3. Dividing larger projects into "sprints" that last about 2 weeks
4. Daily "Scrum" sessions for status reports
5. Pair programming, or two programmers taking turns coding or "navigating"
6. Test-driven development, or writing test cases before the code is written
7. Specialized agile metrics such as velocity, burn down, burn up, and so on
8. Agile "coaches" to help introduce agile concepts

Agile has developed a number of variations, including Agile Light, Agile Unified Process (AUP), and the Disciplined Agile Delivery (DAD) variations. The latter two are by Scott Ambler.

Other methodologies that don't use "Agile" as part of their name share common themes with Agile: Extreme Programming (XP), test-driven development, Crystal development, and the older iterative development methodology and spiral development methodology all have some common characteristics with Agile.

There are also hybrid approaches, such as Agile with CMMI, Agile with DevOps, Agile with TSP, and Agile with waterfall, in alphabetic order. The existence of so many Agile variations shows that Agile is not a panacea that works equally well with all sizes and types of software applications.

Several Agile terms require definitions.

An Agile *sprint* is a small work package that can be finished within a few weeks and when finished, provides users with the functionality that they have requested. The full project will unroll over a series of multiple sprints. In the 1000 function point example at the end of this chapter, there would be one initial planning sprint and then five development sprints each of 200 function points. This is an arbitrary division, and 10 sprints of 100 function points might be used for some Agile projects.

The second term is *Scrum*, which is derived from the sport of rugby. In rugby, the term means a kind of jam where the teams are crowded together. In Agile, the term *Scrum* refers to a daily status meeting chaired by a Scrum master. The purpose is to discuss progress, problems, and interesting technical topics as needed.

Agile requirements are often gathered in the form of *user stories* or scenarios of how specific features will operate when the software is working. Multiple stories are called an *epic.*

Another term and concept is that of *embedded user(s).* Instead of application users being separate from the development team, one or more users works side by side with the development team on a daily basis and helps create requirements for each sprint.

It is obvious that both the Agile Manifesto and the common Agile practices are aimed at fairly small internal projects in which both the user community and the developers are co-located in the same building.

For larger projects with geographic separation of teams (especially across many time zones), the Agile approach loses effectiveness. Agile is also a bit tricky for projects that involve multiple sub-contractors.

Agile also loses effectiveness for projects with over 1000 users who are geographically dispersed and who work in many different companies. There are some available methods for scaling Agile up to larger sizes of perhaps 5000 function points. But there are also quality-strong methods such as team software process (TSP) that might be a better choice.

In other words, Agile was intended for projects that probably have fewer than 100 users and are small enough to require fewer than 10 developers, hopefully in the same building. This usually means projects <500 function points in size and often <200 function points in size. An average Agile project among the author's clients is about 275 function points.

Agile metrics include story points, velocity, burn down, burn up, and several others. No Agile metric has any International Organization for Standardization (ISO) or Object Management Group (OMG) standards, and none has certification exams to ensure consistency. In fact, story points are among the least consistent of all known software metrics and vary widely from company to company and project to project. Applications of exactly the same size in function point metrics have varied by about 400% using story points.

Also, story points only work for applications that use "user stories." This means that story points are worthless for comparing Agile with other methods, such as TSP or waterfall, that don't use user stories. The Agile community would have been wiser to select function point metrics, which do have ISO standards and certification exams and have been proved to be both accurate and effective.

In this book, none of the Agile metrics are useful at all, so the data is presented using International Function Point Users Group (IFPUG) function points version 4.3. This same metric is used for all the methodology comparisons. Quality comparisons use the metric of defect removal efficiency (DRE), which was developed by IBM in 1973 and is widely used among technology companies.

Agile is not an optimal choice for large systems with over 10,000 users and 100 development personnel, although some scaling up is possible. Nor is Agile optimal for distributed development, and especially for international development, where there may be multiple time zone differences that make even Skype or web Scrum meetings inconvenient.

Some Agile concepts are useful and are expanding to other methods, such as test-driven development and Scrum. Other concepts, such as pair programming, are expensive and not as effective as alternates such as inspections and static analysis.

Also, with 126 software occupation groups, the pair concept might be considered for business analysts, architects, designers, and testers, all of whom can introduce errors into software deliverable items. Pair programming exists primarily because the software industry does not measure well and does not know effective software quality control techniques.

It is not obvious, but it is also true, that Agile is aimed primarily at new projects and is not optimal for enhancing legacy applications or for maintenance and defect repairs.

It is also obvious that Agile is not a good choice for projects that need government certification, such as medical devices and avionics packages. The casual Agile attitude about documentation is not suitable for the mandatory documentation forced on projects by government policies.

Agile requires extensive modifications for projects that need certification by the FDA or the FAA. Nor is Agile the best choice for financial applications that require Sarbanes–Oxley governance.

The numerous sprints and the Agile readiness to absorb frequent requirements changes have made Agile planning and estimating somewhat more difficult than for other methods such as Rational Unified Process (RUP) and TSP. Agile is also weak in measuring results, so quality and productivity data is sparse. The unique structure of Agile projects in terms of multiple sprints makes productivity comparisons with other methods difficult.

This book and the author's Software Risk Master (SRM) estimating tool convert Agile sprints into a standard chart of accounts and convert Agile productivity and quality results into IFPUG function point metrics. This allows us to carry out side-by-side comparisons of Agile with all other methodologies. These comparisons cannot be done with actual Agile measures and metrics such as story points, velocity, burn up, burn down, and so on.

Agile concepts are not always intuitive. As a result, Agile has created a new software occupation called the *Agile coach*. Agile coaches are normally employed when companies first decide to adopt Agile. They teach both managers and team members the fundamental concepts of Agile and often participate in the initial sprints and planning sessions.

Several methods of planning, estimating, and measuring Agile have been developed. One of these was developed by Namcook Analytics and consists of converting Agile story points into function points and converting Agile sprints into a standard chart of accounts. In other words, Namcook consolidates all the Agile sprints into one set of activities such as requirements, design, coding, testing, documentation, and so on.

These two Namcook techniques allow Agile to be compared with other methods using a side-by-side format. The Namcook SRM tool can predict Agile sprints but convert overall Agile results into a standard chart of accounts.

Agile principles can be combined with other methods, and a number of "hybrid" methods exist in 2016, including, but not limited, to Agile/waterfall, Agile/RUP, and Agile/TSP. In general, hybrids seem slightly superior to pure methods.

The overall demographics of Agile in 2016 and the results of a typical Agile project of 1000 function points using the Java programming language are shown in Table 5.1. It should be noted that the project itself would be comprised of six agile

Table 5.1 Overall Demographics of Agile and Results of a Typical Agile Project

Project Size and Class Scores		Agile
1	Small projects <500 FP	10
2	New projects	10
3	IT projects	10
4	Web/cloud projects	9
5	Medium projects	7
6	Systems/embedded projects	5
7	Enhancement projects	4
8	High-reliability, high-security projects	4
9	Maintenance projects	1
10	Large projects > 10,000 FP	-1
	Subtotal	59
Project Activity Strength Scores		**Agile**
1	Requirements changes	10
2	Requirements	9
3	Coding	8
4	Change control	8
5	Defect prevention	8
6	Testing defect removal	8
7	Design	6
8	Integration	6
9	Models	5
10	Test case design/generation	5
11	Pre-test defect removal	4
12	Pattern matching	3
13	Reusable components	3

(Continued)

Table 5.1 (Continued) Overall Demographics of Agile and Results of a Typical Agile Project

14	Architecture	3
15	Planning, estimating, measuring	-5
	Subtotal	81
	Total	140
Regional Usage (Projects in 2014)		**Agile**
1	North America	135,000
2	Pacific	110,000
3	Europe	133,000
4	South America/Central America	45,000
5	Africa/Middle East	12,000
	Total	435,000
Defect Potentials per Function Point (1000 function points; Java language)		**Agile**
1	Requirements	0.30
2	Architecture	0.15
3	Design	0.70
4	Code	1.25
5	Documents	0.45
6	Bad fixes	0.30
	Total	3.15
	Total defect potential	3,150
Defect Removal Efficiency (DRE)		**Agile**
1	Pre-test defect removal (%)	50.00
2	Testing defect removal (%)	85.00
	Total DRE (%)	92.50
	Delivered defects per function point	0.236
	Delivered defects	236

(Continued)

Table 5.1 (Continued) Overall Demographics of Agile and Results of a Typical Agile Project

	High-severity defects	35
	Security defects	3
Productivity for 1000 Function Points		**Agile**
	Work hours per function point 2016	11.40
	Function points per month	11.58
	Schedule in calendar months	13.80

Data source: Namcook Analysis of 600 companies.

Agile scoring
Best = +10
Worst = −10

sprints, each of 200 function points in size: one sprint for initial planning and five for actual development.

In conclusion, Agile is a useful methodology that seems best suited for smaller internal projects. Within its proper domain, Agile has better productivity and quality than the older waterfall method. However, Agile does not show significant advantages against some of the other and newer methodologies, such as DevOps and container development. Agile is not the best choice for larger systems >10,000 function points or for projects needing FDA, SEC, or FAA certification.

Chapter 6

Animated 3D Full Color Software Design Methodology

Executive summary: *The software industry has built powerful animated design tools for many other industries, such as aircraft production, ship building, automotive construction, and medical devices. But software itself still depends on text and fairly primitive black and white diagrams.*

Software is among the most dynamic products ever built by the human species. Software is the fastest known moving product when it is operating, and it continues to change fairly rapidly during development at more than 1% per month and after release at about 8% per year. If any industry needs animated full color 3D design tools, it is software.

Custom designs and manual coding are intrinsically slow and error prone. The software industry needs to move away from custom-built software and toward construction from standard reusable components. But in order to achieve high-efficiency construction from reusable materials, preliminary tasks are needed:

1. *Feature lists of common software functions need to be prepared.*
2. *Collections of standard functions need to be certified for quality and security.*
3. *Both internal and commercial feature libraries must be stocked.*
4. *Software applications and features need to be placed on a standard taxonomy.*

5. *Software architecture and design need more effective graphical representations.*
6. *Animated 3D simulations are needed of software when it is executing.*
7. *Animated 3D simulations are needed of software evolution over time.*
8. *Animated 3D simulations are needed of various forms of cyber-attack.*

Custom designs for software applications and manual coding by human programmers are intrinsically expensive, error prone, and slow regardless of which programming languages and which development methodologies are used. Agile may be a bit faster than waterfall, but it is still slow compared with actual business needs.

User stories and the Unified Modeling Language (UML) are useful for specific projects, but they are also fairly primitive static representations and fail to capture the dynamic features of software, such as requirements growth and performance problems. Worse, they are not standardized, have no certification exams for accuracy, and vary by up to 400% from company to company. They are useless for benchmarks. Only function point metrics are useful for benchmarks and economic analysis. This is why 35 or 37 benchmark companies only support function point metrics.

In the modern world, deployed software is at great risk from external cyber-attacks of various kinds: hacking, data theft, viruses, worms, and denial of service. This needs to be modeled dynamically before software is deployed and at risk.

For many years, the first stage of conventional software development, whether Agile, waterfall, the Rational Unified Process (RUP), or something else, is to gather user requirements. This is the wrong starting place and only leads to later problems.

The first stage of development should be *pattern matching*, or using a formal taxonomy to identify exactly what kind of software application will be needed. Pattern matching is possible because in 2016 the majority of applications are not new and novel, but either replacements for existing applications or variations based on existing applications.

The elements of the formal taxonomy used with the author's Software Risk Master (SRM) tool show what a taxonomy looks like, as seen in Table 6.1.

Software Risk Master (SRM) Application Taxonomy

1. Project nature: New or enhancement or package modification, and so on
2. Hardware platform: Cloud, smartphone, tablet, personal computer, mainframe, and so on
3. Operating system: Android, Linux, Unix, Apple, Windows, and so on
4. Scope: Program, departmental system, enterprise system, and so on
5. Class: Internal, commercial, open-source, military, government, and so on
6. Type: Telecom, medical device, avionics, utility, tool, social, and so on
7. Problem complexity: Very low to very high
8. Code complexity: Very low to very high
9. Data complexity: Very low to very high

Table 6.1 Software Risk Master™ Multi-Year Sizing

	Nominal application size in IFPUG function points	**10,000**	**1,369**	
		Function Points	**SNAP Points**	
1	Size at end of requirements	10,000	1,389	
2	Size of requirement creep	2,000	278	
3	Size of planned delivery	12,000	1,667	
4	Size of deferred functions	–4,800	(667)	
5	Size of actual delivery	7,200	1,000	
6	Year 1	12,000	1,667	
7	Year 2	13,000	1,806	
8	Year 3	14,000	1,945	
9	Year 4	17,000	2,361	Kicker
10	Year 5	18,000	2,500	
11	Year 6	19,000	2,639	
12	Year 7	20,000	2,778	
13	Year 8	23,000	3,195	
14	Year 9	24,000	3,334	Kicker
15	Year 10	25,000	3,473	

10. Country of origin: Any country or combination of countries (using telephone codes)
11. Geographic region: State or province (using telephone codes)
12. Industry: North American Industry Classification (NAIC) codes
13. Project dates: Start date; delivery date (if known)
14. Methodology: Agile, RUP, team software process, TSP, Waterfall, Prince2, and so on
15. Languages: C, C#, Java, Objective C, combinations, and so on
16. Reuse volumes: 0% to 100%
17. Capability maturity model integrated (CMMI) level: 0 for no use of CMMI; 1–5 for actual levels
18. Software Engineering Method and Theory (SEMAT) use: Using or not using software engineering methods and practices

19. Start date: Planned start date for application construction
20. Project age: Calendar months from first deployment to current date
21. Growth rate Function points from first deployment to current date
22. Custom features Percentage of total features
23. Reused features Percentage of total features
24. Custom parts lists (added in 2016)
25. Reused parts list (added in 2016)

Once an application is defined by the taxonomy, the next step is to dispatch intelligent agents to find out exactly how many applications exist that have the same patterns and what their results have been. Assuming that the intelligent agents find 50 similar existing applications, using data from benchmark sources and web searches, it is then possible to aggregate useful information derived from the set of similar projects.

Starting with prior results is a normal practice in other kinds of work outside software. If you want to hire a contractor to build a 3000 square foot home, it is normal to check references and consider the results of previous homes built by the same contractor.

If the contractor has done well in the past, the odds are that this will continue for your house. If the contractor has been sued or denied certificates of occupancy, then some other contractor is urgently needed. When commissioning an architect, it is also normal to check references and examine past building designed by the architect being considered.

Contractors are also knowledgeable in and follow local building codes for structural strength, wiring, plumbing, and other aspects of urban home construction. We have no building codes for software as of 2016.

The useful information collected from similar projects will include rates of change, development schedules, staffing, costs, languages, methodologies, and other relevant factors. Further, post-deployment information would show maintenance, the probability and kinds of cyber-attacks that might have occurred for similar projects, and the security precautions used to recover or prevent such attacks.

The data would not be displayed merely in text, spreadsheets, or static diagrams but rather, by using a moving, dynamic picture of the application. This animated plan would show rates of requirements change, staff coming and going by occupation group, bugs or defects from various origins entering the software, defect removal methods and effectiveness, and other key factors, more or less as shown in Figure 6.1.

Additional useful information would be architecture patterns, design patterns, code patterns, and defect removal patterns derived from the most successful applications that share the same taxonomy as the application under development.

The basic idea is a continuous stream of information starting with predictions and estimates before starting, shifting to collection of actual data as the project unfolds, and eventually collection of maintenance, support, and enhancement data for as long as the application is deployed and in use: possibly 25 years or more.

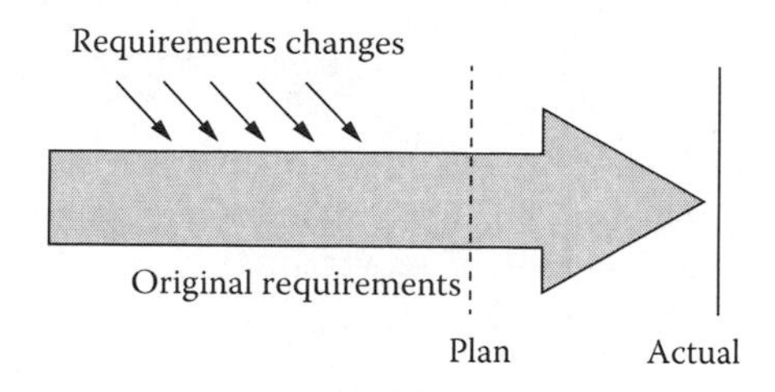

Figure 6.1 Changing requirements impact on delivery schedules.

Table 6.2 Software Graphics in Document Sets (Sample for 10,000 function points)

	Document Sizes	*Pages*	*Words*	*Percentage Complete*	*Graphics and Illustrations*
1	Requirements	2,126	850,306	73.68	425
2	Architecture	376	150,475	78.63	150
3	Initial design	2,625	1,049,819	68.71	656
4	Detail design	5,118	2,047,383	75.15	1,280
5	Test plans	1,158	463,396	68.93	0
6	Development plans	550	220,000	76.63	46
7	Cost estimates	376	150,475	79.63	0
8	User manuals	2,111	844,500	85.40	282
9	HELP text	1,964	785,413	86.09	25
10	Courses	1,450	580,000	85.05	290
11	Status reports	1,249	499,782	78.63	0
12	Change requests	2,067	826,769	78.68	0
13	Bug reports	11,467	4,586,978	81.93	0
	Total	32,638	13,055,298	78.24	3,154

Software applications, once deployed, tend to keep growing for as long as they are used. Table 6.1 shows typical annual software growth patterns.

Table 6.2 is based on IBM commercial software applications such as operating systems, the RUP tool suites, and the Information Management System (IMS) database product. Note that commercial vendors such as IBM tend to add "mid-life kickers" about every 4 years to stay competitive. These kickers are collections of many new features intended to stay ahead of competitive software applications. Of course, these kickers add more than the usual annual number of function points.

Because software is dynamic and changes over time, as well as changing rapidly when it operates, animated 3D visual representations would be helpful in understanding the dynamic nature of software.

- Flow charts were an early attempt to simulate the flow of dynamic information through software. Although flow charts were static, they at least provided a useful idea of the dynamic nature of software when it is in operation.
- Another early form of software diagram was a technique developed by IBM called "HIPO," which stood for "hierarchy, plus input, processing, output."
- Another early graphical method was Nassi-Shneiderman charts, which were useful in showing decision logic among multiple choices, which often occurs in software applications.
- Another early method to use visual data is that of the classic "decision table," which is still very useful in 2017 and especially for test planning and test case designs.
- Another common graphic element used for reporting topics such as software effort over time is a Pareto diagram, named for Vilfredo Pareto. These combine bar graphs of individual values and a curve that shows the accumulated results of all the bars.
- As of 2016, there are at least 50 software design graphics packages available, and some are open source and available for free. For example, Fusion Charts is an open-source package of graphs widely used for dashboards and project management displays.
- The IBM-developed graphics tool called "Many Eyes" can extract and visualize data from existing files, including even text files.
- JavaScript also offers a wide variety of visualization methods that are already applied to software.
- UML has a long history in software and is among the most common and popular of software diagrams. However, software is dynamic, so here, too, animation, multiple colors, and 3D would aid in visual understanding of complex and dynamic software processes.
- As software projects develop, factors such as changing requirements would trigger either extra staffing, overtime, deferred features, or delayed schedules (or all of these), but the data would be clearly visible in time to take effective steps. Every single day, both time-to-complete and cost-to-complete data would be available for clients, managers, and other stakeholders.
- Ordinary Gantt charts are also helpful for visualizing software progress. Somewhat more complex than Gantt charts, Program Evaluation Review Technique (PERT) diagrams are also of interest for visualizing software progress.

- Graphs and diagrams in 3D would also add value to software representations and could be used to show simultaneous development of documents, code, and test materials over time.

The metaphor that comes to mind is that software development would perhaps resemble the flow of a river from its origin point to its destination, with changes occurring like tributary streams entering the main flow. After release, various kinds of cyber-attacks might be visualized as attempts to dam the stream by diverting some of the water from its normal channel, introducing pollutants, or seine netting to extract valuable fish from the flow of the river: that is, data theft and identify theft.

Another possible metaphor is the unloading, setup, and reloading of a major circus that travels by train from one city to another giving daily shows. A full circus crew can top 1300 people, and the logistics of circuses are so good that both the U.S. Army and the German Army sent officers to travel with the Ringling Brothers to find out how they could move so much equipment so rapidly and so efficiently. Part of the secret of both software and circuses is knowing where to start and the optimal sequence of activities from beginning to end. Animated 3D graphics would be of considerable value in visualizing software project logistics and also software project architectures and designs.

Use of 3D Images, 3D Printing, and 3D Holographs

Today in 2016, there are over a dozen graphics packages that can produce 3D graphics, such as Wolfram Alpha, MATLAB®, and many others. The widespread use of 3D printers and the emerging use of holographic displays make it possible to envision both animation and three dimensions as future software design tools.

This would obviously be useful as a technique for visualizing software progress across multiple dimensions. It would also be useful for visualizing multi-tier architectural layers. Already, Shutterstock sells 3D design icons for topics such as wireless.

A minor issue of moving to 3D animation is that it does not work with an ordinary flat screen monitor. Probably, a special device such as Google Glasses or one of the virtual reality game headsets will be needed to achieve realistic visual perceptions of animated 3D models.

Of course, 3D printers already exist, and probably holographic displays will soon be available, although likely to be expensive at first.

It would also be possible to use 3D printing to create simulations of various software features that could then be assembled more or less like Lego blocks to show a full application. Already, 3D printing is becoming a major tool in the design of dozens of technical activities ranging from automotive and aircraft design to dentistry using 3D printers to create crowns for patients that exactly match the patient's teeth. The sample in shows that modern 3D printers can recreate almost anything.

Although holographic displays cannot be duplicated on paper such as this document, holographic displays are common and could easily be used in the production of animated 3D project plans and the creation of 3D architectural models for software. An example of an IBM logo using a holographic display is shown in Figure.

In 2016, software has among the most primitive and ambiguous design methods of any technical product and among the least effective scheduling and planning tools of any industry. The use of dynamic 3D animation should benefit both software architecture and design and also software project logistical planning. There are many kinds of software "patterns" that need effective visual representations: architecture, design, schedule planning, quality control, cyber-attacks, and many more.

Once the project taxonomy is firm and empirical results have been analyzed from similar projects, another set of patterns come into play, as shown by Table 6.3.

Elements of Pattern-Based Software Development

1. Taxonomy patterns
2. Risk patterns
3. Size patterns
4. Size change patterns
5. Schedule patterns
6. Staffing and occupation patterns
7. Departmental organization patterns (matrix, hierarchical)
8. Cost patterns
9. Value and return on investment (ROI) patterns
10. Intellectual property and patent patterns
11. Governance patterns
12. Security patterns
13. Cyber-attack patterns
14. Hardware platform patterns
15. Software platform patterns
16. Algorithm patterns
17. Requirements patterns
18. Requirements creep patterns
19. Architectural patterns
20. Design patterns
21. Code structure patterns
22. Defect removal patterns

Table 6.3 Sample of 1000 function point Animated 3D Methodology with Java Language

Project Size and Nature Scores		**3D Graphics**
1	Small projects <100 FP	10
2	Medium projects	10
3	Large projects >10,000 FP	10
4	New projects	10
5	IT projects	10
6	Systems/embedded projects	10
7	Web/cloud projects	10
8	High-reliability, high-security projects	10
9	Enhancement projects	7
10	Maintenance projects	1
	Subtotal	88
	Project Activity Scores	**3D Graphics**
1	Pattern matching	10
2	Reusable components	10
3	Planning and estimating	10
4	Requirements	10
5	Architecture	10
6	Models	10
7	Design	10
8	Coding	8
9	Defect prevention	10
10	Test case design/generation	10
11	Testing defect removal	8
12	Requirements changes	10
13	Change control	10
14	Integration	8

(*Continued*)

Table 6.3 (Continued) Sample of 1000 function point Animated 3D Methodology with Java Language

15	Pre-test defect removal	10
	Subtotal	144
	Total	232
Regional Usage (3D designs 2016)		**3D Graphics**
1	North America	10
2	Pacific	0
3	Europe	0
4	South America/Central America	0
5	Africa/Middle East	0
	Total	10
Defect Potentials per Function Point (1000 function points; Java language)		**3D Graphics**
1	Requirements	0.15
2	Architecture	0.05
3	Design	0.20
4	Code	0.80
5	Documents	0.20
6	Bad fixes	0.05
	Total	1.45
	Total defect potential	1450
Defect Removal Efficiency (DRE)		
1	Pre-test defect removal (%)	90.00
2	Testing defect removal (%)	92.00
	Total DRE (%)	99.17
	Delivered defects per function point	0.12
	Delivered defects	12
	High-severity defects	2

(Continued)

Table 6.3 (Continued) Sample of 1000 function point Animated 3D Methodology with Java Language

	Security defects	0
Productivity for 1000 Function Points		**3D Graphics**
	Work hours per function point 2016	7.5
	Function points per month 2016	17.6
	Schedule in calendar months	9.26

Data source: Namcook Analysis of 600 companies.

Animated 3D scoring
Best = +10
Worst = –10

23. Test patterns
24. Data patterns
25. Deployment patterns
26. Maintenance patterns
27. Enhancement patterns
28. Customer support patterns
29. Usage patterns and use cases
30. Litigation patterns (if any)

These patterns, expressed visually using 3D images, combined with growing libraries of standard reusable components, should be able to increase application productivity rates from today's average of below 10 function points per staff month to more than 100 function points per staff month. In some cases, for fairly small applications, productivity could approach or even exceed 200 function points per staff month. The software industry should not be satisfied with custom design and manual coding, because these are intrinsically expensive, slow, and error prone.

Many people will be surprised to know that even today in 2016, producing paper documents costs more than coding for large systems! Table 6.2 shows the approximate numbers of documents created for a telephone switching system of 10,000 function points in size.

In round numbers, describing a software application requires about 1300 English words per function point, or roughly 26 English words per line of code assuming a mid-level language such as Java. (For U.S. military software documents, sizes and costs can be more than twice those of civilian software projects of the same size and type.)

Today, graphics and illustrations are somewhat sparse, at only about one diagram for every three function points. None of the diagrams are animated, and very few have color or use 3D representation.

If you consider the common problems of major software projects, such as outright cancellation, schedule slips, and cost overruns, it is clear that project managers and developers need better tools and better data than are common today.

Within 10 years, these fully integrated capabilities will probably be available, which combine predictive analytics, full project measures, and advanced development from certified reusable materials:

1. Early sizing and estimating prior to full requirements via pattern matching (available today)
2. Formal taxonomies of both applications and feature sets
3. Automated metrics conversion between function point methods, story points, and so on (available today)
4. Early predictions of probable requirements creep (available today)
5. Early predictions of defects in requirements, design, code, and other sources (available today)
6. Early predictions of probable security problems in software (available today)
7. Early predictions of defect removal efficiency (DRE) against all defect types (available today)
8. Early production of animated graphical project plans integrating requirements creep
9. Early acquisition and analysis of benchmarks of similar projects already completed
10. Early analysis of all similar projects completed within the past 5 years
11. Early acquisition of data on security attacks against similar projects already completed
12. Early security plan to avoid security flaws and optimize data security
13. Early predictions of necessary test cases and test scripts (available today)
14. Early predictions of all necessary documents, sizes, and costs (available today)
15. Early predictions of full staffing needs including specialists such as project office staff (available today)
16. Early risk analysis of all potential application risks prior to full requirements (available today)
17. Early analysis of both available reusable components and novel components needing custom development
18. Dynamic, interactive, animated architecture and design documents for performance and security analysis
19. Acquisition of certified reusable components from commercial sources
20. Nomination of selected new features to become new certified reusable components
21. Semi-automatic construction of applications from certified reusable components

22. Semi-automatic construction of test scripts and test plans
23. Semi-automatic testing from unit test through system test
24. Semi-automatic tracking of project effort and costs as they accumulate (available today)
25. Semi-automatic logging and tracking of defects found via inspection, static analysis, and testing (available today)
26. Daily refreshed cost-to-complete and time-to-complete estimates
27. Daily refreshed quality and DRE estimates
28. Running FOG and Flesch indices on all text documents to ensure clarity and readability
29. Formal inspections of all critical requirements and design features
30. Running text static analysis on all text documents to find errors and contradictions
31. Running code static analysis on all code changes on a daily basis
32. Daily runs of cyclomatic and essential complexity of all code segments
33. Daily runs of test coverage tools when testing begins
34. Daily status reports to clients and executives that highlight any critical issues
35. Achieving full licensing and board certification for software engineering specialists
36. Achieving defect potentials <2.00 per function point
37. Achieving test coverage of >95.0% for risks and paths
38. Achieving DRE >99.5%
39. Achieving security flaw removal efficiency (SRE) >99.9%
40. Achieving productivity rates >100 function points per month
41. Achieving development schedules <24 months for 10,000 function points
42. Achieving cancellation rates of <1.0% for 10,000 function points
43. Achieving maintenance assignment scopes >5000 function points
44. Achieving certified reuse volumes >75%
45. Achieving mean time to failure of >365 days
46. Achieving schedule and cost overruns of <1.0% of plan for all applications
47. Achieving blockage of cyber-attacks of >99.99%

Of these 47 needs, about a third are already available in 2016, even if not yet widely deployed. The toughest needs are those associated with fully certifying the reusable materials and making sets of reusable materials available as needed. The cyber-security goals are also tough, in part because cyber-security during development is still not a critical-path topic for many companies and government agencies.

Paul Starr's excellent book *The Social Transformation of American Medicine* (Pulitzer Prize in 1982) shows the path followed by the American Medical Association (AMA) to improve medical education and weed out malpractice. This book should be on the shelf of every software engineer and software engineering academic, because we are still far behind medicine in terms of professionalism, licensing, and board certification for recognized specialties. As of 2016, the Department of Commerce still ranks software as a craft and not as an engineering field.

By contrast, the current quantitative results of software projects circa 2016 are much worse than future results anticipated for the topics in the list of 47 items. The industry achieved the following in 2016:

1. Software listed as a craft and not as an engineering profession by the Department of Commerce
2. Defect potentials >4.00 per function point
3. Test coverage of <85.0% for risks and paths
4. Average DRE <93.0%
5. SRE <85.0%
6. Average productivity rates <10.00 function points per month
7. Development schedules >48 months for 10,000 function points
8. Canceled projects >32.0% at 10,000 function points
9. Maintenance assignment scopes <1500 function points
10. Certified reuse volumes <15%
11. Mean time to failure of <10 days
12. Schedule and cost overruns of >25.0% for most large applications
13. Blockage of cyber-attacks <85.0%

These mediocre results for 2016 are unfortunately industry norms, in spite of Agile, TSP, and other new methods and in spite of new languages such as C#, Go, R, and Swift.

Custom designs and manual coding are intrinsically slow, expensive, and error prone, and nothing will change that fact. Software construction from certified reusable components is the only known method of elevating productivity, quality, and security at the same time.

The labor content of software is much higher than it should be. Quality and reliability are much worse than they should be. Security flaws and cyber-attacks represent one of the major business threats in all history, and improvements are urgently needed. Certified reuse has the potential for improving speed, costs, quality, and security at the same time. But software also needs better requirements, design, and architecture methods based on animated 3D full color representations. Text and static black and white diagrams are fundamentally inadequate for software.

It is technically possible that everything in this chapter could be done by 2020. But probably, another 10 years will pass before a synergistic combination of formal taxonomies, pattern matching, dynamic animated software planning tools, animated design engines, certified reusable components, intelligent agents, formal risk analysis, and clearing houses for reusable components becomes widespread or common.

Table 6.3 shows the scores, productivity, and quality results for animated 3D full color architecture and design methods.

Software reuse augmented by full color 3D animated graphics have a great potential for reducing software defect potentials, shortening development schedules, and reducing work hours per function point even for major applications. In fact the benefits should go up with application size. Small projects can be designed and built using almost any design method and any development methodology. Large systems need clear and effective designs and quality-strong methodologies.

Chapter 7

Anti-Pattern-Based Development

Executive summary: *The term "anti-pattern-based development" was coined in 1995 by software researcher Arthur Koenig. The term is the opposite of "pattern-based development." True patterns are beneficial and have tangible benefits for quality and productivity. Anti-patterns are harmful and slow down productivity and lower quality. Unfortunately, in 2014, anti-patterns seem to be more common in the software industry than beneficial patterns.*

The concepts of patterns and anti-patterns are opposite sides of the same coin. Patterns are collections of methods and practices that are beneficial. Anti-patterns are collections of methods and practices that are harmful and degrade both quality and productivity.

From working as an expert witness in over a dozen lawsuits for canceled projects or projects that did not work when delivered, the author can testify that harmful anti-patterns are far too common in the software industry even in 2014.

There are several web images that illustrate software anti-patterns, including Figure 7.1, which shows how many areas are impacted by poor practices.

In order to illustrate the concept of software anti-patterns, Table 7.1 shows 75 anti-patterns ranked in order of severity based on observations from Namcook clients.

Table 7.2 shows the same set of 75 anti-patterns ranked in order of frequency. Unfortunately, the software industry in 2016 tends to operate more like a religious cult than a technical industry. That is, projects and companies either keep making the same mistakes over and over or adopt methodologies based on nothing more

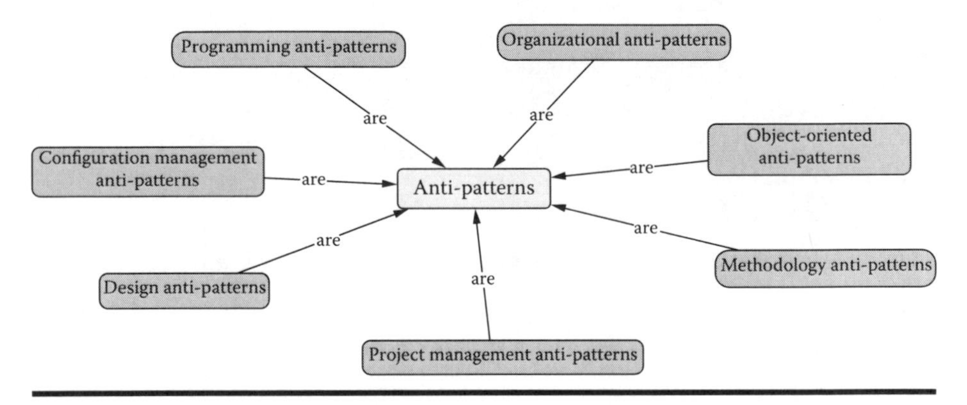

Figure 7.1 Illustration of software anti-patterns.

than popularity without performing any due diligence or evaluating the suitability of the methods to local projects.

An important topic when it comes to process improvements or rejections of better methods is "cognitive dissonance," first studied by Dr. Leon Festinger. He found that entrenched beliefs are cemented in the brain and difficult to change until the evidence is overwhelming. This is why so many innovations are rejected. Cognitive dissonance is also why methods are adopted due to popularity rather than effectiveness or technical merits.

Cognitive dissonance is a major factor in issues such as the initial rejection of sterile surgical instruments; John Ericsson's bankruptcy before he could convince navies of the merits of screw propellers and iron-clad ships; Samuel Colt's bankruptcy before armies and police recognized the value of his six-gun revolver; Christie being unable to sell modern tank treads to the U.S. Army until the Russian Army bought them; and the rejection of Wegener's theory of continental drift and Darwin's theory of evolution. Instead of evaluating the actual technologies or theories, most people merely look around at how many people use the technology. If it is truly new with few users, they assume it won't work. If it has many users, they assume it is the best.

An actual letter by the British Admiralty to the inventor of self-leveling naval cannon exists; "We have examined your proposal and found that none of the navies of the world use such cannon mounts, so we must assume that it is not effective." However, due to the inventor having a relative in Parliament, the Admiralty was forced to build a working prototype, which changed naval warfare forever. This discussion was taken from the naval historian Samuel Eliot Morrison, courtesy of Dr. Gerald Weinberg.

Today in 2016, many companies adopt Agile as their primary development method because it is the most popular. Alternate methods such as team software process (TSP) for large systems or model-based development are not included in the

Table 7.1 Software Anti-Patterns Circa 2016 by Severity

	Software Anti-Patterns	*Scoring*	*Frequency (%)*
1	Inadequate removal of security flaws before release	–10.50	95.00
2	Undiscovered security flaws in software	–10.50	98.00
3	Defect potentials > 6.00 per function point	–10.25	20.00
4	Excessively optimistic manual cost estimates	–10.00	75.00
5	Death-march projects with 100% failure odds	–10.00	15.00
6	Reuse of hazardous materials without validation	–10.00	70.00
7	Status reports inaccurate and concealing problems	–10.00	80.00
8	Inadequate risk analysis before starting	–10.00	90.00
9	Failure to use parametric cost estimates > 500 function points	–10.00	75.00
10	Excessive schedule pressure	–10.00	70.00
11	Defect removal efficiency < 85%	–10.00	35.00
12	Inadequate sizing before starting project	–10.00	70.00
13	Missing or inadequate inspections for critical features	–10.00	60.00
14	Methodology cults (insisting on one method for all projects)	–9.75	55.00
15	Rejection of estimates for business reasons	–9.70	50.00
16	Missing or bypassing static analysis	–9.60	45.00
17	Friction/antagonism among management	–9.52	15.00
18	Friction/antagonism among stakeholders	–9.52	15.00
19	Inadequate communications with stakeholders	–9.52	20.00
20	Friction/antagonism among team members	–9.52	10.00
21	Litigation (intellectual property theft)	–9.50	0.10
22	Litigation (patent violation)	–9.50	0.10
23	Litigation (poor quality/damages)	–9.50	3.00
24	Litigation (security flaw damages)	–9.50	3.00

(Continued)

Table 7.1 (Continued) Software Anti-Patterns circa 2016 by Severity

	Software Anti-Patterns	*Scoring*	*Frequency (%)*
25	Litigation (breach of contract)	−9.50	5.00
26	Litigation (software causes injury)	−9.50	0.10
27	Litigation (non-compete violation)	−9.50	0.10
28	Omitting quality from outsource contracts	−9.27	60.00
29	Omitting scope creep from outsource contracts	−9.27	45.00
30	Peter principle (promotion to incompetence levels)	−9.00	25.00
31	Inadequate security controls	−9.00	80.00
32	Error-prone modules in applications	−9.00	85.00
33	Bad fixes (new bugs in bug repairs)	−9.00	95.00
34	Leakage from historical quality data	−8.50	70.00
35	Inadequate governance	−8.50	50.00
36	Micro management	−8.00	15.00
37	High complexity: Code	−8.00	30.00
38	Excessive defects: Architecture	−8.00	15.00
39	Inadequate cost tracking	−7.75	70.00
40	Excessive defects: Source code	−7.67	80.00
41	Excessive defects: Requirements	−7.33	65.00
42	Layoffs/loss of key personnel	−7.33	10.00
43	Inadequate defect tracking methods	−7.17	50.00
44	Test case errors	−7.00	80.00
45	Failure to estimate requirements changes	−6.85	70.00
46	Excessive defects: Design changes	−6.83	60.00
47	Excessive defects: Data errors	−6.67	97.00
48	Inadequate problem reports	−6.67	60.00
49	Inadequate measurement of quality	−6.50	97.00
50	Leakage from historical cost data	−6.50	90.00

(Continued)

Table 7.1 (Continued) Software Anti-Patterns circa 2016 by Severity

	Software Anti-Patterns	*Scoring*	*Frequency (%)*
51	Requirements toxic or harmful that should not be in software	−6.50	45.00
52	Inadequate testing	−6.38	55.00
53	Government mandates (short lead times)	−6.33	10.00
54	Low-level languages (obsolete)	−6.25	10.00
55	Excessive requirements creep (> 5% per calendar month)	−6.00	30.00
56	Stovepipe organizations	−6.00	30.00
57	Defects: Bad test cases	−6.00	50.00
58	Complexity: Requirements	−5.83	50.00
59	Inadequate value analysis	−5.50	70.00
60	Inadequate change control	−5.42	60.00
61	High complexity: Architecture	−5.33	45.00
62	High complexity: Cyclomatic (> 25)	−5.33	20.00
63	High complexity: Essential	−5.33	10.00
64	Inadequate communications among team	−5.33	15.00
65	High-level languages (obsolete)	−5.00	15.00
66	Analysis paralysis	−5.00	20.00
67	High complexity: Data	−4.83	75.00
68	High complexity: Organizational	−4.67	35.00
69	Lines of code metrics (physical LOC)	−4.50	45.00
70	Partial productivity measures (coding)	−4.40	75.00
71	Lines of code metrics (logical LOC)	−4.50	45.00
72	Complexity: Problem	−4.33	45.00
73	Requirements deferrals	−4.17	30.00
74	Design by committee	−3.00	20.00
75	Pair programming	−2.00	1.00
	Averages	**−7.52**	45.06

Table 7.2 Software Anti-Patterns Circa 2016 by Frequency

	Software Anti-Patterns	*Scoring*	*Frequency (%)*
1	Undiscovered security flaws in software	−10.50	98.00
2	Excessive defects: Data errors	−6.67	97.00
3	Inadequate measurement of quality	−6.50	97.00
4	Inadequate removal of security flaws before release	−10.50	95.00
5	Bad fixes (new bugs in bug repairs)	−9.00	95.00
6	Inadequate risk analysis before starting	−10.00	90.00
7	Leakage from historical cost data	−6.50	90.00
8	Error-prone modules in applications	−9.00	85.00
9	Status reports: Inaccurate and concealing problems	−10.00	80.00
10	Inadequate security controls	−9.00	80.00
11	Excessive defects: Source code	−7.67	80.00
12	Test case errors	−7.00	80.00
13	Excessively optimistic manual cost estimates	−10.00	75.00
14	Failure to use parametric cost estimates >500 function points	−10.00	75.00
15	High complexity: Data	−4.83	75.00
16	Partial productivity measures (coding)	4.40	75.00
17	Reuse of hazardous materials without validation	−10.00	70.00
18	Excessive schedule pressure	−10.00	70.00
19	Inadequate sizing before starting project	−10.00	70.00
20	Leakage from historical quality data	−8.50	70.00
21	Inadequate cost tracking	−7.75	70.00
22	Failure to estimate requirements changes	−6.85	70.00
23	Inadequate value analysis	−5.50	70.00

(*Continued*)

Table 7.2 (Conitnued) Software Anti-Patterns Circa 2016 by Frequency

	Software Anti-Patterns	*Scoring*	*Frequency (%)*
24	Excessive defects: Requirements	−7.33	65.00
25	Missing or inadequate inspections for critical features	−10.00	60.00
26	Omitting quality from outsource contracts	−9.27	60.00
27	Excessive defects: Design changes	−6.83	60.00
28	Inadequate problem reports	−6.67	60.00
29	Inadequate change control	−5.42	60.00
30	Methodology cults (insisting on one method for all projects)	−9.75	55.00
31	Inadequate testing	−6.38	55.00
32	Rejection of estimates for business reasons	−9.70	50.00
33	Inadequate governance	−8.50	50.00
34	Inadequate defect tracking methods	−7.17	50.00
35	Defects: Bad test cases	−6.00	50.00
36	Complexity: Requirements	−5.83	50.00
37	Missing or bypassing static analysis	−9.60	45.00
38	Omitting scope creep from outsource contracts	−9.27	45.00
39	Requirements toxic or harmful that should not be in software	−6.50	45.00
40	High complexity: Architecture	−5.33	45.00
41	Lines of code metrics (physical LOC)	−4.50	45.00
42	Lines of code metrics (logical LOC)	−4.50	45.00
43	Complexity: Problem	−4.33	45.00
44	Defect removal efficiency <85%	−10.00	35.00
45	High complexity: Organizational	−4.67	35.00
46	High complexity: Code	−8.00	30.00

(Continued)

Table 7.2 (Conitnued) Software Anti-Patterns Circa 2016 by Frequency

	Software Anti-Patterns	*Scoring*	*Frequency (%)*
47	Excessive requirements creep (>5% per calendar month)	−6.00	30.00
48	Stovepipe organizations	−6.00	30.00
49	Requirements deferrals	−4.17	30.00
50	Peter principle (promotion to incompetence levels)	−9.00	25.00
51	Defect potentials >6.00 per function point	−10.25	20.00
52	Inadequate communications with stakeholders	−9.52	20.00
53	High complexity: Cyclomatic (>25)	−5.33	20.00
54	Analysis paralysis	−5.00	20.00
55	Design by committee	−3.00	20.00
56	Death-march projects with 100% failure odds	−10.00	15.00
57	Friction/antagonism among management	−9.52	15.00
58	Friction/antagonism among stakeholders	−9.52	15.00
59	Micro management	−8.00	15.00
60	Excessive defects: Architecture	−8.00	15.00
61	Inadequate communications among team	−5.33	15.00
62	High-level languages (obsolete)	−5.00	15.00
63	Friction/antagonism among team members	−9.52	10.00
64	Layoffs/loss of key personnel	−7.33	10.00
65	Government mandates (short lead times)	−6.33	10.00
66	Low-level languages (obsolete)	−6.25	10.00
67	High complexity: Essential	−5.33	10.00
68	Litigation (breach of contract)	−9.50	5.00
69	Litigation (poor quality/damages)	−9.50	3.00

(Continued)

Table 7.2 (Conitnued) Software Anti-Patterns Circa 2016 by Frequency

	Software Anti-Patterns	*Scoring*	*Frequency (%)*
70	Litigation (security flaw damages)	−9.50	3.00
71	Pair programming	−2.00	1.00
72	Litigation (intellectual property theft)	−9.50	0.10
73	Litigation (patent violation)	−9.50	0.10
74	Litigation (software causes injury)	−9.50	0.10
75	Litigation (non-compete violation)	−9.50	0.10
	Averages	−7.52	45.06

evaluation or selection, which is merely a popularity contest rather than an actual technical due diligence.

The existence of more than 60 software development methods is an interesting sociological issue. Most companies want a "one size fits all" method and tend to adopt one method for all projects, rather than a suite of methods aimed at various sizes and types of projects.

In spite of more than 60 methodologies, security flaws and cyber-attacks are increasing; the majority of large software systems run late and exceed their budgets, if they are finished at all; and quality remains much worse than it should be.

Cognitive dissonance is one of the reasons why so few people use inspections, why static analysis is frequently omitted, and why inaccurate manual estimates are more common than accurate parametric estimates. Anti-patterns and cognitive dissonance are both part of the poor performance of the software industry. Table 7.2 shows a variety of anti-patterns.

Table 7.3 shows the quantitative results for anti-patterns on a project of 1000 function points and the Java programming languages. Readers may find it interesting to compare the anti-pattern results with the pattern-based results, since they are both for the same size and type of software project. It is dismaying that as of 2016, anti-patterns are more widely deployed by the software industry than beneficial patterns.

As can be seen by the frequency of anti-patterns in software, the industry is careless in selecting effective technologies and even worse in measuring results. Barely 1 project in 100 starts with accurate size information and a realistic estimate based on a combination of parametric estimation tools and historical data from similar projects. In spite of years of proof of damages, anti-patterns are still distressingly common even in 2016.

Table 7.3 Software Size and Nature Scoring

Project Size and Nature Scores		Anti-Patterns
1	Large projects >10,000 FP	−10
2	Web/cloud projects	−10
3	High-reliability, high-security projects	−10
4	Medium projects	−3
5	Systems/embedded projects	−3
6	Small projects <100 FP	−1
7	New projects	−5
8	Enhancement projects	−5
9	Maintenance projects	−5
10	IT projects	−5
	Subtotal	−57
		Anti-Patterns
1	Change control	−10
2	Planning, estimating, measurement	−10
3	Requirements changes	−7
4	Coding	−5
5	Requirements	−1
6	Models	−1
7	Design	−1
8	Defect prevention	−1
9	Pre-test defect removal	−1
10	Testing defect removal	−1
11	Pattern matching	−1
12	Reusable components	−5
13	Architecture	−5

(*Continued*)

Table 7.3 (Continued) Software Size and Nature Scoring

14	Integration	−1
15	Test case design/generation	−8
	Subtotal	−58
	Total	−115
Regional Usage (Projects in 2016)		**Anti-Patterns**
1	North America	3,000
2	Pacific	2,000
3	Europe	2,000
4	South America/Central America	1,000
5	Africa/Middle East	3,000
	Total	11,000
Defect Potentials per Function Point (1000 function points; Java language)		**Anti-Patterns**
1	Requirements	1.20
2	Architecture	0.80
3	Design	2.00
4	Code	1.80
5	Documents	0.60
6	Bad fixes	1.00
	Total	7.40
	Total defect potential	7,400
Defect Removal Efficiency (DRE)		**Anti-Patterns**
1	Pre-test defect removal (%)	40.00
2	Testing defect removal (%)	60.00
	Total DRE (%)	76.00
	Delivered defects per function point	1.776

(*Continued*)

Table 7.3 (Continued) Software Size and Nature Scoring

	Delivered defects	1,776
	High-severity defects	275
	Security defects	25

Data source: Namcook Analysis of 600 companies.

Anti-Pattern Scoring
Best = +10
Worst = −10

Chapter 8

CASE Software Development

Executive summary: *The term "computer-aided software engineering" (CASE) was a very hot topic in the 1980s, when dozens of companies were intrigued by the concept. Many start-up companies entered a growing market for CASE tools. Some survived, but many failed. CASE stumbled due to the endemic software industry problems of poor measurement practices and making vast claims for productivity and quality without actual empirical data to back up the claims. That being said, CASE still exists in 2016, even though the name is no longer popular. Many useful tools were created, and that process was beneficial.*

As we all know, before any code can be written, there must be an agreed-to set of user requirements, a written or graphical design, and for large systems, a formal architecture. As applications grew from <500 to >10,000 function points, these pre-code efforts grew beyond easy human control.

The essential idea of CASE was to automate the front-end of software development as well as code creation. In 1968, the University of Michigan started an academic project called the *Information System Design and Optimization System* (ISDOS) to help in analyzing software requirements.

There were also interesting papers by Daniel Teichroew and the creation of a formal language for software requirements called the *problem statement language* (PSL) combined with a kind of theorem prover called the *problem statement analyzer* (PSA).

The use of project data dictionaries and the use of meta-data for encapsulating requirements and design issues would eventually feed into model-driven development, which is active today in 2016, even though it is no longer called CASE.

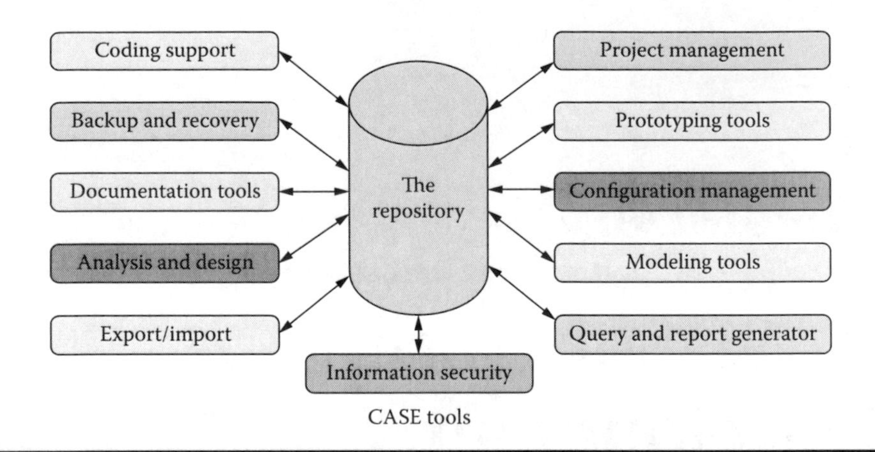

Figure 8.1 CASE tools.

A newer, related field is that of computer-aided method engineering (CAME), which is active in 2016. The CAME concept is to use meta-models for creating useful families of software development methods tailored for specific kinds of problem areas: real time, data rich, or something else.

The web has many illustrations of CASE and CAME tools, meta-models, and other relevant images such as Figure 8.1. It can be seen that CASE covers a wide range of topics.

Eventually, lack of empirical data and lack of proof of success dampened enthusiasm for CASE, which gradually fell out of popularity when Agile development appeared in 1997.

Among the permanent concepts that originated with the CASE concept was a family of interesting tools for requirements, analysis, design, and architecture, many of which still exist, such as the IBM Rational DOORS tool suite for requirements.

Other related useful concepts include, but are not limited to,

- Meta-data and meta-models
- Data dictionaries
- Central repositories of requirements and design information

In 2016, software is supported by a fairly rich collection of tools that encompass project management, pre-code analysis, coding, integration, and quality control. Some of the useful tool categories include the following:

Representative Software Tools circa 2016

1. Project sizing tools
2. Automated function point tools

3. Project cost and schedule estimating tools
4. Project quality estimation tools
5. Project tracking and cost reporting tools
6. Requirements gathering and analysis tools
7. Graphical design tools for visual models
8. Code creation support tools
9. Debugging tools
10. Design support tools
11. Architecture support tools
12. Defect tracking tools
13. Test library support tools
14. Automated test tools
15. Inspection support tools
16. Static analysis tools
17. Prototyping tools
18. Web-application support tools
19. Cloud-application support tools
20. Complexity analysis tools
21. Project office support tools
22. Verification and validation support tools
23. Development work benches
24. Maintenance work benches
25. Customer support tools

In spite of numerous tools, software is way behind other engineering fields, such as aeronautical engineering, marine engineering, and automotive engineering. Software engineering builds wonderful dynamic animated tools for other engineering fields, but continues to rely on text and monochrome static diagrams such as Unified Modeling Language (UML) diagrams.

Nothing moves faster than an operating software application. Nothing changes more often than a software project. It is obvious that software engineering has an urgent need for dynamic, animated, full color, and three-dimensional meta-models.

With animation, topics could be modeled that are totally invisible to the current generation of meta-models:

1. Requirements creep during development
2. Speed and load characteristics of software in use
3. Improved responses to threat vectors such as hacking, viruses, worms, and so on
4. Evolution of software over a multi-year life cycle
5. Bug injection and removal activities
6. Growth of entropy and cyclomatic complexity over time
7. Impact of refactoring and renovation of applications

8. Impact of varying quantities of reusable components
9. Visual highlights of software differences across multiple platforms
10. Multi-tier architecture in three dimensions

Without multiplying examples, software engineering is at least 10 years behind aeronautical and automotive engineering in the sophistication of design tools and meta-modeling. Software needs the equivalent of a "wind tunnel" to examine the characteristics of software under real-world usage conditions. Primitive UML diagrams, static user stories, and static monochrome meta-model diagrams are like antique automobiles trying to compete against a modern Formula 1 racing car.

Tools such as Mathematica10 can actually build dynamic animated meta-models that would show software attributes in real time under varying conditions.

The bottom line is that software engineering in 2016 is far behind other engineering fields in the power and sophistication of its design and planning tools. Even current meta-models such as International Organization for Standardization (ISO) and Object Management Group (OMG) are primitive. In fact, they are essentially useless for representing dynamic topics such as requirements growth, cyber-attacks after deployment, software performance under varying numbers of users, and other important technical and operational topics.

Unfortunately, the whole CASE/CAME meta-model concept has a more serious flaw besides primitive static graphical representation. Custom designs and manual coding are always expensive and error prone, no matter how they are designed, what methodology is used, or what meta-model is used.

The only effective software engineering method that can solve quality, security, and productivity problems at the same time is to move away from custom designs and manual coding toward construction of software from at least 85% reusable materials.

In theory, CASE and CAME tools might be used to construct suites of standard reusable components, but today, they seem to be mainly aimed merely at speeding up custom designs and making minor improvements in code development.

Also, with cyber-crime and cyber-warfare escalating to alarming levels, CASE and CAME seem to have intrinsic gaps in dealing with security issues. In fact, the whole software world continues to use the von Neumann computer architecture and casual permission lists, which make both hardware and software intrinsically vulnerable to attack, even though alternative secure methods exist.

From working as an expert witness in a number of software lawsuits for poor quality, canceled projects, or other disasters, it soon became obvious that project management failures outnumber technical software engineering failures by a wide margin.

It also became obvious that lack of effective project management tools is a major source of canceled projects, poor quality, and other disasters.

The differences in project management tool usage are both significant and striking. The lagging projects typically use only three general kinds of project management tools, while the leading projects use 18. Indeed, the project management tool family is one of the key differentiating factors between lagging and leading projects.

In general, managers on the lagging projects typically use manual methods for estimating project outcomes, although quite a few may use schedule planning tools such as Microsoft Project. However, project managers on lagging projects tend to be less experienced in the use of planning tools and to use fewer of the available features. The sparseness of project management tools does much to explain why so many lagging software projects tend to run late, to exceed their budgets, or to behave in more or less unpredictable fashion. Table 8.1 shows project management tool size ranges.

It is interesting that project managers on successful projects tend to use tools for about 4 h per work day. On average projects, use is about 2.5 h per work day. On lagging projects, barely 1 h per day is devoted to tool use.

By contrast, the very significant use of project management tools on the leading projects results in one overwhelming advantage: "No surprises." The number of on-time projects in the leading set is far greater than in the lagging set, and all measurement attributes (quality, schedules, productivity, etc.) are also significantly better.

There is very high use of parametric software cost estimation tools, quality estimation tools, and risk analysis tools for successful projects. Manual estimates seldom occur on successful projects, because they are too optimistic and too unreliable.

Differences in the software project management domain are among the most striking in terms of the huge differential in tool use between the laggards and the leaders. The variance in the number of tools deployed is about 7 to 1 between the leaders and the laggards, while the variance in the tool capacities expressed in function points has a ratio of approximately 17 to 1 between the leaders and the laggards. These differences are far greater than for almost any other category of tool.

While project management tools are not part of academic CASE and CAME research, their importance is as great as that of traditional software engineering tools.

Quality tools are also important and need additional study.

When the software quality assurance tool suites are examined, one of the most striking differences of all springs into focus. Essentially, the projects and companies in the "laggard" set have no software quality assurance function at all except for rudimentary defect tracking, usually not even starting until post release (Table 8.2).

Table 8.1 Numbers and Size Ranges of Software Project Management Tools

	Project Management Tools	*Lagging*	*Average*	*Leading*
1	Project planning	1,000	1,250	3,000
2	Project cost estimating			3,000
3	Statistical analysis			3,000
4	Methodology management		750	3,000
5	Reusable feature analysis			2,000
6	Quality estimation			2,000
7	Assessment support		500	2,000
8	Project office support		500	2,000
9	Project measurement			1,750
10	Portfolio analysis			1,500
11	Risk analysis			1,500
12	Resource tracking	300	750	1,500
13	Governance tools			1,500
14	Value analysis		350	1,250
15	Cost variance reporting	500	500	1,000
16	Personnel support	500	500	750
17	Milestone tracking		250	750
18	Budget support		250	750
19	Function point analysis		250	750
20	Backfiring: LOC to FP			300
21	Earned value analysis		250	300
22	Benchmark data collection			300
	Subtotal	1,800	4,600	30,000
	Tools	4	12	22

Note: Tool sizes are expressed in terms of International Function Point Users Group (IFPUG) function points, version 4.2.

FP: function point.

Table 8.2 Numbers and Size Ranges of Software Quality Assurance Tools

	Quality Assurance Tools	*Lagging*	*Average*	*Leading*
1	Quality estimation tools	–		2,500
2	Reliability modeling tools	–		1,000
3	Quality measurement tools	–	750	1,500
4	Six Sigma analysis tools	–		1,250
5	Data quality analysis tools	–		1,250
6	QFD analysis tools	–		1,000
7	COQ analysis tools	–		1,000
8	Inspection support tools	–		1,000
9	Static analysis: text	–		3,000
10	Static analysis: code	–	1,500	3,000
11	Complexity analysis tools	–	500	1,000
12	Defect tracking tools	250	500	1,000
	Subtotal	250	3,250	18,500
	Tools	1	4	12

Note: Size is expressed in terms of IFPUG function points version 4.2.

Quality assurance provides the greatest contrast between tool use and lack of tool use. Quality assurance personnel on leading projects use tools for almost 5 h per day; use is only about 2 h per day on average projects. For lagging projects, tools might not even be used during development at all, but only to track bugs after release. If quality tools are used on lagging projects, use is seldom more than 1 h per day.

By contrast, the leaders in terms of delivery, schedule control, and quality all have well-formed independent software quality assurance groups that are supported by powerful and growing tool suites. Leaders are far more likely than laggards to use static analysis tools and quality estimation tools.

Unfortunately, even leading companies are sometimes understaffed and underequipped with software quality assurance tools. In part, this is due to the fact that so few companies have software measurement and metrics programs in place, so that the significant business value of achieving high levels of software quality is often unknown to the management and executive community.

Several tools in the quality category are identified only by their initials, and need to have their purpose explained. The set of tools identified as *QFD support* are those that support the special graphics and data analytic methods of the quality function deployment methodology.

The set of tools identified as *COQ support* are those that support the reporting and data collection criteria of the cost of quality concept.

The other tools associated with the leaders are the tools of the trade of the software quality community: tools for tracking defects, tools to support design and code inspections, quality estimation tools, reliability modeling tools, and complexity analysis tools.

Complexity analysis tools are fairly common, but their use is much more frequent among the set of leading projects than among either average or lagging projects. Complexity analysis is a good starting point prior to beginning complex maintenance work such as error-prone module removal.

Another good approach prior to starting major maintenance tasks would be to run static analysis tools on the entire legacy application. However, a caveat is that static analysis tools only support about 25 languages out of the approximately 2500 programming languages in existence. Static analysis is available for common languages such as C, C++, C#, Java, COBOL, FORTRAN, ASP.NET, and some others. Static analysis is not available for the less common languages used for legacy applications, such as JOVIAL, CMS2, CHILL, or CORAL.

Unfortunately, since the laggards tend to have no quality assurance tools at all, the use of ratios is not valid in this situation. In one sense, it can be said that the leading projects have infinitely more software quality tools than the laggards, but this is simply because the lagging set often have zero quality tools deployed.

Table 8.3 shows the results of a fairly well-equipped software project of 1000 function points and the Java programming language. Assume the use of about 5 project management tools, 25 software development tools, 10 quality assurance tools, and 15 testing tools.

CASE and CAME are still in use in 2016, but today, they are used in a secondary role to provide tools for more recent and popular methodologies such as Agile, container development, and DevOps. However, the number and usefulness of software engineering tools are steadily improving. Whether or not the terms CASE and CAME are popular in 2016, tool use is steadily increasing and gradually adding value as tools become more sophisticated.

Table 8.3 Results of a Fairly Well-Equipped Software Project

Project Size and Nature Scores		**CASE**
1	Small projects < 100 FP	9
2	New projects	9
3	Medium projects	8
4	Large projects > 10,000 FP	7
5	Enhancement projects	7
6	Maintenance projects	7
7	IT projects	7
8	Systems/embedded projects	7
9	Web/cloud projects	7
10	High-reliability, high-security projects	7
	Subtotal	75
		CASE
1	Design	9
2	Coding	9
3	Change control	9
4	Integration	9
5	Defect prevention	9
6	Pre-test defect removal	8
7	Test case design/generation	8
8	Reusable components	7
9	Planning, estimating, measurement	7
10	Requirements changes	7
11	Testing defect removal	7
12	Architecture	5
13	Requirements	4
14	Models	4

(Continued)

Table 8.3 (Continued) Results of a Fairly Well-Equipped Software Project

15	Pattern matching	2
	Subtotal	104
	Total	179
Regional Usage (Projects in 2016)		**CASE**
1	North America	4000
2	Pacific	3000
3	Europe	3000
4	South America/Central America	2000
5	Africa/Middle East	1000
	Total	13000
Defect Potentials per Function Point (1000 function points; Java language)		**CASE**
1	Requirements	0.30
2	Architecture	0.35
3	Design	0.40
4	Code	1.35
5	Documents	0.40
6	Bad fixes	0.25
	Total	3.05
	Total defect potential	3050
Defect Removal Efficiency (DRE)		**CASE**
1	Pre-test defect removal (%)	65.00
2	Testing defect removal (%)	85.00
	Total DRE (%)	94.75
	Delivered defects per function point	0.160
	Delivered defects	160
	High-severity defects	25

(Continued)

Table 8.3 (Continued) Results of a Fairly Well-Equipped Software Project

	Security defects	2
Productivity for 1000 Function Points		**CASE**
	Work hours per function point 2016	13.90
	Function points per month 2016	9.50
	Schedule in calendar months	14.60

Data source: Namcook Analysis of 600 companies.

CASE Scoring
Best = +10
Worst = −10

Chapter 9

Cleanroom Software Engineering (CSE)

Executive summary: *The term "cleanroom" alludes to the highly controlled rooms used in semiconductor manufacturing to prevent introduction of dust or defects during chip fabrication. The software flavor of cleanrooms was developed by the late Dr. Harlan Mills of IBM. Software CSE is based on five principles: (1) formal specification; (2) incremental development; (3) structured programming; (4) static verification using rigorous inspections (no testing at the unit or module level); and (5) statistically based testing at the system level based on an operational profile. CSE is not widely used, as many believe it is too theoretical, too mathematical, and too radical because it relies on correctness verification and statistical quality control rather than unit testing. Successful deployment requires highly motivated and experienced teams. The CSE process was developed by Harlan Mills and others at IBM during the 1980s. Demonstration projects in the U.S. military began during the 1990s.*

CSE embeds software development within a statistical quality control framework—it is intended to produce software that is correct by design. Statistical usage-based testing processes enable inferences about software reliability and hence provide a basis for certification of fitness for use at delivery.

The development life cycle starts with formal specification of functional and performance requirements as well as the intended operational use of the software and a nested sequence of user function subsets that can be implemented and "tested" (including formal inspections) to produce a final system in a series of increments.

Correctness verification (formal inspections) is used to identify and correct defects prior to the execution of the delivered software.

CSE has been applied to both new systems and the evolution of legacy systems (which generally entails reengineering). The CSE process involves language and development environment. CSE is application independent and has been used in real-time, embedded, host, distributed, client-server, and microcode systems. CSE enables reuse through the precise definition of common services and certification of component reliability.

Substantial improvements over traditional practices have been reported, for example:

- Quality: IBM developed an embedded, real-time, bus architecture, multiprocessor device controller that had no field failures in 3 years' use at over 300 customer locations.
- Productivity: An Ericsson Telecom project to develop a 374 thousand lines of code (KLOC) operating system reported a 1.7-fold gain over former methods. (However, LOC is an unreliable metric.)
- Life cycle costs: IBM developed a COBOL restructuring product that had seven minor defects in 3 years of field use with a corresponding drop in maintenance compared with similar products.

CSE Teams

Each team is typically composed of five to eight individuals. On small projects, some individuals may participate in more than one team.

- Specification team: Develops and maintains the system specification
- Development team: Constructs software and verifies using formal inspections (software is not complied or tested)
- Certification team: Develops a set of statistical tests to exercise the software after development; uses reliability growth models to assess reliability

CSE Technology

CSE includes a number of technology topics:

- Incremental development—in common with other incremental/iterative/"agile" approaches, CSE offers opportunities for early customer feedback and hence facilitates the avoidance of risks inherent in late integration. Unlike other incremental approaches, however, CSE entails a degree of rigor and formality not found in other approaches. Specifically, the technical basis for incremental development in CSE is the mathematical property of "referential transparency." In a software context, this property requires that a specification and its design decomposition define the same mapping of inputs to valid outputs so that a design can be shown to be a correct expression of the specification. This requires that the order of design decomposition is organized for development into a sequence of verifiable and executable increments, each delivering additional functions. When this condition is satisfied, referential transparency is maintained throughout development.

- Precise specification and design—in CSE, programs are regarded as rules for mathematical functions or relations; that is, programs carry out transformations of input to output that can be specified as function mappings. This concept can be applied at any desired level of abstraction from large systems down to individual control structures. Three forms of "box structures," all of which map stimuli (inputs) and stimulus histories (previous inputs) to responses (outputs), are commonly used to accomplish the decomposition process:
  - Black box: Defines the required external behavior of a system or system element in all possible circumstances of use. The transition function for a black box specification takes the following form: ((current stimulus, stimulus history) response).
  - State box: Derived from and verified against a corresponding black box. The state box specification takes the following form: ((current stimulus, current state) (response, new state)). Note that in the state box, the stimulus history of the black box is replaced by retained state data required to achieve black box behavior.
  - Clear box: Defines procedures required to implement the state box transition, possibly introducing new black boxes for further decomposition. A clear box is a program that implements the state box and introduces and connects operations in a program structure for decomposition to the next level. The clear box specification takes the following form: ((current stimulus, current state) (response, new state)) by procedures.

Box structures can be applied to functional decomposition, object-oriented (OO) approaches, and so on. In the OO approach, the black box defines the behavior of an object, the state box defines its encapsulation, and the clear box defines its procedural services or methods. Box structures provide a systematic framework for reuse.

- Correctness verification: All CSE-developed software is subject to function-theoretic correctness verification prior to release to the certification test team. A correctness theorem defines the conditions to be met for achieving correct

software: These conditions are verified in mental/verbal proofs of correctness during formal team inspections. The correctness theorem is based on verifying individual control structures (if-then-else, etc.) rather than tracing paths. This step has proved to be extremely effective in finding defects (commonly more than 10–20 times better than testing).
- Statistical testing and software certification: No testing process, no matter how extensive, can sample more than a minute fraction of all the potentially executable paths. CSE accepts this immutable fact and creates usage models that include scenario probabilities using, for example, Markov models. Test cases are randomly generated from the usage model(s), so every case represents a potential use. This approach tends to detect errors with high failure rates—it is an efficient approach.

Comparisons with Other Approaches

Cleanroom and the Capability Maturity Model Integrated (CMMI)

The CMMI is fundamentally about management, and Cleanroom is fundamentally about methodology. There is considerable overlap between the scopes of the two, and each has areas that are not addressed by the other.

The CMMI, for example, has key process areas (KPAs) at Level 2 (repeatable) that are outside the scope of Cleanroom. Configuration management and sub-contractor management, for example, are important management issues not addressed by Cleanroom ideas. Cleanroom, on the other hand, enforces the mathematical basis of software development and the statistical basis of software testing, while the CMMI is silent on the merits of various methods. In general, the CMMI and Cleanroom are compatible and complementary. The combination of CMMI management and organizational capabilities and Cleanroom technical practices represents a powerful process improvement paradigm.

Cleanroom and the Object-Oriented Approach

Most people who have studied the relationship between Cleanroom and OO regard the two as complementary, each having strengths that can enhance the practice of the other.

Common Characteristics

With respect to the life cycle, Cleanroom follows the incremental development of the project, while the OO approach follows the iterative development. The pitfalls of the waterfall approach are recognized in both practices. Cleanroom incremental development and OO iterative development are both intended to provide opportunity for user feedback and to accommodate changing requirements.

The usage scenario involves the application of use case in OO and the usage model in Cleanroom. The OO use case and the Cleanroom usage model are both techniques for characterizing the user's view early in development. Artifacts from these activities are used in both design and testing. The state machine representation is common to both the methods for describing the behavior of a design entity.

Both the methods share a common methodology concerning reuse, which is an explicit objective in both practices. The OO class and the Cleanroom common service are the units of reuse.

An example of a cleanroom project of 1000 function points using Java is shown in Table 9.1.

Cleanroom software engineering started in IBM under the auspices of Dr. Harlan Mills, who worked in IBM's Federal Systems Division. Cleanroom has had some limited success, primarily with systems software, embedded software, and defense software. These sectors are probably more comfortable with the rigor and mathematical procedures associated with Cleanroom than are web applications, information technology applications, or ordinary commercial software and open-source software.

The Cleanroom methodology is at the opposite end of the spectrum from the Agile methodology. The Cleanroom method asks for very rigorous requirements and specifications, while Agile prefers to get these in a less formal manner. Cleanroom works best for applications with stable requirements and is not optimal for dynamic changing projects. Quality is good, but schedules and productivity are somewhat sluggish.

Table 9.1 Example of a Cleanroom Project of 1000 Function Points Using Java

Project Size and Nature Scores		Cleanroom
1	Systems/embedded projects	10
2	High-reliability, high-security projects	10
3	Small projects < 100 function points	9
4	Medium projects	8
5	New projects	8
6	Large projects > 10,000 function points	5
7	IT projects	5
8	Enhancement projects	4
9	Web/cloud projects	4
10	Maintenance projects	3
	Subtotal	66
		Cleanroom
1	Coding	10
2	Defect prevention	10
3	Pre-test defect removal	10
4	Test case design/generation	10
5	Testing defect removal	10
6	Planning, estimating, measurement	9
7	Requirements	9
8	Pattern matching	8
9	Reusable components	8
10	Models	8
11	Change control	8
12	Integration	7
13	Architecture	6

(Continued)

Table 9.1 (Continued) Example of a Cleanroom Project of 1000 Function Points Using Java

14	Design	5
15	Requirements changes	–1
	Subtotal	117
	Total	183
Regional Usage (Projects in 2016)		**Cleanroom**
1	North America	2500
2	Pacific	1500
3	Europe	1000
4	South America/Central America	500
5	Africa/Middle East	250
	Total	5750
Defect Potentials per Function Point (1000 function points; Java language)		**Cleanroom**
1	Requirements	0.20
2	Architecture	0.10
3	Design	0.30
4	Code	1.15
5	Documents	0.40
6	Bad fixes	0.20
	Total	2.35
	Total defect potential	2350
Defect Removal Efficiency (DRE)		**Cleanroom**
1	Pre-test defect removal (%)	85.00
2	Testing defect removal (%)	90.00
	Total DRE (%)	98.50
	Delivered defects per function point	0.035

(*Continued*)

Table 9.1 (Continued) Example of a Cleanroom Project of 1000 Function Points Using Java

	Delivered defects	35
	High-severity defects	5
	Security defects	0
Productivity for 1000 Function Points		**Cleanroom**
	Work hours per function point 2016	18.50
	Function points per month 2016	7.14
	Schedule in calendar months	16.20

Data source: Namcook Analysis of 600 companies.

Cleanroom scoring
Best = +10
Worst = –10

Chapter 10

CMMI Level 3 (Capability Maturity Model Integrated) Software Development

Disclosure: *Capers Jones, the author of this report, worked with two of the creators of the CMMI: Watts Humphrey and Dr. Bill Curtis. Watts and the author were colleagues at IBM. Dr. Curtis worked in the same department as the author at the ITT Programming Technology Center in Stratford, CT. The author also had a contract from the U.S. Air Force to examine the benefits of ascending from Level 1 to Level 5 of the CMMI.*

Executive summary: *The capability maturity model is not a "methodology" in the usual sense of the term. Rather, the CMMI points out and quantifies a diverse collection of "key process indicators" that have been observed to improve quality and productivity. The CMMI approach can be used with several development methods, but it is most often used with the team software process (TSP), since both the CMMI and TSP were developed by Watts Humphrey. The CMMI concepts are not well known in the civilian sector, such as banks and insurance companies. However, achieving Level 3 on the CMMI is mandatory for defense and many government software contracts. Since Level 3 is mandatory, it is included in this book.*

The Software Engineering Institute (SEI) was created on the campus of Carnegie Mellon University in 2005. It was funded in part by the Defense Advanced Research Projects Agency (DARPA) and chartered to find ways of improving software development for military and defense software projects.

About 5 years after the SEI was incorporated, the first version of the capability maturity model (CMM) was released in 1987. The capability maturity model was eventually replaced in 2002 by the capability maturity model integrated (CMMI).

The best-known feature of the CMMI is the examination or "appraisal" of the way companies build software. The appraisers are certified by the SEI and use standard questions with all companies.

One of the results of the CMMI appraisal process is to be placed on one of the five levels of the famous CMMI scoring system first developed by Watts Humphrey. There are several forms of appraisal of varying degrees of rigor: Class A (most rigorous), Class B, and Class C. Only Class A leads to a specific score.

There are many illustrations of the CMMI scoring system on the web. Indeed, images of the five levels of the CMMI are probably the most common software image on the web after the spiral development image (see Figure 10.1).

Because the CMMI deals with process structures in a fairly comprehensive and rigid fashion, it is obvious that some CMMI concepts do not blend well with the more flexible Agile concepts. However, a number of attempts have been made to merge CMMI concepts with Agile, and some are fairly successful.

However, CMMI is a very good fit for many other methods, and especially for the team software process (TSP), since both the maturity model itself and TSP were developed by the famous software researchers Watts Humphrey.

Because the CMMI uses the word "maturity," that term has become popular in a host of other concepts, such as human factors maturity, database maturity, estimating maturity, measurement maturity, and quite a few others.

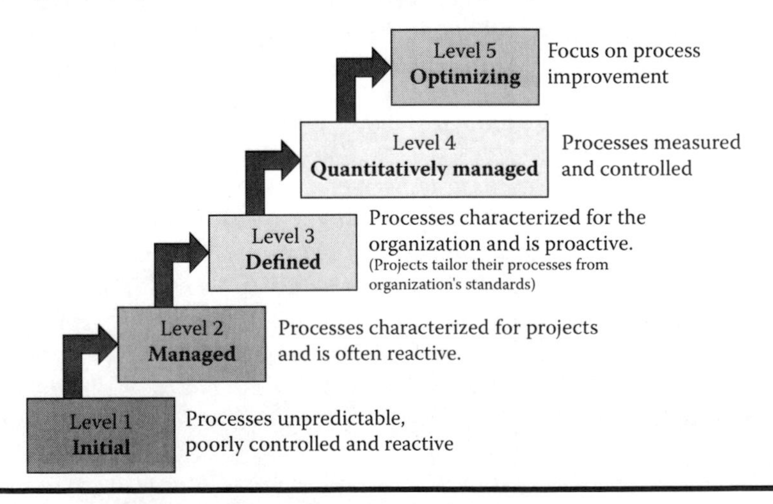

Figure 10.1 Characteristics of the maturity level.

Since the author of this report had a contract with the U.S. Air Force to study the impact of the various CMMI levels, quite a lot of quantitative data is available. Table 10.1 shows approximate quality results for the five CMMI levels for large systems of 10,000 function points in size.

It can be seen that the higher levels of the CMMI show very significant quality improvements compared with the lowest level or CMMI 1.

A related chart (Figure 10.2) shows a sample of CMMI levels and also development methodologies plotted against the same scale. The vertical axis shows defect potentials, and the horizontal axis shows defect removal efficiency (DRE).

This particular chart can be used to show any methodology, any CMMI level, or any other interesting variable. The idea is to stay away from the upper left quadrant and get as close as possible to the lower right quadrant, which is where CMMI 5 can be found.

Because the CMMI is mandatory for defense contracts, there are hundreds of certified appraisers, dozens of books, and hundreds of journal articles. A Google search on "CMMI" will turn up more than 39,000,000 web citations. By contrast, a Google search on "Agile" will turn up only about 21,000,000 citations. Both are well documented, of course, but the CMMI has a much larger set of documents and web citations than Agile does, probably because the CMMI is almost 20 years older than Agile.

A study of companies that used or did not use the CMMI found some interesting anomalies. For example, quality in the best CMMI Level 1 companies was better than the quality of the worst CMMI Level 3 companies. This seemed to be due to the fact that they were small companies and could not afford some of the rather high overhead costs associated with the CMMI appraisals.

Table 10.1 Software Quality and the SEI Capability Maturity

Model Integrated (CMMI) for 2500 Function Points				
CMMI Level	*Defect Potential per Function Point*	*Defect Removal Efficiency (%)*	*Delivered Defects per Function Point*	*Delivered Defects*
SEI CMMI 1	4.50	87.00	0.585	1463
SEI CMMI 2	3.85	90.00	0.385	963
SEI CMMI 3	3.00	96.00	0.120	300
SEI CMMI 4	2.50	97.50	0.063	156
SEI CMMI 5	2.25	99.00	0.023	56

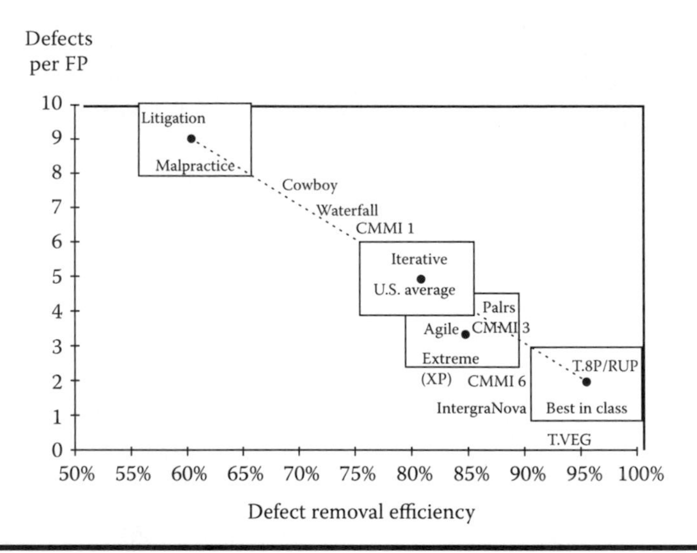

Figure 10.2 Defect potentials and defect removal efficiency.

We also found that some companies that did not even use the CMMI at all had quality even better than CMMI level 5. These were primarily technology companies, such as those building medical devices, where high quality is mandatory due to very large liabilities.

The CMMI is not a formal method per se, but it does recommend a significant suite of "best practices." Unlike Agile, pair programming, and some of the other software cults, the CMMI approach does have empirical data proving that quality and productivity increase with CMMI levels.

A sample project of 1000 function points using the Java language is shown in Table 10.2.

The SEI's CMM was an interesting and fairly successful way of bringing rigor and standardization to software development practices. Because funding came from DARPA, the focus of the CMMI has always been on military and defense software projects. Indeed, in 2016, achieving Level 3 on the CMMI is mandatory for U.S. defense software projects.

In spite of reasonable successes for projects at Level 3 and above, the CMMI never became popular in the civilian sector. Indeed, the author has met C-level executives in both Fortune 500 companies and state governments who have never even heard of the SEI or the CMMI.

Table 10.2 Sample CMMI 3 Project of 1000 Function Points Using the Java Language

Project Size and Nature Scores		CMMI 3
1	Large projects > 10,000 function points	10
2	New projects	10
3	Systems/embedded projects	10
4	High-reliability, high-security projects	10
5	Medium projects	8
6	Web/cloud projects	8
7	Enhancement projects	7
8	Maintenance projects	5
9	IT projects	5
10	Small projects < 100 function points	3
	Subtotal	76
		CMMI 3
1	Defect prevention	10
2	Pre-test defect removal	10
3	Test case design/generation	10
4	Testing defect removal	10
5	Coding	9
6	Change control	9
7	Pattern matching	8
8	Planning, estimating, measurement	8
9	Requirements	8
10	Requirements changes	8
11	Architecture	8
12	Integration	8
13	Models	7
14	Design	7

(Continued)

Table 10.2 (Continued) Sample CMMI 3 Project of 1000 Function Points Using the Java Language

15	Reusable components	6
	Subtotal	126
	Total	202
Regional Usage (Projects in 2016)		**CMMI 3**
1	North America	10,000
2	Pacific	8,000
3	Europe	9,000
4	South America/Central America	6,000
5	Africa/Middle East	2,000
	Total	35,000
Defect Potentials per Function Point (1000 function points; Java language)		**CMMI 3**
1	Requirements	0.35
2	Architecture	0.25
3	Design	0.60
4	Code	1.15
5	Documents	0.45
6	Bad fixes	0.20
	Total	3.00
	Total defect potential	3,000
Defect Removal Efficiency (DRE)		**CMMI 3**
1	Pre-test defect removal (%)	70.00
2	Testing defect removal (%)	85.00
	Total DRE (%)	96.00
	Delivered defects per function point	0.135
	Delivered defects	135
	High-severity defects	21
	Security defects	2

(*Continued*)

Table 10.2 (Continued) Sample CMMI 3 Project of 1000 Function Points Using the Java Language

Productivity for 1000 Function Points		CMMI 3
	Work hours per function point 2016	17.50
	Function points per month 2016	7.54
	Schedule in calendar months	15.10

Data source: Namcook Analysis of 600 companies.

CMMI Level 3 scores
Best = +10
Worst = −10

Chapter 11

Commercial Off-the-Shelf Software (COTS) Customization

Executive summary: *In 2016, most large companies own about as many COTS packages as they own internal software applications. Small companies often come close to 100% COTS packages combined with open-source packages. Many companies and government agencies find that vanilla COTS packages need customization. This can be a very difficult task, based in part on whether the COTS vendors support or discourage modifications. Even if modifications are supported, they can still jeopardize future COTS updates. COTS packages include accounting software, financial software, human resource software, engineering and architectural design tools, enterprise resource planning packages, office suites, operating systems, and many more.*

All corporations and government agencies have certain business functions in common that now use software. They need to support accounts payable and accounts receivable, personnel or human resource data, payrolls for fulltime personnel, payrolls for consulting personnel, records of customers, purchasing, shipping and receiving, marketing, sales, and inventory management, among others.

Commercial software (the developers of COTS packages) began in the 1960s and has expanded to become one of the largest and most profitable industries in world history. Some of the major companies in this sector include, in alphabetical order, Apple, Cisco, Computer Aid, Computer Associates, Facebook, Google, IBM (where the author worked on commercial applications), Microsoft, Oracle, SAP, Symantec, and hundreds of others.

Although the acronym "COTS" is the most widely used, there are several variations that define different forms of commercial software. MOTS means "modifiable off-the-shelf software," which means that the source code is available to clients and can be changed at will. GOTS means "government off-the-shelf software," which, as the name implies, supports various government functions such as taxation, child support, motor vehicles, property taxes, and the like.

Commercial software had a slow start, because companies such as IBM "bundled" applications or gave away software for free when customers purchased computers. As a result of an anti-trust suit, IBM "unbundled" in 1969, and that gave commercial software a major impetus to grow and expand.

Commercial software started in the mainframe era with applications such as database packages and sorts, which were widely used. Today, in 2017, commercial applications run on all platforms and under all operating systems. While there are still some mainframe commercial packages, PC applications, tablet applications, and smartphone applications are dominant. There are not many embedded commercial packages, although there are some.

The largest commercial COTS applications are enterprise resource planning (ERP) systems such as SAP and Oracle at over 250,000 function points. Big operating systems such as Windows 10 and IBM's operating systems can top 100,000 function points. There are also hundreds of smaller commercial packages, such as static analysis and test tools, of perhaps 3000 function points. Some smartphone packages are only a few hundred function points in size.

Commercial software is usually constructed by employees of the software companies. Requirements come from market analysis or from individual inventors. To be successful, commercial software needs high quality and reliability, so the larger commercial software vendors tend to be quite sophisticated in quality control, but they are not perfect, as clients find to their dismay. (The author has worked as an expert witness in litigation for poor quality against several major COTS vendors. Most cases are filed by disgruntled clients, but one case was actually filed by the company's own shareholders, who claimed that quality was so bad that stock values were depressed.)

As of 2017, the major cost drivers for commercial software are the following.

2017 Commercial Software Cost Drivers

1. The cost of finding and fixing bugs
2. The cost of producing paper documents
3. The cost of marketing and sales
4. The costs of code development
5. The costs of requirements changes
6. The costs of training and learning
7. The costs of project management
8. The costs of meetings and communication

(The data on commercial software in this book comes from benchmarks carried out by the author and colleagues in over 160 corporations, including about 15 COTS vendors. The author also worked on COTS applications at IBM, and, of course, has owned two software companies that market COTS estimation tools.)

The COTS industry is a fairly effective form of software reuse. COTS packages support all of the many generic functions that are used by thousands of companies in hundreds of industries. The COTS vision is that since generic functions are common across all industries and all companies, it is wasteful for each company to create its own custom software for business functions that are generic and widely used.

Even though major COTS and ERP packages are large and top 250,000 function points, most corporate software portfolios are larger than 1,000,000 function points. The corporate portfolio for Fortune 500 corporations may top 10,000,000 function points. Most state governments own software in excess of 5,000,000 function points, as do major cities such as New York and Los Angeles. Even smaller towns have portfolios that top 500,000 function points. The U.S. federal government, when civilian and military organizations are viewed as a whole, is the world's largest user of computer software, with a national portfolio that probably tops 100,000,000 function points. Russia and China are not far behind and probably have national portfolios that approach 75,000,000 function points.

Because even the largest and most extensive COTS and ERP packages can't support 100% of corporate software functions, the majority of companies customize some COTS packages during or after initial deployment.

Some COTS packages support customization, and many even have collections of variables that users can modify to meet local needs. Other COTS packages, including office suites such as Microsoft Office, Open Office, and Libre Office, are usually deployed and used straight out of the box with no customization at all.

The author's clients (mainly Fortune 500 companies) use about 40% vanilla COTS packages and customize about 60%. Small companies may use 100% vanilla COTS packages and do no customization at all, because they don't employ any software personnel. Table 11.1 shows the distribution of internal, COTS, and open-source software in a Fortune 500 client.

This client had a software organization with more than 10,000 personnel and built a lot of its own software for internal use, as well as embedded software in many of its products. But even a company with thousands of software personnel still owned over 1200 COTS packages. It is interesting that in this company, COTS packages outnumbered internal software for two software categories: tools and web applications.

COTS packages have pros and cons. If the COTS vendor is a major corporation with a proven track record, such as IBM or Microsoft, most COTS packages will probably have acceptable quality and security and good customer support.

On the other hand, COTS packages from smaller or newer vendors without a track record may have marginal quality and unknown security levels. It is always

Table 11.1 Software Owned by a Fortune 500 Company

Portfolio Application Types	*Internal*	*COTS*	*Open Source*	*Total*
Information systems	1460	96	78	1634
Systems software (telecommunications mainly)	913	125	0	1038
Tools and support software	365	433	0	798
Web applications	335	345	25	705
Embedded applications (switches mainly)	548	125	12	685
Manufacturing applications	365	20	0	385
Cloud applications	205	65	20	290
End-user applications	200	0	0	200
Total	4390	1209	135	5734

wise to check out smaller COTS vendors and contact some of their clients before making a big purchase.

Modifying COTS packages, assuming the vendor supports modifications, is usually a bit more difficult than internal development. This is due to the need to understand the inner workings of the COTS package being modified. If it is a large package in the 10,000 function point range or an ERP package in the 200,000 function point range, just figuring out where the changes need to go will be troublesome and expensive. Testing will be more costly, since the COTS package may need some test cases. Static analysis is possible for the COTS modifications, but probably not for the COTS package itself.

Some of the software methodologies used with COTS modifications include DevOps, Iterative, team software process/personal software process (TSP/PSP), test-driven, and waterfall. Some of the methodologies that may be difficult to use with COTS modifications include Agile, containers, feature-driven, Rational Unified Process (RUP), and lean. Agile is difficult because the COTS package itself does not fit well with the Agile philosophy. There are no live users to discuss requirements with, and only the changes themselves can be included.

Table 11.2 shows the results of a 1000 function point COTS modification project in the Java programming language. This might be a project that adds function point calculations, software cost estimates, and benchmark support to a project office tool such as Microsoft Project to customize it for software projects.

In terms of sheer numbers of projects, COTS modifications rank as #2 in the world, and today in 2016, may outnumber new internal applications for information technology projects. Of course, custom development is still dominant

Table 11.2 Results of a COTS Modification Project in the Java Programming Language

Project Size and Nature Scores		COTS Changes
1	Small projects <100 function points	8
2	Medium projects	7
3	Large projects >10,000 function points	5
4	New projects	4
5	IT projects	10
6	Systems/embedded projects	7
7	Web/cloud projects	8
8	High-reliability, high-security projects	6
9	Enhancement projects	9
10	Maintenance projects	5
	Subtotal	69
Project Activity Scores		**COTS Changes**
1	Pattern matching	5
2	Reusable components	10
3	Planning and estimating	5
4	Requirements	5
5	Architecture	5
6	Models	3
7	Design	6
8	Coding	7
9	Defect prevention	6
10	Test case design/generation	7
11	Testing defect removal	8
12	Requirements changes	8
13	Change control	9
14	Integration	9

(Continued)

Table 11.2 (Continued) Results of a COTS Modification Project in the Java Programming Language

15	Pre-test defect removal	6
	Subtotal	99
	Total	168
Regional Usage (COTS Changes 2016)		**COTS Changes**
1	North America	125,000
2	Pacific	110,000
3	Europe	115,000
4	South America/Central America	75,000
5	Africa/Middle East	65,000
	Total	490,000
Defect Potentials per Function Point (1000 function points; Java language)		**COTS Changes**
1	Requirements	0.50
2	Architecture	0.25
3	Design	0.75
4	Code	1.05
5	Documents	0.20
6	Bad fixes	0.65
	Total	3.40
	Total defect potential	3,400
Defect Removal Efficiency (DRE)		
1	Pre-test defect removal (%)	65.00
2	Testing defect removal (%)	75.00
	Total DRE (%)	91.20
	Delivered defects per function point	0.30
	Delivered defects	299
	High-severity defects	46

(Continued)

Table 11.2 (Continued) Results of a COTS Modification Project in the Java Programming Language

	Security defects	4
Productivity for 1000 Function Points		**COTS Changes**
	Work hours per function point 2016	16.70
	Function points per month 2016	7.90
	Schedule in calendar months	16.10

Data source: Namcook Analysis of 600 companies.

COTS modification scoring
Best = +10
Worst = −10

for systems and embedded software such as medical devices, where there are few COTS packages available, to say nothing of embedded weapons software.

The Department of Defense has actively pursued the expanded use of COTS packages. In 1994, Secretary of Defense William Perry wrote "the Perry memorandum," which asked for expanded defense COTS use to lower costs. This recommendation became a Federal law in 1994 in the Federal Acquisition Streamlining Act and also the Clinger–Cohen Act. Today in 2016, the Department of Defense uses both COTS packages and a smaller number of open-source packages. (Open-source is a special subset of the large COTS concept. Security and quality are more uncertain with open-source packages, because the numerous independent developers are not in one organization, and some of them may even be moles planted by hostile governments.)

COTS modifications are fairly cost-effective, but productivity is difficult to measure. Actual changes to the COTS packages are slow, but of course, the COTS packages themselves are already built. It would take longer and cost much more to duplicate their features.

If you include the size of the COTS package itself in the productivity calculations, the true economic value of COTS can be seen. It might take 16 work hours per function point to add 100 function points to a COTS package of 1000 function points. But if you consider the whole combination of 1100 function points, the net productivity is only 1.45 work hours per function point. In other words, what is important for COTS productivity is not *development* productivity but rather, *delivery* productivity; that is, how many function points are actually delivered to users.

Chapter 12

Container Software Development

Executive summary: *Container development is a fairly new method. The most popular current form of containers, from Docker, came out in 2013. Container development is not a full life-cycle methodology that includes requirements, design, architecture, and a full sequence of tasks. Container development focuses on using "containers" that allow software applications to run easily on many platforms such as Android, Linux, and Windows. The container development methodology can be used in conjunction with other methods such as Agile, DevOps, team software process (TSP), and so on.*

Container software development is one of the newer methodologies in this book. It originated in 2013 from Docker for Linux containers, but older versions can be traced back more than 10 years, although they did not use the name "container."

Many container development projects use proprietary "containers" developed by the Docker Corporation. These allow software applications to be easily shifted from platform to platform, such as from Windows to Linux to Android.

Container technology originated in the Linux environment, but Microsoft has announced Windows container support. Eventually, containers will probably be supported on all major operating systems.

Containers are similar to virtual machines, but with important differences. With virtualization, an application package has a version of the primary operating system that it needs as well as the application itself. This means a very bulky package.

With containers, only the applications and selected services are included in the package, so containers are much smaller and use fewer resources than virtualization.

Container development is not a full life-cycle methodology that includes requirements, architecture, design, quality assurance, and so on. Rather, the container development method is mainly a way of allowing software to run easily on multiple platforms. This speeds things up, because application developers are reusing many shared services instead of needing to develop them.

For example, if an application of 1000 function points of user features is going to operate on multiple platforms and multiple operating systems, probably 150 function points of special connectors would be needed for each environment. Assuming that Android, Linux, and Windows would be used, that would mean 450 function points of connection features. With containers, the developers don't need these special connecting features, because they are part of the container package.

Container development is methodology agnostic and can work with other methodologies, such as Agile, TSP, DevOps, and many others.

Containers are probably not as secure as virtualization, because they are sharing operating system features pretty much in the open.

Today in 2016, the Docker Company is a main source for containers, and it is usually cited in the literature. The Docker containers are open source. But similar container technology was available with Linux over 10 years ago. There are also Docker competitors, such as CorOs. To be effective, containers need to follow the guidelines in the App Container Image Specification.

This section is not a tutorial on container technology but rather, a quantitative comparison of development with containers versus other popular methodologies. Container development can be used in tandem with many other software methodologies, such as Agile, DevOps, iterative, TSP, Rational Unified Process (RUP), and so on.

Table 12.1 shows the scores and quality and productivity results for container software for 1000 function points and the Java language.

Container development is an effective technical response to cloud computing and to the modern need to have software applications run on a variety of hardware and software platforms without needing expensive customization for each platform. Containers and virtualization are similar in concept, but containers are much smaller and require fewer system resources than virtualization.

Table 12.1 Sample of a 1000 Function Point Project with the Java Language

Project Size and Nature Scores		Container
1	Small projects <100 function points	10
2	Medium projects	7
3	Large projects >10,000 function points	8
4	New projects	10
5	IT projects	10
6	Systems/embedded projects	6
7	Web/cloud projects	10
8	High-reliability, high-security projects	7
9	Enhancement projects	3
10	Maintenance projects	0
	Subtotal	71
Project Activity Scores		**Container**
1	Pattern matching	8
2	Reusable components	10
3	Planning and estimating	8
4	Requirements	7
5	Architecture	5
6	Models	3
7	Design	6
8	Coding	9
9	Defect prevention	7
10	Test case design/generation	5
11	Testing defect removal	7
12	Requirements changes	7
13	Change control	7

(Continued)

Table 12.1 (Continued) Sample of a 1000 Function Point Project with the Java Language

14	Integration	9
15	Pre-test defect removal	8
	Subtotal	106
	Total	177
Regional Usage (Container development 2016)		**Container**
1	North America	25,000
2	Pacific	20,000
3	Europe	18,000
4	South America/Central America	10,000
5	Africa/Middle East	3,500
	Total	76,500
Defect Potentials per Function Point (1000 function points; Java language)		**Container**
1	Requirements	0.40
2	Architecture	0.25
3	Design	0.35
4	Code	0.85
5	Documents	0.20
6	Bad fixes	0.30
	Total	2.35
	Total defect potential	2,350
Defect Removal Efficiency (DRE)		
1	Pre-test defect removal (%)	65.00
2	Testing defect removal (%)	85.00
	Total DRE (%)	94.75
	Delivered defects per function point	0.189
	Delivered defects	189

(*Continued*)

Table 12.1 (Continued) Sample of a 1000 Function Point Project with the Java Language

	High-severity defects	29
	Security defects	3
Productivity for 1000 Function Points		**Container**
	Work hours per function point 2016	12.80
	Function points per month 2016	10.31
	Schedule in calendar months	11.20

Data Source: Namcook Analysis of 600 companies.

Container scoring
Best = +10
Worst = −10

Chapter 13

Continuous Development

Executive summary: *The term "continuous development" is fairly recent. See also the DevOps chapter, which also features continuous development. Container development also can be used with continuous development. The concept of continuous development expanded from continuous integration. Continuous development requires a set of supporting tools and methods, including automated testing and an automated integration work bench. Continuous development is somewhat controversial and not always the best choice. For example, it has no place in the development of medical device software such as cochlear implants. Nor is it suitable for embedded software such as automobile brakes and navigation. Continuous development is aimed at internal IT projects and web projects where everything is done at one location, and all personnel are in frequent contact.*

The concept of continuous development is an extension of continuous integration, which was first proposed by Grady Booch, the chief scientist of IBM's Research Division.

Continuous integration and continuous development were soon popular with the agile methodology called Extreme Programming (XP), which also pioneered test-driven development, in which test cases are written before code. Empirical data shows that there are benefits from this approach.

For continuous integration and continuous development to work, a variety of supporting tools and sub-methods are needed. These support tools include, but are not limited to,

1. Automated testing
2. Automated, or at least frequent, static analysis
3. Automated build servers

4. Automated defect tracking
5. Automated defect routing
6. Cloning of the production environment

In ordinary development with teams of multiple programmers, each programmer is making changes individually. These changes need to be integrated into the overall application. Obviously, there will be instances of two or more programmers making changes that interfere with each other. There will be instances of changes containing bugs and causing regressions. The longer the time line between the introduction of conflicting or buggy changes and their discovery, the greater the difficulty of restoring the application to fully workable form.

With continuous integration, all code changes are submitted at least several times a day, and sometimes more often. Integration also occurs several times a day, as does automated testing. Thus, in general, the gap between possibly conflicting code changes and their discovery and correction is only a few hours, as opposed to several weeks or even longer with conventional build and test cycles.

The essential elements of continuous integration include, but are not limited to,

1. Maintain a formal code repository
2. Run static analysis on all changes
3. Automate unit testing
4. Automate the build process
5. Unit test prior to submitting code changes
6. Make the builds self-testing
7. Make the builds fast—minutes not hours
8. Do changes to clones of the production system

It is fairly obvious that continuous development concentrates only on the code base. Continuous development has little to say and little control over other software deliverables, including, but not limited to,

1. Requirements
2. Requirements creep
3. Reusable materials from certified sources
4. Reusable materials from legacy software
5. Commercial off-the-shelf (COTS) packages that are being modified
6. Data errors
7. Architecture
8. Design
9. HELP screens
10. User manuals

Because continuous development only deals with code, it is theoretically possible to build imperfect systems with missing requirements and a bad architecture and have the software developed and released under continuous development without anyone being the wiser until it is too late.

In other words, don't think of continuous integration and continuous development as panaceas for quality issues. You still need to use formal inspections of major deliverables before integration and testing even start. Static analysis should also be part of your quality strategy.

Unfortunately, the software industry tends to regard many technologies as panaceas, and continuous development is one of them.

There are many web images of continuous development, including the one shown in Figure 13.1.

Continuous development is most widely deployed on applications where the whole team is co-located. It can be used with distributed development, and even with multiple countries. However, the timing of continuous development becomes tricky when work locations are more than 8 hours apart.

Table 13.1 shows the results of continuous development for an internal IT project of 1000 function points and the Java programming language.

Continuous development is more or less a niche methodology that works mainly on internal web applications and information technology products. It is not optimal for systems or embedded software, for commercial software, or for open-source software. Nor is it recommended for software needing Food and Drug Administration or Federal Aviation Authority certification or subject to the Sarbanes–Oxley law. Since the focus of continuous development is primarily on source code, there are no safeguards against faulty requirements or bad design issues. The security of continuous development may also be an issue that needs additional steps.

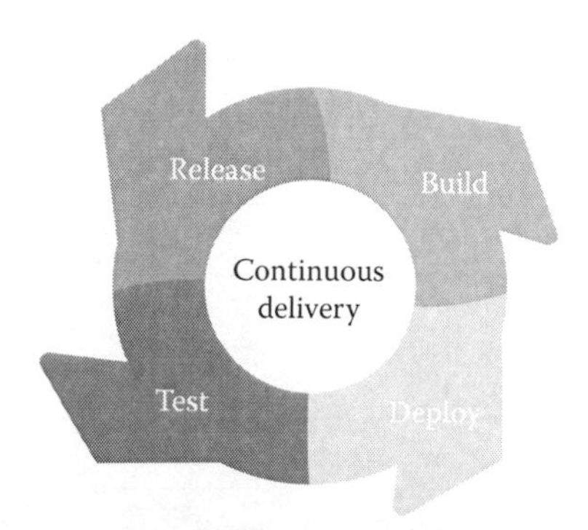

Figure 13.1 Example of continuous development.

Table 13.1 Example of Continuous Development for an Internal IT Project of 1000 Function Points and the Java Programming Language

Project Size and Nature Scores		**Continuous Development**
1	New projects	10
2	IT projects	10
3	Web/cloud projects	10
4	Small projects <100 function points	9
5	Medium projects	8
6	Large projects >10,000 function points	6
7	Enhancement projects	5
8	Maintenance projects	2
9	Systems/embedded projects	–3
10	High-reliability, high-security projects	–3
	Subtotal	54
Project Activity Scores		**Continuous Development**
1	Change control	10
2	Integration	10
3	Test case design/generation	10
4	Testing defect removal	10
5	Reusable components	8
6	Architecture	8
7	Defect prevention	8
8	Requirements	7
9	Requirements changes	7
10	Coding	7
11	Pre-test defect removal	7
12	Models	6
13	Design	6
14	Pattern matching	5

(*Continued*)

Table 13.1 (Continued) Example of Continuous Development for an Internal IT Project of 1000 Function Points and the Java Programming Language

15	Planning, estimating, measurement	5
	Subtotal	114
	Total	168
Regional Usage (Projects in 2016)		**Continuous Development**
1	North America	4,000
2	Pacific	4,000
3	Europe	3,000
4	South America/Central America	2,000
5	Africa/Middle East	1,000
	Total	14,000
Defect Potentials per Function Point (1000 function points; Java language)		**Continuous Development**
1	Requirements	0.45
2	Architecture	0.40
3	Design	0.60
4	Code	1.25
5	Documents	0.40
6	Bad fixes	0.20
	Total	3.30
	Total defect potential	3,330
Defect Removal Efficiency (DRE)		
1	Pre-test defect removal (%)	70.00
2	Testing defect removal (%)	75.00
	Total DRE (%)	92.50
	Delivered Defects Per Function Point	0.248
	Delivered defects	248

(*Continued*)

Table 13.1 (Continued) Example of Continuous Development for an Internal IT Project of 1000 Function Points and the Java Programming Language

	High-severity defects	38
	Security defects	3
Productivity for 1000 Function Points		**Continuous Development**
	Work hours per function point 2016	13.00
	Function points per month 2016	10.15
	Schedule in calendar months	12.70

Data Source: Namcook Analysis of 600 companies.

Continuous development scoring
Best = +10
Worst = −10

Chapter 14

Cowboy Development

Executive summary: *This method is the oldest method of all in software. It was the primary method in the 1950s, when computers were in their infancy, and applications were small and usually developed by one person, who handled the whole project from end to end. With the cowboy method, applications were usually less than 100 function points in size or less than 5000 source code statements in total. With cowboy development, individual programmers did the entire project. There were no development "teams" like those that began to occur when software applications grew in size and complexity. In spite of a complete lack of formal methods and rigor, a few brilliant "cowboy" developers did create some amazing software.*

Note: this chapter uses excerpts from the author's book The Technical and Social History of Software Engineering, *Addison Wesley, 2014.*

The "cowboy" development method originated in the 1950s, when computers were new, and software was even newer. This was long before waterfall, long before structured development, and almost 50 years before Agile.

The 1950s witnessed the migration of computers from military and academic purposes into the business domain. They also witnessed the evolution from custom-built, special-purpose computers to commercial computers such as the Ferranti Mark 1, LEO 1, Univac 1, IBM 701, and IBM 650. Even more important, transistors and integrated circuits were patented during this decade and began to be used on production computers by the end of the decade. By the end of the 1950s, computers were being built globally in China, Russia, Poland, Japan, and many other countries.

Programming became a significant occupation, and the early "high-level" languages of FORTRAN, COBOL, and LISP were created by the end of the decade. Until then, basic assembly language and the slightly more useful macro-assembly language were the only languages available other than actual machine language. The term "software" as we know it was coined during this decade.

At the start of the decade, mathematically trained human "computers" had been employed in their thousands to perform both sophisticated scientific mathematical calculations and also mundane accounting math for billing, salaries, taxes, and the like.

By the end of the decade, the term "computer" had morphed into meaning a digital computer, and significant numbers of people were employed in the new occupation of computer programmer, while the older occupation of human mathematical "computers" was in rapid decline and would soon disappear.

In 1950, universities with strong engineering and science departments began to teach courses on computers and software. For example, University of California, Los Angeles (UCLA) had its initial courses in 1950 taught by Douglis Pfister and Willis Ware. This was an extension of an earlier Institute for Numerical Analysis, formed at UCLA to work in conjunction with RAND and military organizations on the use of computers.

Other universities, such as Princeton, Harvard, and MIT, also began to incorporate computer-related courses into engineering curricula.

During this decade, several major inventions began the expansion of computers and software from massive and complex laboratory instruments to global commercial products.

Two critical background inventions, among the most important in the history of science, were the development of transistors to replace vacuum tubes and the invention of integrated circuits to replace discrete electronic components.

In 1947, William Bardeen and Walter Brattan of AT&T Bell Labs developed a prototype semiconductor based on germanium. The group leader, William Shockley, participated in expanding the idea. In 1956, Bardeen, Brattan, and Shockley received the Nobel Prize in Physics for the discovery of the transistor effect.

In 1949, a German scientist from Siemens AG filed a patent for an "integrated circuit" that he envisioned would use transistors. Later, in 1952, Geoffrey Dummer from the British Royal Radio Establishment gave a public lecture on the need for integrated circuits.

In 1950, the Japanese Union of Scientists and Engineers (JUSE) invited W. Edwards Deming of the United States to teach a 30-day seminar on statistical quality control. This seminar also paved the way for the later Deming Prize for quality. Later in the decade, Deming's work would help Japan become a major vendor of computers and computer components such as dynamic random-access memory (DRAM), chips, and screens.

It is widely known that Japanese industrial companies were among the first to apply the concepts of W. Edwards Deming and Joseph Juran to industrial quality control. Deming's contributions started in the 1940s and early 1950s; Juran's started in early 1954. The high quality of Japanese products benefited their market share in a variety of products, including computer chips, computer memory, televisions and portable radios, and automobiles. Later, Japanese quality methods, such as Kaizen, Kanban, and Poke Yoke, would also become important for software engineering.

In 1958, a researcher named Jack Kilby at Texas Instruments demonstrated a working integrated circuit. This idea was patented in 1959 and was soon used by the U.S. Air Force. Later, in 2000, Kilby won the Nobel Prize in Physics for the invention of the integrated circuit.

Without transistors and integrated circuits, software would probably be a small niche industry supporting a few dozen mainframe computers that used vacuum tubes. Neither personal computers nor embedded devices would be possible without low-power microscopic transistors and integrated circuits.

In 1958, a mathematician and statistician from Bell Labs named John Wilder Tukey used the word "software" in a paper. This was in the context of being a separate entity from "hardware." This is the first known use of the word "software" in a computer context.

Three key "high-level" programming languages were developed in the mid-1950s, and their use expanded during the 1960s: FORTRAN in 1955; LISP in 1958; and COBOL in 1959. Ideas and seeds for other languages such as ALGOL also started prior to 1960. However, for much of the decade, assembly language was the most common language in use.

While COBOL, LISP, and FORTRAN were developed in the United States, dozens of other languages were developed in other countries and had different names, even though some were variations on older languages. ALGOL was developed jointly between U.S. and European researchers.

Not only do languages have many dialects or variations, but they also change and add features over time. For example, there were ALGOL versions called ALGOL 58, ALGOL 60, and ALGOL 68, and another revision in 1973. There are also dozens of "ALGOL-like" languages based on some of the defined features, such as SIMULA.

Early computers were coded in "machine language," which is so highly complex that errors were rampant and hard to find and fix. Assembly languages introduced mnemonic source instructions that were somewhat easier to understand than binary numbers or machine language.

Originally, assembly languages had a one-to-one correspondence with machine languages, in that each source statement was translated into a single machine instruction. Later, the concept of "macro-assembly languages" expanded the scope of assembly source code statements.

A "macro instruction" was a method that allowed a number of statements to be created and named separately. These were called "macros," and a macro instruction was a quick way of adding reusable features to an application.

The 1950s saw the first signs of the future explosion of programming languages that would eventually top 3000 by 2016. The reasons for so many different programming languages are uncertain and probably sociological. There seems to be no technical need for so many, especially considering that more than 2500 out of 3000 are either dead, dying, or never used.

In these early days, applications were small, and coding was the dominant activity. This is the background of "cowboy programming." Table 14.1 shows the evolution of software from the 1950s through 2010 with projections to 2020.

When software started in the 1950s, there was only one occupation (programming) and one programming language (basic assembly). Every decade since then, programming languages and software occupation groups have grown larger. In 1950, the largest software application in the world was barely 100 function points in size. Today in 2016, the largest software applications top 300,000 function points.

In addition to programmers, software development teams now include architects, business analysts, configuration control, designers, function point counters, integration, quality assurance, technical writers, testers, and many more. A survey by the author funded by AT&T found a total of 126 software occupation groups. A few very large systems (Windows 10, SAP, Oracle, etc.) may use as many as 50 occupation groups concurrently. Software in 2016 is a team activity, and cowboy programming was mainly an individual activity.

Cowboy programming really only existed for about a decade when applications were small and done by one person, although even in 2016 some developers of smartphone applications and a few computer games still work by themselves and use cowboy programming.

In the era of cowboy programming, coding comprised 95% of the total job of creating software, and this job was done by one skilled programmer using only one language. Today in 2016, coding comprises only about 33% of the total job, and there are 126 different occupation groups involved with software, based on a

Table 14.1 Programming Evolution by Decade

Decade	*Coding (%)*	*Non-Code (%)*	*Programming Languages*	*Average Team*	*Maximum Team*	*Software Occupations*
1950s	95.00	5.00	1	1	1	1
1960s	90.00	10.00	5	1.5	10	5
1970s	85.00	15.00	25	4.5	100	25
1980s	75.00	25.00	90	8.5	500	75
1990s	65.00	35.00	750	9	1000	100
2000s	45.00	55.00	2250	8	1500	125
2010s	35.00	65.00	3000	7.5	2500	130
2020s	25.00	75.00	3300	6.5	3000	150

study by the author commissioned by AT&T. There are over 3000 programming languages, and most applications use at least two, such as Java and HTML. Some use as many as 15 programming languages.

It is a matter of historical interest that when software consisted of small applications where coding made up 95% of the work and only one occupation was needed, the "lines of code" metric was reasonably effective for measuring both productivity and quality.

IBM funded the development of function point metrics in the 1970s for showing the value of newer high-level programming languages such as PL/I and APL, which could not be done using lines of code.

The year 1952 marked the introduction of the IBM 701, which is claimed to be the first successful commercial computer, although only 19 were made. The IBM 701 was also delivered after the Univac 1 in the United States, so it was not the first chronologically. However, it did have the greatest long-range impact on the industry, since it evolved through several generations.

The IBM 701 was among the first computers used by the Department of Defense, and it alerted all defense organizations in every country to the fact that computers were a vital tool for logistics support and consolidation of defense information in real time. (Today, as this book is being written, the Department of Defense owns more computers than most countries.) The 701 was a vacuum tube computer that used magnetic tape for storage. It was intended for scientific and defense calculations.

The IBM 701 triggered an interesting and important sociological event. In 1955, a group of IBM customers in the Los Angeles area founded a user association called SHARE Inc. This association later moved to Chicago and became a non-profit corporation.

The idea of computer and software user groups was beneficial to both customers and manufacturers. SHARE members influenced IBM in terms of future features and needs. SHARE members also helped each other with technical advice and even programs and source code.

Another important invention occurred in 1953, when engineers at IBM's San Jose research facility created the first disk drive, which allowed "random access" to data instead of sequential access, which was provided by tape drives. The first commercial disk drive was the IBM RAMAC 350 in 1956.

Without disk drives and random access to data, computers would have very limited functionality, and later database technologies would not have occurred.

In 1953, IBM released the 650 computer, which was aimed squarely at business customers (by contrast, the earlier 701 had been aimed at scientific customers). The IBM 650 was a market success, and between the initial release and 1963, more than 2000 of these were sold.

The IBM 650 featured a rotating magnetic drum for memory. Programmers had to be sensitive to drum rotation speed to optimize performance. If a read instruction missed a piece of data, the next opportunity would be when the data rotated under the read head again. There was also a small amount of magnetic core memory, used as a buffer between the drum and the processing unit.

For external storage, at first, the IBM 650 used only punched cards, but later, tape drives were added. Disks were not originally used on the IBM 650. After RAMAC was invented, disk drives would be added to the later versions of the IBM 650.

The early versions of the IBM 650 were programmed in machine language. But in 1954, Stan Poley of IBM's TJ Watson Research Center added the "Symbolic Optimal Assembly Program" (SOAP). Eventually, more than a dozen programming languages would become available for the IBM 650, including FORTRAN in 1957.

Up till the IBM 650, universities had built a number of computers, but they were used only for a limited range of scientific studies and had no connection to day-to-day university studies. Columbia University, for example, had about 200 users of its IBM 650 computers. The IBM 650 computer started a trend of using digital computers for academic business activities and not just as research tools.

The hallmark of cowboy programming is that the work is done mainly by a single individual without any team support. This was common in the 1950s and even in the early 1960s.

Although cowboy programming is older than structured development, that does not mean that it was uniformly bad. There were many gifted cowboy programmers, and some remarkable applications were created by individuals, such as the first sorting program, the first compilers, and even the first rudimentary operating system and the APL programming language. More recent successes include the Google search algorithm and the original Facebook algorithms, both of which have led to multi-billion-dollar companies.

Table 14.2 shows the results of cowboy development for a project of 1000 function points using the Java programming language. Incidentally, a size of 1000 function points is about as large as it is technically possible for one person to develop,

Table 14.2 Example of Cowboy Development for 1000 Function Points and the Java Language

Project Size and Nature Scores		Cowboy
1	Small projects <100 function points	6
2	Web/cloud projects	6
3	Maintenance projects	5
4	New projects	4
5	Enhancement projects	4
6	IT projects	4
7	Medium projects	2
8	Systems/embedded projects	–3

(Continued)

Table 14.2 (Continued) Example of Cowboy Development for 1000 Function Points and the Java Language

9	Large projects >10,000 function points	–10
10	High-reliability, high-security projects	–10
	Subtotal	8
Project Activity Scores		**Cowboy**
1	Reusable components	9
2	Requirements	8
3	Requirements changes	8
4	Coding	8
5	Testing defect removal	5
6	Pattern matching	3
7	Architecture	3
8	Design	3
9	Change control	3
10	Integration	3
11	Pre-test defect removal	3
12	Test case design/generation	3
13	Models	–1
14	Defect prevention	–1
15	Planning, estimating, measurement	–7
	Subtotal	50
	Total	58
Regional Usage (Projects in 2016)		**Cowboy**
1	North America	6,000
2	Pacific	1,500
3	Europe	2,000
4	South America/Central America	4,000
5	Africa/Middle East	3,000
	Total	16,500

(*Continued*)

Table 14.2 (Continued) Example of Cowboy Development for 1000 Function Points and the Java Language

Defect Potentials per Function Point (1000 function points; Java language)		**Cowboy**
1	Requirements	0.30
2	Architecture	0.50
3	Design	0.80
4	Code	1.70
5	Documents	0.60
6	Bad fixes	1.10
	Total	5.00
	Total defect potential	5,000
Defect Removal Efficiency (DRE)		
1	Pre-test defect removal (%)	40.00
2	Testing defect removal (%)	80.00
	Total DRE (%)	92.00
	Delivered defects per function point	0.40
	Delivered defects	400
	High-severity defects	62
	Security defects	6
Productivity for 1000 Function Points		**Cowboy**
	Work hours per function point 2016	20.30
	Function points per month 2016	6.50
	Schedule in calendar months	19.40

Data source: Namcook Analysis of 600 companies.

Cowboy development scoring
Best = +10
Worst = −10

which is why the schedule is so long. Staffing is not shown here, but would probably consist of at least three programmers.

Cowboy programming has a bad reputation, but some top-gun programmers created outstanding applications by themselves when using it. Even today in 2016, a few top-gun software engineers have developed powerful and useful programs, and especially smartphone applications. Of course, Facebook also started with cowboy-style programming.

However, cowboy programming is mainly a one-person activity, and the majority of modern software applications are created by teams. The existence of teams is actually an unstated reason for all the more recent software methodologies, such as Agile and DevOps. If programs were still written by one person, then the 60 methodologies in this book would not have been created.

For example, the reason for the slow productivity in this cowboy example is that 1000 function point projects normally require a team of five developers. While individual cowboys can do a good job, getting five of them to cooperate and coordinate on the same project is beyond the normal range of cowboy development.

Chapter 15

Crystal Methods

Executive summary: *Crystal methods are a family of Agile-related methodologies developed by Alistair Cockburn in the mid-1990s. All are considered to be "lightweight" methods. These methods focus on people, interaction, community, skills, talents, and communication—Cockburn claims that process (methodology), while important, should be considered only in the context of the foregoing. Hence, his methods emphasize flexibility and leave much to team discretion. Eight members of the Crystal family have been described and are identified by colors (Clear, Yellow, Orange, Orange Web, Red, Maroon, Diamond, and Sapphire). They differ by "weight." Clear, Orange, and Yellow might be used for smaller, less critical projects, and Diamond and Sapphire for mission-critical projects. As with other Agile-related methods, little empirical evidence of efficacy is available. It is fair to say that these methods are more philosophy than prescription—tailoring for each project is expected.*

All the Crystal variants have the following common properties, each followed by the corresponding Agile Manifesto principle in italics:

- Frequent delivery: *"Our highest priority is to satisfy the customer through early and continuous delivery of valuable software"*
- Reflective improvement: *"At regular intervals, the team reflects on how to become more effective, then tunes and adjusts its behavior accordingly"*
- Close or osmotic communication: *"Business people and developers must work together daily throughout the project"*
- Personal safety: *"Build projects around motivated individuals. Give them the environment and support they need, and trust them to get the job done"*

- Focus: Refers to minimizing interruptions and multi-tasking—no manifesto equivalent
- Easy access to expert users: *"Business people and developers must work together daily throughout the project"*
- Technical environment with automated tests, configuration management, and frequent integration: *"Continuous attention to technical excellence and good design enhances agility"*

Note that there are other principles included in the Manifesto that are not addressed by these properties (e.g., *"Agile processes promote sustainable development. The sponsors, developers, and users should be able to maintain a constant pace indefinitely"*).

What Is a "Methodology"?

According to Cockburn, a methodology consists of (defines) the following elements. Most methodologists broadly agree with this list.

- Roles: Essentially, job descriptions. For example, requirements engineer, software architect, designer, coder, project manager, configuration management specialist, and so on.
- Skills: Define what each role is expected to know. For example, modeling techniques, design patterns, specific language(s), critical-path planning tools, and so on.
- Teams: How people are grouped, how people are assigned roles and possibly assigned to organizational units. For example, business analyst team, project office, and so on.
- Techniques: For example, Unified Modeling Language (UML), formal software inspections, facilitation.
- Activities: Meetings, reviews, milestones, construction of specific deliverable artifacts (work products). These may be specified in a particular sequence and may specify particular roles and techniques to be employed.
- Work products: Defines specific deliverables, their content, quality assurance required, and who creates/reviews/approves/receives each work product. For example, project plans, requirements specifications, use cases, test plans/cases/scripts, interface definitions, and so on.
- Standards: What is or is not permitted in a work product. For example, various national standards, style guides, delivered quality requirements, and so on.
- Quality: Defines expectations for delivered quality. For example, x defects per function point.
- Values: May define the sociological norms to be promoted. For example, Agile methods generally embrace the Manifesto principle: *"The best architectures,*

requirements, and designs emerge from self-organizing teams." Cockburn's approach certainly supports this view, whereas other approaches (e.g., capability maturity model integrated [CMMI], Cleanroom) may place higher value on conformance to process.

A methodology may also be characterized by its scope—that is, the extent to which it explicitly addresses roles (all or only some), life-cycle coverage (where it starts and ends—for example, covers requirements or not, covers maintenance or not), and activity coverage (i.e., whether or not all activities for all roles are explicitly defined). Crystal takes a "minimalist" view—just barely enough definition to satisfy the needs of a project of a given scope and scale associated with the colors in the Crystal family.

- A larger, "heavier" methodology is required when more people are involved.
- More publicly visible "correctness" is called for when risk is high (e.g., mission-critical systems). This implies greater emphasis on quality assurance and metrics. Cockburn defines four levels of criticality:
  - "Loss of comfort": Users are inconvenienced, efficiency is compromised.
  - "Loss of discretionary money": Uncomfortable, but strategically significant.
  - "Loss of irreplaceable money": Bankruptcy may result.
  - "Loss of life": For example. avionics, medical equipment.
- "Weight" is cost: A relatively small increase in methodology weight may add a relatively large cost. This is the basis for preferring "lightweight" approaches.
- The most effective form of communication for transmission of ideas is face-to-face interaction.

Tailoring Methodology to the Project

Cockburn believes, we think correctly, that there is no "one size fits all" methodology. As indicated in the previous section, he believes that methods should be scaled to the size of the team and also to the criticality of the project. So far, we agree.

Unfortunately, he goes further, asserting that the methodology should be tailored for each project, taking into account the preferences, experiences, and fears of the team doing the project. In one of his papers, Cockburn quotes Kent Beck as saying "All methodology is based on fears" and goes on to state: "Each element in the methodology is a stop against a bad experience some project has had. Afraid that programmers make coding mistakes? Hold code reviews. Afraid that users don't know what they really want? Create prototypes. Afraid that designers will leave in the middle of the project? Have them write extensive design documentation as they proceed."

Project-specific tailoring is problematic for large projects. Who will do the tailoring? According to what criteria? We believe a more sound approach is to use one of the well-proven methods such as Rational Unified Process (RUP) (which permits tailoring within well-defined limits), team-specific process (TSP), or Cleanroom. Industry data clearly establishes that a focus on quality in larger projects always leads to the lowest attainable cost and schedule.

The Crystal development methodologies are used more widely in Europe than in the United States, where several other varieties and hybrids of Agile are more widely deployed. However, as of 2016, newer methodologies such as container development and DevOps are expanding rapidly into the Crystal space. In the United States, Disciplined Agile Delivery (DAD) also competes with Crystal.

Table 15.1 shows an example of a 1000 function point Crystal project, assuming Java as the programming language and "orange web" as the flavor of Crystal.

Table 15.1 Example of 1000 Function Point Crystal Project with Java Language

Project Size and Nature Scores		**Crystal**
1	Small projects <100 function points	10
2	New projects	10
3	IT projects	10
4	Web/cloud projects	9
5	Medium projects	8
6	Systems/embedded projects	6
7	High-reliability, high-security projects	5
8	Large projects >10,000 function points	4
9	Enhancement projects	4
10	Maintenance projects	1
	Subtotal	67
Project Activity Scores		**Crystal**
1	Requirements changes	10
2	Requirements	9
3	Coding	9
4	Change control	8
5	Defect prevention	8
6	Testing defect removal	8

(Continued)

Table 15.1 (Continued) Example of 1000 Function Point Crystal Project with Java Language

7	Design	6
8	Integration	6
9	Models	5
10	Test case design/generation	5
11	Pre-test defect removal	4
12	Pattern matching	3
13	Reusable components	3
14	Architecture	3
15	Planning, estimating, measurement	–1
	Subtotal	86
	Total	153
Regional Usage (Projects in 2016)		**Crystal**
1	North America	1500
2	Pacific	500
3	Europe	5000
4	South America/Central America	750
5	Africa/Middle East	500
	Total	7500
Defect Potentials per Function Point (1000 function points; Java language)		**Crystal**
1	Requirements	0.20
2	Architecture	0.15
3	Design	0.35
4	Code	1.25
5	Documents	0.40
6	Bad fixes	0.30
	Total	2.65
	Total defect potential	2650

(*Continued*)

Table 15.1 (Continued) Example of 1000 Function Point Crystal Project with Java Language

Defect Removal Efficiency (DRE)		
1	Pre-test defect removal (%)	70.00
2	Testing defect removal (%)	85.00
	Total DRE (%)	95.50
	Delivered defects per function point	0.119
	Delivered defects	119
	High-severity defects	17
	Security defects	2
Productivity for 1000 Function Points		**Crystal**
	Function points per month 2016	11.50
	Work hours per function point 2016	11.48
	Schedule in calendar months	13.65

Data Source: Namcook Analysis of 600 companies.

Crystal scoring
Best = +10
Worst = −10

Chapter 16

DevOps

Executive summary: *The term "DevOps" is a combination of "development" and "operations." DevOps is one of the newer methodologies and is based on collaboration between the development team and the operations and maintenance teams. DevOps uses modern methods such as continuous integration and continuous delivery. The goal of DevOps is not only faster development and higher quality than waterfall, but more stable and reliable operations and reduced maintenance. Clearly, DevOps is aimed at software projects with the operations personnel in the same location as the development personnel. DevOps is not aimed at commercial software that will be installed at unknown numbers of distant locations in hundreds or perhaps thousands of companies.*

The DevOps methodology is one of the newest methodologies, having originated in Belgium in 2009. Since then, DevOps has moved quickly through the world, with DevOps conferences in India, Brazil, the United States, Sweden, Germany, and Australia.

DevOps concepts have also been quickly adopted by many of the world's major software producers, including IBM, Computer Associates, Netflix, Facebook, Amazon, Google, and Twitter, as well as hundreds of medium to small companies.

DevOps has the flavor of a start-up company where developers go beyond pure development and work as quality assurance (QA) experts and testers, and even have operations support roles. The new term *full stack developer* has come to be associated with these multi-faceted roles.

DevOps also has a strong Agile flavor and tends to use many Agile principles, including embedded user personnel for requirements, test-driven development, and sometimes Scrum sessions.

The growing popularity of DevOps has led to the creation of dozens of specialized new tools and even specialized new companies that offer tutoring and support for new DevOps groups. In 2014, DevOps was growing faster than Agile.

Like Agile, DevOps suffers from poor measurement practices and a dearth of reliable quantitative data on productivity, schedules, quality, and customer satisfaction.

Not everyone likes the DevOps concept, and some say it dilutes the effectiveness of good programmers by forcing them into test and QA roles for which they may lack training, skills, or both.

The web has many images of DevOps concepts. A Google search of "DevOps images" turns up more than 50 diagrams and images.

DevOps is a new methodology and is not yet "stable," since DevOps concepts are still growing and maturing. The essential features of DevOps include

1. Constant contact between development and operations personnel
2. Constant contact between stakeholders and developers
3. Continuous integration
4. Continuous delivery on a daily basis
5. Early QA
6. A larger quality role for developers
7. Early testing by developers
8. A lack of standard productivity metrics such as function points
9. A lack of standard quality metrics
10. Little data on defect potentials and defect removal efficiency (DRE)

As the name implies, DevOps requires access to the data center and operations personnel who run the software. This means that DevOps is aimed directly at internal information technology (IT) or web projects. It is clearly not appropriate for systems and embedded software that operate untended once deployed. For example, DevOps has no role in automobile navigation packages or brake systems. DevOps is also clearly not appropriate for medical devices such as pacemakers, cochlear implants, magnetic resonance imaging devices, computed tomography scans, and other modern computer-controlled medical equipment. The same is true for weapons systems and many kinds of military software used under combat conditions.

Because DevOps concepts are in flux, it is desirable for those interested in DevOps concepts to do frequent Google searches in order to gain access to the latest information, which seems to be changing on a weekly basis.

DevOps and container development are two of the fastest-growing recent software development methodologies. Both are somewhat specialized and not suited for all types and sizes of software applications.

Within its main niche of IT applications with high transaction rates and frequent or continuous runs, DevOps has shown respectable results through 2016. However, it is not a suitable methodology for systems and embedded applications. DevOps' main targets are business software applications with high transaction rates and frequent runs; possibly continuous 24/7 run.

Table 16.1 Example of 1000 Function Point DevOps Project with Java Language

Project Size and Nature Scores		DevOps
1	Medium projects	10
2	New projects	10
3	Enhancement projects	10
4	Maintenance projects	10
5	IT projects	10
6	Small projects <100 function points	9
7	Large projects >10,000 function points	9
8	Web/cloud projects	8
9	High-reliability, high-security projects	4
10	Systems/embedded projects	–10
	Subtotal	70
Project Activity Scores		**DevOps**
1	Coding	10
2	Change control	10
3	Integration	10
4	Planning, estimating, measurement	8
5	Requirements	8
6	Requirements changes	8
7	Testing defect removal	8
8	Pattern matching	7
9	Reusable components	7
10	Design	7
11	Defect prevention	7
12	Pre-test defect removal	7
13	Architecture	6
14	Test case design/generation	6

(*Continued*)

Table 16.1 (Continued) Example of 1000 Function Point DevOps Project with Java Language

15	Models	4
		113
	Total	183
Regional Usage (DevOps 2016)		**DevOps**
1	North America	25,000
2	Europe	20,000
3	Pacific	18,000
4	South America/Central America	12,500
5	Africa/Middle East	3,000
	Total	53,500
Defect Potentials per Function Point (1000 function points; Java language)		**DevOps**
1	Requirements	0.30
2	Architecture	0.15
3	Design	0.35
4	Code	1.30
5	Documents	0.40
6	Bad fixes	0.35
	Total	2.85
	Total defect potential	2,850
Defect Removal Efficiency (DRE)		
1	Pre-test defect removal (%)	75.00
2	Testing defect removal (%)	90.00
	Total DRE (%)	92.60
	Delivered defects per function point	0.074
	Delivered defects	211
	High-severity defects	33
	Security defects	3

(*Continued*)

Table 16.1 (Continued) Example of 1000 Function Point DevOps Project with Java Language

Productivity for 1000 Function Points		DevOps
	Work hours per function point 2016	13.61
	Function points per month 2016	9.70
	Schedule in calendar months	12.40

Data source: Namcook Analysis of 600 companies.

DevOps scoring
Best = +10
Worst = −10

Chapter 17

Disciplined Agile Development (DAD)

Executive summary: *Disciplined agile development (DAD) is one of about half a dozen Agile variations. DAD was created by Scott Ambler while he was at IBM. Scott and Mark Lines wrote a book of the same name, which is widely cited and used. (IBM has developed a number of software innovations, including function point metrics and the defect removal efficiency [DRE] metric, both of which are used in this book.) DAD uses basic Agile concepts but also includes risk analysis and techniques for scaling Agile up to larger applications. The original Agile with Scrum was not optimal for larger applications above about 1000 function points. DAD can scale up above 5000 function points and supports teams of specialists, which are normal for large software applications.*

The original Agile is partly an evolutionary method based on iterative development and partly a new approach based on the famous "Agile Manifesto." In the year 2001, some 17 well-known software experts met at the Snowbird resort in Utah to discuss software development problems and the potential for solving the problems. The result of this meeting was the Agile Manifesto, published in February, 2001. The main principles of the Agile Manifesto are

- Individuals and interactions are better than formal processes and tools.
- Working software is better than comprehensive documentation.
- Customer collaboration is better than comprehensive contracts.
- Responding to change is better than following a rigid plan.

As of 2016, Agile is one of the world's most popular software development methods. It has come to include a number of practices and techniques, although these are not always followed on every project, and some are used with other methodologies:

1. Embedded users who provide requirements.
2. User stories for requirements analysis.
3. Dividing larger projects into "sprints" that last about 2 weeks.
4. Daily "Scrum" sessions for status reports.
5. Pair programming, or two programmers taking turns coding or "navigating."
6. Test-driven development, or writing test cases before the code is written.
7. Specialized agile metrics such as velocity, burn down, burn up, and so on.
8. Agile "coaches" to help introduce Agile concepts.
9. The DAD version of Agile uses many original Agile principles but expands to support cross-functional teams and some expanded risk and value analysis tasks.

Software project team members with DAD include

1. Stakeholders
2. Product owner
3. Team leader
4. Team member(s)
5. Architecture owner
6. Integrator (as needed for large projects)
7. Specialists (as needed)
8. Testers (independent)
9. Domain experts (as needed)
10. Technical experts (as needed)

Among the author's clients, original Agile projects are fairly small, averaging about 275 function points. The author does not have enough DAD clients in 2016 for a statistical average of DAD size, but DAD projects up to 5000 function points have been reported.

A speculative average size for DAD software projects might be in the 1000 function point size range. This size range normally uses teams with specialists, while smaller original Agile projects normally use teams of generalists.

(A study carried out by the author's company and funded by AT&T explored software occupation groups in large companies and large government organizations. IBM participated in the study, as did the U.S. Navy, Texas Instruments, and a dozen other large organizations, all of which employed over 25,000 software personnel. The study identified a total of 126 software occupation groups in total. No single company employed all 126, but IBM employed about 50 kinds of software specialists.

Some large systems in the 100,000 function point size range have total staff of about 350 personnel and use more than 25 kinds of specialists of various occupation groups, such as business analysts, quality assurance, technical writers, database specialists, integration specialists, software engineers, test personnel, function point counters, project office staff, and many more. Table 17.1 illustrates 20 occupation group staffing patterns for 10,000 function points.)

Table 17.1 Twenty Common Software Occupations

	20 Software Occupations	*Normal Staff*
1	Programmers	43
2	Testers	38
3	Designers/architects	18
4	Business analysts	18
5	Technical writers	7
6	Quality assurance	6
7	First line managers	7
8	Database administration	4
9	Project office staff	3
10	Administrative support	4
11	Configuration control	2
12	Project librarians	2
13	Second line managers	1
14	Estimating specialists	1
15	Architects	1
16	Security specialists	1
17	Performance specialists	1
18	Function point counters	1
19	Human factors specialists	1
20	Third line managers	1
		160

It can be seen from Table 17.1 that large software projects are multi-disciplinary. The DAD approach recognizes this fact and provides cohesion for disparate occupations so that their work becomes an effective collaboration.

As of 2016, DAD is not as popular as the original Agile, but it is growing rapidly and has been used successfully for many applications. Banking and insurance are very well suited for DAD, and in fact, Barclay's Bank has standardized on DAD. A Google search on "disciplined agile" will turn up many interesting studies and an extensive set of citations.

It is not obvious, but it is true, that original Agile is aimed primarily at new projects, and it is not optimal for enhancing legacy applications or for maintenance and defect repairs. DAD, on the other hand, does support enhancement and maintenance projects.

One caveat about both original Agile and DAD is that while story point metrics may be effective locally for individual projects, they are essentially worthless for software benchmark analysis. Some 35 of 37 software benchmark organizations only support function point metrics. Not a single benchmark group supports story points as a primary metric, although several allow story points as an optional metric.

All the commercial parametric estimation tools support function point metrics, but few support story point metrics, although the author's Software Risk Master (SRM) estimating tool supports 23 size metrics, including story points and use-case points. However, International Function Point Users Group (IFPUG) function point metrics are the SRM default and the metric of choice for all software projects in the United States.

As of 2016, story points have no International Organization for Standardization (ISO) standards and no certification exams. Among the author's clients, story points have varied by over 400% for projects with identical sizes based on IFPUG function point metrics.

The author's SRM tool avoids story point and Agile measurement problems by converting story points to function points and by converting sprint data into a standard chart of accounts, which allows side-by-side comparisons with other methodologies such as team software process, DevOps, container development, waterfall, and so on.

Many Agile projects in 2016 use function point metrics as well as story point metrics. A rule of thumb for converting story point metrics to IFPUG function point metrics is that one story point is roughly equivalent to three function points. However, due to the random variances of story points, it is better to get local rules for converting story points to function points.

Original Agile was not an optimal choice for state government software or the Federal government due to their burdensome oversight requirements and mandatory documentation. DAD would probably be a good choice for state and government software projects.

However, there is a major caveat: any company that wants to get a government software contract with the governments of Brazil, Italy, Japan, Mexico, Malaysia, South Korea, and soon Poland must use function point metrics, because a growing number of national governments have seen the power of function points and the hazards of older unstandardized metrics such as lines of code and story points. Eventually, all national governments will require function point metrics.

IFPUG function point metrics have ISO standards and also certification exams, so size expressed in function point metrics is quite accurate. Story points lack both standards and certification exams, so they vary widely. It would probably be best to size applications directly in function points rather than converting from story points, assuming that clients have an interest in seeing benchmarks from similar applications.

In this book, none of the Agile metrics are useful at all, so the data is presented using IFPUG function points version 4.3. This same metric is used for all the methodology comparisons. Quality comparisons use the metric of defect removal efficiency (DRE), which was developed by IBM in 1973 and is widely used among technology companies.

Original Agile requires extensive modifications for projects that need certification by the Food and Drug Administration (FDA) or the Federal Aviation Authority (FAA). Nor is original Agile the best choice for financial applications that require Sarbanes–Oxley governance. However, DAD projects are rigorous enough that certification and Sarbanes–Oxley governance requirements are easily accomplished.

Agile concepts are not always intuitive. As a result, Agile has created a new software occupation called the *Agile coach*. Agile coaches are normally employed when companies first decide to adopt Agile. They teach both managers and team members the fundamental concepts of Agile and often participate in the initial sprints and planning sessions. The DAD version of Agile also has coaches and offers training, as well as a textbook.

The overall demographics of DAD in 2016 and the results of a typical Agile project of 1000 function points using the Java programming language are shown in Table 17.2. It should be noted that the project itself would be comprised of six

Table 17.2 Example of 1000 Function Point DAD Project with Java Language

Project Class and Size Scores		Disciplined Agile
1	Small projects <500 function points	10
2	New projects	10
3	IT projects	10
4	Web/cloud projects	10
5	Medium projects	10
6	Systems/embedded projects	7
7	Enhancement projects	7

(Continued)

Table 17.2 (Continued) Example of 1000 Function Point DAD Project with Java Language

8	High-reliability, high-security projects	7
9	Maintenance projects	5
10	Large projects >10,000 function points	9
	Subtotal	85
Project Activity Strength Impact		**Disciplined Agile**
1	Requirements changes	10
2	Requirements	10
3	Coding	10
4	Change control	8
5	Defect prevention	8
6	Testing defect removal	8
7	Design	6
8	Integration	6
9	Models	5
10	Test case design/generation	5
11	Pre-test defect removal	4
12	Pattern matching	9
13	Reusable components	3
14	Architecture	7
15	Planning, estimating, measuring	5
	Subtotal	104
	Total	189
Regional Usage (Projects in 2016)		**Disciplined Agile**
1	North America	4000
2	Pacific	500
3	Europe	1500
4	South America/Central America	750
5	Africa/Middle East	250

(*Continued*)

Table 17.2 (Continued) Example of 1000 Function Point DAD Project with Java Language

	Total	7000
Defect Potentials per Function Point (1000 function points; Java language)		**Disciplined Agile**
1	Requirements	0.25
2	Architecture	0.10
3	Design	0.30
4	Code	1.25
5	Documents	0.40
6	Bad fixes	0.20
	Total	2.50
	Total defect potential	2500
Defect Removal Efficiency (DRE)		
1	Pre-test defect removal (%)	65.00
2	Testing defect removal (%)	85.00
	Total DRE (%)	96.00
	Delivered defects per function point	0.113
	Delivered defects	100.0
	High-severity defects	16.0
	Security defects	1.0
Productivity for 1000 Function Points		**Disciplined Agile**
	Work hours per function point 2016	12.40
	Function points per month 2016	10.65
	Schedule in calendar months	13.30

Data source: Namcook Analysis of 600 companies.

Disciplined Agile Development (DAD) scoring
Best = +10
Worst = −10

agile sprints, each 200 function points in size: one sprint for initial planning and five for actual development.

In conclusion, DAD is a useful methodology that keeps the virtues of the original Agile concept but expands the range of software projects from small applications up to large systems. The improved rigor of DAD also makes it useful for projects that need FAA or FDA certification or are subject to Sarbanes–Oxley governance.

Like the original Agile, the DAD version has better productivity and quality than the older waterfall method. Disciplined Agile might also be used in conjunction with other and newer software methodologies, such as DevOps and container development, where the DAD discipline would add value. In Europe, DAD competes with Crystal, also an Agile variation.

Chapter 18

Dynamic Systems Development Method (DSDM)

Executive summary: *DSDM claims to be another flavor of Agile methods, yet it is "heavier" than most alternatives. Originally released in 1994, it sought to provide discipline to rapid application development (RAD). DSDM, as with other Agile approaches, inverts the "iron triangle"—the functionality to be delivered is a variable, while cost, time, and (theoretically) quality are fixed. DSDM differs from many other Agile approaches in that it addresses the complete life cycle from concept through to post-project maintenance. DSDM is managed by a consortium, which now has more than 1000 members. Many regard this approach as the de facto standard for use of RAD. As with other Agile methods, empirical evidence of effectiveness is lacking. The latest version of DSDM (The DSDM Agile Project Framework, 2014) is an evolution of the 2008 "Atern" release 4.2. The description in this chapter is based on 4.2, as the details of the 2014 version are not yet fully available. An available summary of changes suggests that they are relatively minor.*

Dynamic system development method (DSDM) Atern differs from some of the other Agile frameworks in that it advocates "just enough design up front," whereas other approaches consider overall design to be an emergent property. Also unlike most other Agile frameworks, the authors of DSDM indicate that it can be used as a complement to project management frameworks such as the PMBoK or Prince2 and that it can be combined with other Agile methods such as Extreme Programming (XP) or Scrum.

DSDM, like Crystal, views every project as different and advocates selecting the "correct level of rigor" by tailoring the approach to each project based on a risk assessment. DSDM espouses a "philosophy" supported by eight principles:

- Focus on the business need
- Deliver on time
- Collaborate
- Never compromise quality*
- Build incrementally from firm foundations
- Develop iteratively
- Communicate continuously and clearly
- Demonstrate control

DSDM identifies the following critical-to-success factors:

- Acceptance of the Atern philosophy at the outset
- Appropriate empowerment of the development team
- Commitment of a business representative to the project
- Incremental delivery
- Team access to business roles
- Team stability
- Adequate skills among the team
- Optimum team size 7 ± 2—potentially multiple teams operating in parallel
- Supportive relationship (facilitate requirements changes)

DSDM Atern Life Cycle

The pre-project phase has the following objectives:

- To describe the business problem to be addressed
- To identify a business sponsor and business visionary
- To confirm that the project is in line with business strategy
- To scope, plan, and resource the feasibility phase

The pre-project phase should be short, sharp, and ideally restricted to the creation of a short statement that has the purpose of justifying and prioritizing a feasibility investigation.

* While certainly a sound principle for larger projects, this is potentially in conflict with the "fail fast" notion prevalent in most Agile approaches—there may be tendency to over-invest in something that may not turn out to be what the customer actually wants.

Objectives of the feasibility phase include

- To establish whether there is a feasible solution to the business problem described in the terms of reference defined during pre-project
- To identify the benefits likely to arise from the delivery of the proposed solution
- To outline possible approaches for delivery, including strategies for sourcing the solution and project management
- To describe the organization and governance aspects of the project
- To state first-cut estimates of timescale and costs for the project overall
- To plan and resource the foundations phase

The feasibility phase should be kept as short and sharp as possible, remembering that its only purpose is to justify progressing to the foundations phase. The detail of the investigation happens in the foundations phase.

The foundations phase has the following objectives:

- To baseline the high-level requirements for the project and describe their priority and relevance to the business need
- To describe the business processes to be supported by the proposed solution (where appropriate)
- To identify the information used, created, and updated by the proposed solution
- To describe the strategies for all aspects of solution deployment
- To detail the business case for the project
- To start designing the solution architecture and identifying the physical or infrastructural elements of the solution
- To define technical implementation standards
- To describe how quality will be assured
- To establish appropriate governance and organization for the project
- To describe the solution development life cycle for the project along with techniques to be applied in managing the project and for demonstrating and communicating progress
- To baseline a schedule for development and deployment activities for the solution
- To describe, assess, and manage risk associated with the project

Foundation products are prepared to the level that allows the project to move into the first exploratory development phase. Regardless of whether the formal business case product is created, the justification for the project must be assessed, and a conscious decision must be taken to continue with the work beyond this phase: stopping a project with a poor business case now (too risky, too costly, low benefits, etc.) should be considered a successful outcome of the foundations phase.

Exploration phase objectives include

- To elaborate on the requirements captured and baselined in the prioritized requirements list during foundations
- To explore the full detail of the business need and provide detailed requirements for the evolving solution
- To create a functional solution that demonstrably meets the needs of the business
- To give the wider organization an early view of the solution that it will eventually operate, support, and maintain
- Where needed, to evolve the business area definition and system architecture definition products of the foundations phase into models that describe how the solution works and how it supports all impacted business processes and systems

Engineering phase objectives:

- To refine the evolving solution from the exploration phase to meet the agreed acceptance criteria
- To expand and refine any products required to successfully operate and support the solution in live operation

Deployment phase objectives:

- To confirm the ongoing performance and viability of the project and re-plan as required
- To deploy the solution (or increment of it) into the live business environment
- Where applicable, to train the end users of the solution and/or provide necessary documentation to support the live operation of the solution in the business environment
- To train and/or provide documentation for operations and support staff who will be responsible for supporting and maintaining technical aspects of the solution
- To assess whether the deployed solution is likely to enable the delivery of intended elements of business benefit described in the business case (where created)

DSDM Roles and Responsibilities

Business Sponsor

- Owning the business case for the project
- Ensuring ongoing viability of the project in line with the business case
- Ensuring that funds and other resources are made available as needed

- Ensuring that the decision-making process for escalated project issues is effective and rapid
- Responding rapidly to escalated issues

Business Visionary

- Owning the wider implications of any business change from an organizational perspective
- Defining the business vision for the project
- Communicating and promoting the business vision to all interested parties
- Monitoring the progress of the project in line with the business vision
- Contributing to key requirements, design, and review sessions
- Approving changes to the high-level requirements in the prioritized requirements list
- Ensuring collaboration across stakeholder business areas
- Ensuring business resources are available as needed
- Promoting the translation of the business vision into working practice
- Acting as a final arbiter of any disagreements between team members

Project Manager

- Communicating with senior management and the project governance authorities (business sponsor, project board, steering committee, etc.) with the frequency and formality that they deem necessary
- High-level project planning and scheduling, but not detailed task planning
- Monitoring progress against the baselined project plans
- Managing risk and any issues as they arise, escalating to senior business or technical roles as required
- Managing the overall configuration of the project
- Motivating the teams to meet their objectives
- Managing business involvement within the solution development teams
- Resourcing specialist roles as required
- Handling problems escalated from the solution development teams
- Coaching the solution development teams when they are handling difficult situations

Technical Coordinator

- Agreeing and controlling the technical architecture
- Determining the technical environments
- Advising on and coordinating each team's technical activities
- Identifying and owning architectural and other technically based risk, escalating to the project manager as appropriate

- Ensuring that the non-functional requirements are achievable and subsequently met
- Ensuring adherence to appropriate standards of technical best practice
- Controlling the technical configuration of the solution
- Managing the technical aspects of the transition of the solution into live use
- Resolving technical differences between technical team members

Team Leader

- Focusing the team to ensure an on-time delivery of the agreed products
- Encouraging full participation of team members within their defined roles and responsibilities
- Ensuring that the iterative development process is properly focused and controlled
- Ensuring that all testing and review activity is properly scheduled and carried out
- Managing risks and issues at the development timebox level, escalating to the project manager or technical coordinator as required
- Monitoring progress on a day-to-day basis for all team activities
- Reporting progress to the project manager
- Running the daily team meetings, ensuring they are timely, focused, and brief

Business Ambassador

- Contributing to all requirements, design, and review sessions
- Providing the business perspective for all day-to-day project decisions
- Providing the detail of business scenarios to help define and test the solution
- Communicating with other users, involving them as needed and getting their agreement
- Providing day-to-day assurance that the solution is evolving correctly
- Organizing and controlling business acceptance testing of the solution
- Developing business user documentation for the ultimate solution
- Ensuring that user training is adequately carried out
- Attending the daily team meetings

Business Analyst

- Ensuring that all communication between business and technical participants in the project is unambiguous and timely
- Managing development, distribution, and baseline approval of all documentation and products related to business requirements and their interpretation
- Ensuring that the business implications of all day-to-day decisions are properly thought through

Solution Developer

- Working with business roles and solution testers to iteratively develop
  - The deployable solution
  - Models required for the properly controlled development of the solution
  - Models and documentation required for the purpose of supporting the solution in live use
- Recording (and later interpreting) the detail of any
  - Changes to the detailed requirements
  - Changes to the interpretation of requirements which result in rework within the solution
  - Information likely to have an impact on the ongoing evolution of the solution
  - Adhering to technical constraints laid out in the system architecture definition
  - adhering to standards and best practice laid out in the technical implementation standards
  - Participating in any quality assurance work required to ensure that the delivered products are truly fit for purpose
  - Testing the output of their own work prior to independent testing

Solution Tester

- Working with business roles to define test scenarios and test cases for the evolving solution

In accordance with the technical testing strategy,

- Carrying out all types of technical testing of the solution as a whole
- Creating testing products, for example, test cases, plans, and logs
- Reporting the results of testing activities to the technical coordinator for quality assurance purposes
- Keeping the team leader informed of the results of testing activities
- Assisting the business ambassador(s) and business advisor(s) to ensure that they can plan and carry out their tests well enough to ensure that the important areas are covered

Business Advisor

- Requirements, design, and review activities
- Day-to-day project decisions
- Business scenarios to help define and test the solution
- Specialist advice on, or help with
  - Organizing and controlling business acceptance testing of the solution
  - Developing business user documentation for the ultimate solution
  - User training

Workshop Facilitator

For each workshop:

- Agreeing the scope of the workshop with the workshop owner
- Planning the workshop
- Familiarization with the subject area of the workshop
- Engaging with participants to
  - Confirm their suitability as a participant (in terms of knowledge and state of empowerment)
  - Ensure their full understanding of the workshop objectives
  - Understand any major areas of interest and concern in the subject area
  - Encourage completion of any required preparation work
- Facilitating the workshop to meet its objectives
- Reviewing the workshop against its objectives

Atern Coach

- Providing detailed knowledge and experience of Atern to inexperienced Atern teams
- Tailoring the Atern process to suit the individual needs of the project and the environment in which the project is operating
- Helping the team use Atern techniques and practices and helping those outside the team appreciate the Atern philosophy and value set
- Helping the team work in the collaborative and cooperative way demanded by Atern and all Agile approaches
- Building Atern capability within the team

General Observations

While in some respects DSDM may be viewed as an Agile approach, it is certainly much more "heavy-weight" than most of the alternatives. This method is likely best suited to intermediate-size projects—it is probably too formal for many small projects yet not sufficiently rigorous for large projects. While quality is ostensibly fixed within the DSDM framework, it is unlikely that this objective can in fact be achieved without specific techniques such as formal inspections and formal quality plans.

A sample DSDM project with Namcook scoring is shown in Table 18.1. The project is 1000 function points and uses the Java programming language.

DSDM in all its flavors is not well known in the United States, but has achieved some popularity in Europe. In 2016, it is also competing with other and newer methods such as container development and DevOps. In the United States, DSDM competes with Disciplined Agile Development (DAD).

Table 18.1 Example of 1000 Function Point DSDM Project with Java Language

Project Size and Nature Scores		**DSDM**
1	Small projects <100 function points	10
2	New projects	10
3	IT projects	10
4	Web/cloud projects	9
5	Medium projects	8
6	Systems/embedded projects	6
7	High-reliability, high-security projects	5
8	Enhancement projects	4
9	Large projects >10,000 function points	2
10	Maintenance projects	1
	Subtotal	65
		DSDM
1	Requirements changes	10
2	Requirements	9
3	Coding	8
4	Change control	8
5	Defect prevention	8
6	Testing defect removal	8
7	Design	6
8	Integration	6
9	Models	5
10	Test case design/generation	5
11	Pattern matching	4
12	Pre-test defect removal	4
13	Reusable components	3
14	Architecture	3

(*Continued*)

Table 18.1 (Continued) Example of 1000 Function Point DSDM Project with Java Language

15	Planning, estimating, measurement	−1
	Subtotal	86
	Total	151
Regional Usage (Projects in 2016)		**DSDM**
1	North America	500
2	Pacific	500
3	Europe	1000
4	South America/Central America	200
5	Africa/Middle East	200
	Total	2400
Defect Potentials per Function Point (1000 function points; Java language)		**DSDM**
1	Requirements	0.20
2	Architecture	0.15
3	Design	0.40
4	Code	1.25
5	Documents	0.40
6	Bad fixes	0.25
	Total	2.65
	Total defect potential	2650
Defect Removal Efficiency (DRE)		
1	Pre-test defect removal (%)	50.00
2	Testing defect removal (%)	85.00
	Total DRE (%)	92.50
	Delivered defects per function point	0.199
	Delivered defects	191
	High-severity defects	31
	Security defects	3

(Continued)

Table 18.1 (Continued) Example of 1000 Function Point DSDM Project with Java Language

Productivity for 1000 Function Points		**DSDM**
	Work hours per function point 2016	11.90
	Function points per month 2016	11.09
	Schedule in calendar months	13.50

Data source: Namcook Analysis of 600 companies.

DSDM scoring
Best = +10
Worst = −10

Chapter 19

Enterprise Resource Planning (ERP) Customization

Executive summary: *ERP packages are large and complex. Many top 250,000 function points in total size if all features are installed and used. Most companies find it necessary to customize ERP packages, which is among the most difficult forms of software development. ERP customization may make it difficult to install later ERP updates from the vendors. There is no safe methodology for ERP customization. ERP customization is made difficult by the fact that ERP packages themselves may be alarmingly buggy. The author was an expert witness in litigation filed against a major ERP vendor by the company's own shareholders, who claimed that software quality was so bad that stock values were being depressed.*

All corporations and government agencies have certain business functions in common that now use software. They need to support accounts payable and accounts receivable, personnel or human resource data, payrolls for fulltime personnel, payrolls for consulting personnel, records of customers, purchasing, shipping and receiving, marketing, sales, and inventory management, among others.

The enterprise resource planning (ERP) industry started in the late 1980s. Gartner Group used the term *enterprise resource planning* in the 1990s. Today in 2016, there are over 100 ERP vendors, including Oracle, SAP, Microsoft, NetSuite, JD Edwards, and many more. Major ERP packages are about the most expensive software that corporations acquire. But there are also open-source ERP packages, which might be used by smaller organizations that balk at the high costs of commercial packages. These open-source ERP packages include ERPNext and Dolibar. Google searches using "ERP..." will turn up many useful websites and documents.

The ERP industry is a fairly effective form of software reuse. ERP packages support all the generic functions that are used by thousands of companies in hundreds of industries. The ERP vision is that since generic functions are common across all industries and all companies, it is wasteful for each company to create its own custom software for business functions that are generic and widely used.

Based on the specific industry, there may be other common functions as well. Some of these include manufacturing support for specific industries, customer resource management (CRM), and market analysis.

Even though major ERP packages are large and top 250,000 function points, most corporate software portfolios are larger than 1,000,000 function points. The corporate portfolio for Fortune 500 corporations may top 10,000,000 function points. Most state governments own software in excess of 5,000,000 function points, as do major cities such as New York and Los Angeles. Even smaller towns have portfolios that top 500,000 function points. The U.S. federal government, when civilian and military organizations are viewed as a whole, is the world's largest user of computer software, with a national portfolio that probably tops 100,000,000 function points. Russia and China are not far behind and probably have national portfolios that approach 75,000,000 function points.

Because even the largest and most extensive ERP packages can't support 100% of corporate software functions, the majority of companies customize ERP packages during or after initial deployment.

Among the author's clients, only about 10% use vanilla ERP packages without any customization. Most companies customize their ERP packages to a greater or lesser degree. The forms of customization include changing default ERP settings using tools provided by the vendor, porting existing applications into the ERP package, and, which is most challenging, actually modifying the code in the ERP package itself. This form of customization is potentially hazardous because of the chance of voiding ERP warranties or making it difficult to install ERP update releases from the vendors.

The author's Software Risk Master (SRM) tool includes an ERP deployment and customization feature. Table 19.1 shows a list of 82 corporate software functions for a Fortune 500 manufacturing client and the number of these functions that require porting to an ERP package. The company had total employment of about 250,000. Of these, ERP users amounted to about 150,000. Therefore, training is a major expense when moving to ERP packages. Table 19.1 illustrates ports to ERP packages from legacy applications.

As can be seen in Table 19.1, corporate software is a large and eclectic group of functions in 2016. Table 19.2 shows the approximate size of the corporation's portfolio using both function points and lines of code metrics. Table 19.2 also identifies the percentage of features that are handled by commercial off-the-shelf (COTS) packages as opposed to internal development.

Table 19.1 Manufacturing Applications and ERP Ports

	Corporate Function	*ERP Port? 1 = yes; 0 = no*
1	Accounts payable	1
2	Accounts receivable	1
3	Advertising	0
4	Advisory boards: Technical	0
5	Banking relationships	1
6	Board of directors	0
7	Building maintenance	0
8	Business intelligence	1
9	Business partnerships	1
10	Competitive analysis	0
11	Consultant management	0
12	Contract management	1
13	Customer resource management	1
14	Customer support	1
15	Divestitures	1
16	Education: Customers	0
17	Education: Staff	0
18	Embedded software	0
19	Energy acquisition	0
20	Energy consumption monitoring	0
21	Engineering	0
22	ERP: Corporate	1
23	Finances (corporate)	1
24	Finances (divisional)	1

(*Continued*)

Table 19.1 (Continued) Manufacturing Applications and ERP Ports

	Corporate Function	*ERP Port? 1 = yes; 0 = no*
25	Governance	1
26	Government certification (if any)	0
27	Government regulations (if any)	0
28	Human resources	1
29	Insurance	1
30	Inventory management	1
31	Legal department	0
32	Litigation	0
33	Long-range planning	1
34	Maintenance: Buildings	0
35	Maintenance: Product	0
36	Manufacturing	1
37	Market research	0
38	Marketing	1
39	Measures: Customer satisfaction	0
40	Measures: Financial	1
41	Measures: Market share	0
42	Measures: Performance	0
43	Measures: Quality	0
44	Measures: ROI and profitability	1
45	Mergers and acquisitions	1
46	Office suites	0
47	Open-source tools: General	0
48	Order entry	1

(*Continued*)

Table 19.1 (Continued) Manufacturing Applications and ERP Ports

	Corporate Function	*ERP Port? 1 = yes; 0 = no*
49	Outside services: Legal	0
50	Outside services: Manufacturing	0
51	Outside services: Marketing	0
52	Outside services: Sales	0
53	Outside services: Terminations	0
54	Outsource management	0
55	Patents and inventions	0
56	Payrolls	1
57	Planning: Manufacturing	1
58	Planning: Products	1
59	Process management	1
60	Product design	0
61	Product nationalization	0
62	Product testing	1
63	Project management	0
64	Project offices	0
65	Purchasing	1
66	Quality control	1
67	Real estate	0
68	Research and development	0
69	Sales	1
70	Sales support	1
71	Security: Buildings	0
72	Security: Computing and software	0

(*Continued*)

Table 19.1 (Continued) Manufacturing Applications and ERP Ports

	Corporate Function	*ERP Port? 1 = yes; 0 = no*
73	Shareholder relationships	1
74	Shipping/receiving products	1
75	Software development	0
76	Standards compliance	0
77	Stocks and bonds	1
78	Supply chain management	1
79	Taxes	1
80	Travel	0
81	Unbudgeted costs: Cyber-attacks	0
82	Warranty support	1
	Portfolio Totals	**38**

As can be seen in Table 19.2, large corporations own and use a huge quantity of software in 2016. Prior to the use of computers and software, the record keeping for these business functions would probably have required over 50,000 people (the total number of personnel employed by the corporation whose data is in Table 19.2 was about 250,000).

Table 19.3 shows the approximate number of personnel involved in ERP deployment and customization. This is not an easy task, and quite a few people are involved, including external ERP consultants. In general, there were about 16 corporate personnel and 10 consultants on board during the deployment and customization process.

Table 19.4 shows ERP effort and expenses. Since the corporate total employment was about 250,000, and over 100,000 would use the ERP package, it can be seen that training is a major cost element.

Migration to an ERP environment from individual independent software applications is a big expense. But, of course, successful ERP deployment will save time and money, reducing overall maintenance expenses. ERP packages also make corporate data more accessible and more useful to C-level executives.

Table 19.5 shows the scores and quantified results of creating an ERP extension of 1000 function points in the Java language. The development methodologies were a hybrid of Agile and DevOps. Since ERP applications tend to have high use and are run often, DevOps is a frequent choice.

Table 19.2 Manufacturing Portfolio Size in Function Points and Lines of Code

	Corporate Functions	*Function Points*	*Lines of Code*	*COTS (%)*
1	Accounts payable	55,902	3,074,593	62.75
2	Accounts receivable	71,678	3,942,271	62.75
3	Advertising	62,441	2,809,867	47.75
4	Advisory boards: Technical	9,678	532,286	42.75
5	Banking relationships	175,557	9,655,657	61.75
6	Board of directors	7,093	390,118	50.75
7	Building intelligence	3,810	209,556	63.75
8	Business maintenance	94,302	5,185,625	7.75
9	Business partnerships	55,902	3,074,593	22.75
10	Competitive analysis	97,799	5,378,919	6.75
11	Consultant management	5,040	277,174	27.75
12	Contract management	124,883	6,868,564	32.75
13	Customer resource management	193,740	10,655,693	63.75
14	Customer support	90,659	4,986,251	33.75
15	Divestitures	18,017	990,928	22.75
16	Education: Customers	13,205	726,262	42.75
17	Education: Staff	7,093	390,118	45.75
18	Embedded software	359,193	30,531,426	39.75
19	Energy acquisition	7,093	390,118	49.75
20	Energy consumption monitoring	8,032	441,749	50.75
21	Engineering	437,500	32,812,500	24.75
22	ERP: Corporate	400,000	28,000,000	62.25
23	Finances (corporate)	335,247	18,438,586	60.75
24	Finances (divisional)	236,931	13,031,213	64.75
25	Governance	30,028	1,651,546	18.75

(Continued)

Table 19.2 (Continued) Manufacturing Portfolio Size in Function Points and Lines of Code

	Corporate Functions	*Function Points*	*Lines of Code*	*COTS(%)*
26	Government certification (if any)	45,764	2,517,025	35.75
27	Government regulations (if any)	24,583	1,352,043	37.75
28	Human resources	13,205	726,262	66.75
29	Insurance	10,298	566,415	68.75
30	Inventory management	90,659	4,986,251	47.75
31	Legal department	45,764	2,517,025	58.75
32	Litigation	62,441	3,434,282	63.75
33	Long-range planning	22,008	1,210,437	10.75
34	Maintenance: Buildings	158,606	8,723,313	29.75
35	Maintenance: Product	9,678	725,844	54.75
36	Manufacturing	470,014	35,251,071	22.75
37	Market research	75,239	4,138,139	17.75
38	Marketing	51,821	2,850,144	55.75
39	Measures: Customer satisfaction	7,093	390,118	19.75
40	Measures: Financial	45,764	2,517,025	49.25
41	Measures: Market share	14,952	822,381	61.25
42	Measures: Performance	16,931	931,220	26.25
43	Measures: Quality	18,017	990,928	7.25
44	Measures: ROI and profitability	62,441	3,434,282	30.25
45	Mergers and acquisitions	76,273	4,195,041	19.75
46	Office suites	34,889	1,918,888	68.75
47	Open-source tools: General	140,068	7,703,748	67.75
48	Order entry	51,821	2,850,144	32.75
49	Outside services: Legal	45,764	2,517,025	42.75
50	Outside services: Manufacturing	86,368	4,750,241	30.25

(Continued)

Table 19.2 (Continued) Manufacturing Portfolio Size in Function Points and Lines of Code

	Corporate Functions	*Function Points*	*Lines of Code*	*COTS(%)*
51	Outside services: Marketing	27,836	1,530,981	41.75
52	Outside services: Sales	31,520	1,733,601	27.75
53	Outside services: Terminations	13,259	729,258	66.75
54	Outsource management	62,441	3,434,282	20.25
55	Patents and inventions	35,692	1,963,038	11.75
56	Payrolls	67,359	3,704,732	24.75
57	Planning: Manufacturing	85,197	4,685,808	17.25
58	Planning: Products	18,017	990,928	23.75
59	Process management	21,709	1,194,018	11.75
60	Product design	193,740	10,655,693	57.75
61	Product nationalization	36,874	2,028,064	10.75
62	Product testing	75,239	4,138,139	27.75
63	Project management	72,848	4,006,662	49.75
64	Project offices	33,031	1,816,701	59.75
65	Purchasing	58,679	3,227,351	67.75
66	Quality control	24,583	1,352,043	23.75
67	Real estate	14,952	822,381	37.75
68	Research and development	537,321	29,552,650	7.75
69	Sales	90,659	4,986,251	33.75
70	Sales support	27,836	1,530,981	39.75
71	Security: Buildings	40,415	2,222,839	8.75
72	Security: Computing and software	145,697	8,013,325	33.75
73	Shareholder relationships	34,889	1,918,888	17.75
74	Shipping/receiving products	86,368	4,750,241	27.75

(Continued)

Table 19.2 (Continued) Manufacturing Portfolio Size in Function Points and Lines of Code

	Corporate Functions	*Function Points*	*Lines of Code*	*COTS(%)*
75	Software development	337,550	18,565,265	51.75
76	Standards compliance	24,583	1,352,043	55.75
77	Stocks and bonds	94,302	5,186,625	59.75
78	Supply chain management	96,472	5,305,959	14.75
79	Taxes	141,994	7,809,679	45.75
80	Travel	30,028	1,651,546	67.75
81	Unbudgeted costs: Cyber-attacks	114,476	6,296,184	33.75
82	Warranty support	11,661	641,378	17.75
	Portfolio Totals	**7,268,513**	**434,263,439**	**38.95**

Table 19.3 ERP Deployment Activities and Personnel

	Deployment Activities	*Internal Team*	*Consultants*	*Total*
1	Business analysis	32.30	8.08	40.38
2	Requirements analysis	16.00	6.46	22.46
3	ERP selection	9.69	6.78	16.48
4	ERP installation	5.33	11.73	17.07
5	ERP customization	13.33	16.00	29.33
6	Legacy application ports	6.00	1.80	7.80
7	Data migration	16.15	3.23	19.38
8	Testing	18.00	9.00	27.00
9	ERP deployment management	3.65	1.97	5.62
10	Administration	2.56	0.64	3.19
11	Cutover	11.83	7.10	18.93
	Subtotal	**12.30**	**6.62**	**18.88**
12	User training instructors	145.10	7.25	152.35
13	User training students	29,019	1,451	30,470

Table 19.4 ERP Deployment Schedules and Costs

	Deployment Activities	*Schedule Months*	*Effort Months*	*Costs ($)*
1	Business analysis	2.58	104.20	1,041,981
2	Requirements analysis	1.57	35.28	352,776
3	ERP selection	0.94	15.53	155,259
4	ERP installation	0.61	10.45	104,541
5	ERP customization	7.88	231.04	2,310,420
6	Legacy application ports	6.5	50.70	507,000
7	Data migration	0.85	16.51	165,121
8	Testing	9.30	251.10	2,511,000
9	ERP deployment management	8.73	49.07	490,695
10	Administration	6.98	22.30	223,037
11	Cutover	1.14	21.53	215,314
	Subtotal	**15.16**	**807.71**	**8,077,144**
12	User training instructors	1.38	210.33	**2,103,293**
13	User training students	0.24	7361.53	**73,615,257**
	Total	**Total**	**Total**	**83,795,694**

Customizing ERP packages is somewhat sluggish in terms of work hours per function point and has a fairly long schedule in calendar months. Quality is also only marginal, due in part to the latent bugs that may reside in the ERP package itself, which are hard for ERP clients to discover or test.

ERP packages are valuable and cost-effective compared with clients trying to build the same features with custom software. But ERP vendors are not as effective at software quality control as they should be. Latent bugs in ERP packages slow down deployment and make customization difficult.

Table 19.5 Sample of 1000 function point ERP modification project with Java Language

Project Size and Nature Scores		ERP Custom
1	Small projects <100 function points	6
2	Medium projects	7
3	Large projects >10,000 function points	8
4	New projects	10
5	IT projects	10
6	Systems/embedded projects	–1
7	Web/cloud projects	6
8	High-reliability, high-security projects	7
9	Enhancement projects	8
10	Maintenance projects	6
	Subtotal	67
Project Activity Scores		**ERP Custom**
1	Pattern matching	9
2	Reusable components	10
3	Planning and estimating	5
4	Requirements	7
5	Architecture	5
6	Models	3
7	Design	7
8	Coding	8
9	Defect prevention	5
10	Test case design/generation	7
11	Testing defect removal	8
12	Requirements changes	8
13	Change control	9

(Continued)

Table 19.5 (Continued) Sample of 1000 function point ERP modification project with Java Language

14	Integration	9
15	Pre-test defect removal	4
	Subtotal	104
	Total	171
Regional Usage (ERP Customization 2016)		**ERP Custom**
1	North America	17,000
2	Pacific	15,000
3	Europe	16,000
4	South America/Central America	8,000
5	Africa/Middle East	4,000
	Total	60,000
Defect Potentials per Function Point (1000 function points; Java language)		**ERP Custom**
1	Requirements	0.65
2	Architecture	0.30
3	Design	0.75
4	Code	1.10
5	Documents	0.35
6	Bad fixes	0.55
	Total	3.70
	Total defect potential	3,700
Defect Removal Efficiency (DRE)		
1	Pre-test defect removal (%)	50.00
2	Testing defect removal (%)	83.00
	Total DRE (%)	91.49
	Delivered defects per function point	**0.352**
	Delivered defects	352

(Continued)

Table 19.5 (Continued) Sample of 1000 function point ERP modification project with Java Language

	High-severity defects	52
	Security defects	8
Productivity for 1000 Function Points		**ERP Custom**
	Work hours per function point 2016	17.2
	Function points per month 2016	7.67
	Schedule in calendar months	16.5

Data source: Namcook Analysis of 600 companies.

ERP customization scores
Best = +10
Worst = −10

Chapter 20

Evolutionary Development (EVO) Methodology

Executive summary: *The evolutionary development (EVO) methodology was developed by the well-known software researcher and author Tom Gilb in the 1970s. Since Tom is British by birth and now lives in Norway, he speaks at many European events. As a result, EVO is more widely used in Europe than elsewhere. However, EVO concepts were found in IBM's Cleanroom development methodology and are also used by Hewlett Packard for their internal software development. Gilb asserts that EVO is embedded into military standard MIL-498 and International Organization for Standardization (ISO) standard ISO-12207. Some features of EVO have found their way into the Agile family, such as designing and building software in small packages with user value rather than as one large application.*

Evolutionary development (EVO) is one of the older software methodologies. It was first formulated in the 1970s and then described more fully in the 1980s. Tom Gilb, the well-known software author and researcher, is the EVO developer.

Some of the EVO concepts migrated into other methodologies, such as Cleanroom and even Agile. EVO was one of the first to articulate the concept of building software in small increments rather than full systems. An image from the web illustrates that EVO was among the first of the later "iterative/agile" methodologies, as shown in Figure 20.1

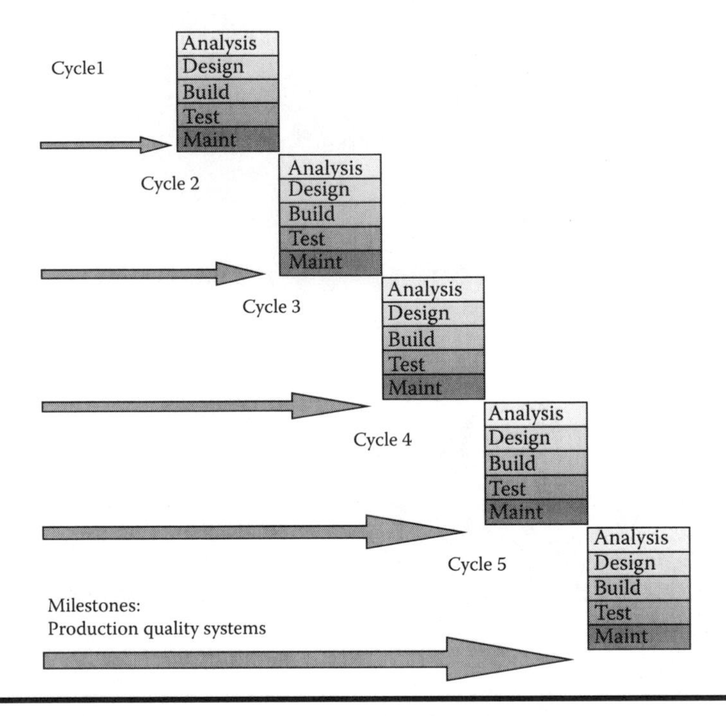

Figure 20.1 Illustration of evolutionary development (EVO).

Tom Gilb has published many papers and given many talks about EVO. In particular, EVO concentrates on 10 core principles:

1. EVO should provide real results of value to stakeholders, and do so frequently.
2. The next EVO delivery should provide maximum value to stakeholders.
3. EVO builds applications by evolution and not as a huge mass.
4. All requirements are not known at first, but they can be discovered by delivering continuous value.
5. EVO is holistic, and all deliverables should be valuable, not just code.
6. EVO projects need an open architecture, since there will be unknown changes.
7. The EVO project team should focus on the current step, not unknown future steps.
8. EVO is about learning from experience and building on successful iterations.
9. EVO should lead to on-time or early product delivery.
10. EVO should let teams prove successful work processes and eliminate bad ones early.

These 10 essential concepts are congruent with Agile, extreme programming (XP), iterative, and others in the Agile family. However, EVO also shares some of the weaknesses of the Agile family, including, but not limited to

1. Difficulty in estimating the full project
2. Poor measurement of quality
3. Poor measurement of productivity
4. No special strength in architecture
5. No special strength in design
6. The patterns and results of similar historical projects being ignored
7. The huge benefits of standard reusable components being ignored

Because Tom is interested in quality and has written books on formal inspections, many EVO projects (unlike other Agile variants) tend to use inspections; this is a good thing.

Under its own name, EVO is more widely used in Europe than in the United States. But by loaning concepts to other Agile methods, EVO has been one of the more influential methodologies.

Table 20.1 shows the results of building a 1000 function point EVO project using the Java programming language.

Table 20.1 Example of 1000 Function Point EVO Project with Java Language

Project Size and Nature Scores		**EVO**
1	IT projects	10
2	Small projects <100 function points	9
3	New projects	9
4	Web/cloud projects	9
5	Medium projects	7
6	Systems/embedded projects	5
7	Enhancement projects	4
8	High-reliability, high-security projects	4
9	Maintenance projects	1
10	Large projects >10,000 function points	−1
	Subtotal	57
Project Activity Scores		**EVO**
1	Requirements	9
2	Requirements changes	9

(*Continued*)

Table 20.1 (Continued) Example of 1000 Function Point EVO Project with Java Language

3	Coding	8
4	Change control	8
5	Defect prevention	8
6	Testing defect removal	8
7	Design	6
8	Integration	6
9	Models	5
10	Test case design/generation	5
11	Pattern matching	4
12	Pre-test defect removal	4
13	Reusable components	3
14	Architecture	3
15	Planning, estimating, measurement	–1
	Subtotal	85
	Total	142
Regional Usage (Projects in 2016)		**EVO**
1	North America	2,000
2	Pacific	1,000
3	Europe	13,000
4	South America/Central America	1,000
5	Africa/Middle East	500
	Total	17,500
Defect Potentials per Function Point (1000 function points; Java language)		**EVO**
1	Requirements	0.30
2	Architecture	0.15

(Continued)

Table 20.1 (Continued) Example of 1000 Function Point EVO Project with Java Language

3	Design	0.45
4	Code	1.30
5	Documents	0.40
6	Bad fixes	0.30
	Total	2.90
	Total defect potential	2,900
Defect Removal Efficiency (DRE)		
1	Pre-test defect removal (%)	85.00
2	Testing defect removal (%)	85.00
	Total DRE (%)	97.80
	Delivered defects per function point	0.065
	Delivered defects	64
	High-severity defects	10
	Security defects	1
Productivity for 1000 Function Points		**EVO**
	Work hours per function point 2016	13.46
	Function points per month 2016	9.80
	Schedule in calendar months	14.30

Data source: Namcook Analysis of 600 companies.

EVO scores
Best = +10
Worst = −10

EVO concepts have found their way into a number of other software development methodologies, such as the Agile family. Because the EVO inventor, Tom Gilb, lives in Norway and speaks at many events in Europe, the EVO method is more popular and better known in Europe than in the United States. EVO has some good concepts and better than average quality results.

Chapter 21

Extreme Programming (XP)

Executive summary: *According to Kent Beck, the author of Extreme Programming, "XP is about social change." It is "a philosophy of software development based on the values of communication, feedback, simplicity, courage, and respect." All of these are very consistent with most other Agile approaches. XP is a "light-weight" method that is focused primarily on developer tasks rather than a full life-cycle perspective. The most significant distinguishing feature of XP is its preference for "pair programming." See the discussion of pair programming earlier in this book.*

Extreme programming (XP) is defined in three dimensions: value, principles, and practices (primary and corollary):

Values

- Communication: Creating a sense of team and effective cooperation
- Simplicity: Finding the simplest approach that can achieve the desired outcome, avoiding unnecessary complexity
- Feedback: Change is inevitable; it is essential to provide opportunities for feedback early and often
- Courage: To speak truth even when unpleasant fosters communication and trust
- Respect: Care about the others on the team; value all contributions

Principles

- Humanity: An environment that meets basic human needs: freedom from fear, a sense of accomplishment, a feeling of belonging, opportunities to grow, a feeling of being understood.
- Economics: Ensuring that development effort delivers real business value, recognizing the time value of money.
- Mutual benefit: Always seek win-win solutions to conflicts.
- Self-similarity: Maximize potential for reuse; re-purpose existing software assets when practical to do so.
- Improvement: Do the best you can today; improve it tomorrow.
- Diversity: Multiple perspectives generate opportunities, but potential for conflict must be managed.
- Reflection: Ensure opportunities to consider how and why work is being done; learn from both successes and failures.
- Flow: Leverage the Lean principle of flow; the daily build is an example of flow in practice.
- Opportunity: View obstacles as opportunities.
- Redundancy: For example, address defects with multiple strategies (pair programming, test-driven development, daily builds, intense customer involvement); don't rely on a single strategy to solve difficult problems.
- Failure: A central tenet of many Agile methods is "fail fast;" if you're not sure, try something, see if it works.
- Quality: According to Beck, "If you don't know a clean way to do a job that has to be done, do it the best way you can. If you know a clean way but it would take too long, do the job as well as you have time for now. Resolve to finish doing it the clean way later." ***Comment:*** *This notion is consistent with many Agile approaches, in which schedule is king, and goes a long way toward explaining why many Agile methods deliver poor quality and have high maintenance costs. This philosophy is ineffective and downright destructive in larger projects.*
- Baby steps: Develop incrementally in small chunks.

Primary Practices

- Sit together: Co-locate the team in an open work area for at least a significant portion of each work day to facilitate communication and cooperation.
- Whole team: Include everyone, create cross-functional teams.
- Informative workspace: Leverage the Lean concept of visible status and potentially Kanban.

- Energized work: Limit work hours to what is productive; excessive overtime is counter-productive. Avoid burnout.
- Pair programming: Write all programs using two people sitting at one machine, but work alone when that is felt to be necessary. Pairs rotate keying and guiding/observing.
- Pairing and personal space: Cultural and potentially gender differences may impact how close is comfortable. Sensitivity in setting up pairs that are comfortable together is very necessary.
- Stories: Use the user stories approach to define small increments of functionality.
- Weekly cycle: Plan work one week at a time. Conduct a weekly meeting to review progress from the prior week, have customers select a week's worth of stories for the upcoming week, and break those stories into tasks for assignment to team members; team members estimate the tasks.
- Quarterly cycle: Plan a quarter at a time at a higher level; select "themes" consisting of sets of related stories to be created ahead of the weekly cycle.
- Slack: Include minor tasks in the plan that can be dropped or deferred if the planned schedule is not being met.
- Ten-minute build: Automate build and test so that the process can run in 10 minutes to ensure that the process is used with great frequency.
- Continuous integration: Run the automated build/test every couple of hours to spot and correct issues early and often.
- Test first programming: Usually known as *test-driven development* (TDD). The test first approach highlights the need for highly cohesive and lightly coupled code.
- Incremental design: Code a little, design a little; orthogonal to the older "design first" paradigm.

Corollary Practices

- Real customer involvement: Customers co-located with the team, ideally a dedicated customer representative.
- Incremental deployment: Avoid "big-bang" deployment, transition incrementally.
- Team continuity: Keep the core team together, avoid task switching.
- Root cause analysis: Use the "five-whys" concept to determine the root cause of identified defects.
- Shared code: All code is "public" and anyone can improve anything at any time.

- Code and tests: These are the only permanent artifacts; generate any other required documentation for the code and tests. According to Beck, "everything else is waste." ***Comment:*** *In certain environments, such as military systems and medical devices, this would not be acceptable.*
- Single code base: Temporary branches for developers never exist for more than a few hours.
- Daily deployment: Move new software into production every evening. ***Comment:*** *This is obviously impractical in many circumstances.*
- Negotiated scope contract: As with other Agile methods, cost and time are fixed, while scope or functionality is variable.

The pair programming aspect of extreme programming (XP) is extremely expensive. After all, two people are being paid to do the work of one. The literature on pair programming is embarrassingly bad. Cowboy programmers are compared with pairs, and there is no discussion of static analysis, testing, or other forms of collaboration.

Also, with a total of 126 different occupation groups, the pair programming concept is only applied to programmers and not to business analysts, testers, architects, database analysts, or any of the other skill positions. The total programming staff on a large system in 2013 comprises less than 30% of the full team. It is sociologically unsound to double up one occupation and ignore 125 other occupations.

Pair programming exists because the software industry does not measure well and does not know how to achieve high quality levels. Both inspections and static analysis are more cost-effective than pair programming and usually better in quality too.

Table 21.1 shows the XP scores and the results of using XP and pair programming on an application of 1000 function points in the Java language. The low productivity rate and long schedule are the results of the unfortunate use of pair programming.

XP has some interesting concepts but also includes the very expensive pair programming concept. Pair programming is a good example of the way software pushes out methodologies with zero validation or proof of success. None of the pair studies bothered to include static analysis, inspections, requirements models, or any other effective quality techniques. All you get is a comparison of one unaided cowboy programmer against two unaided cowboy programmers.

Also, with a total of 126 occupations and programming being less than 30% of large system costs, XP is only a partial methodology centering on coding but not dealing with topics such as business analysis, architecture, and other technical issues of large systems. It is OK for small projects but not recommended for large systems, where more holistic approaches such as Rational Unified Process (RUP) and team software process (TSP) handle a wider range of issues.

Table 21.1 Example of 1000 Function Point XP Project with Java Language

Project Size and Nature Scores		XP
1	Small projects <100 function points	8
2	Enhancement projects	8
3	Maintenance projects	8
4	New projects	7
5	Web/cloud projects	7
6	IT projects	5
7	High-reliability, high-security projects	5
8	Medium projects	4
9	Systems/embedded projects	4
10	Large projects >10,000 function points	–3
	Subtotal	53
Project Activity Scores		**XP**
1	Reusable components	9
2	Coding	7
3	Defect prevention	7
4	Change control	6
5	Integration	5
6	Pre-test defect removal	5
7	Test case design/generation	5
8	Testing defect removal	5
9	Requirements changes	4
10	Pattern matching	3
11	Requirements	3
12	Architecture	3
13	Models	3
14	Design	3

(*Continued*)

Table 21.1 (Continued) Example of 1000 Function Point XP Project with Java Language

15	Planning, estimating, measurement	–1
	Subtotal	67
	Total	120
Regional Usage (Projects in 2016)		**XP**
1	North America	4,000
2	Pacific	3,000
3	Europe	4,000
4	South America/Central America	2,000
5	Africa/Middle East	1,000
	Total	14,000
Defect Potentials per Function Point (1000 function points; Java language)		**XP**
1	Requirements	0.3
2	Architecture	0.3
3	Design	0.6
4	Code	1.1
5	Documents	0.4
6	Bad fixes	0.2
	Total	2.9
	Total defect potential	2,900
Defect Removal Efficiency (DRE)		
1	Pre-test defect removal (%)	70.00
2	Testing defect removal (%)	85.00
	Total DRE (%)	95.50
	Delivered defects per function point	0.131
	Delivered defects	131
	High-severity defects	20
	Security defects	2

(*Continued*)

Table 21.1 (Continued) Example of 1000 Function Point XP Project with Java Language

Productivity for 1000 Function Points		**XP**
	Work hours per function point 2016	16.50
	Function points per month 2016	8.00
	Schedule in calendar months	15.20

Data Source: Namcook Analysis of 600 companies.

XP scoring
Best = +10
Worst = −10

Chapter 22

Feature-Driven Development (FDD)

Executive summary: *The methodology of "feature-driven development" is usually placed in the Agile family. However, the early identification of features also overlaps pattern-based development and model-based development. Feature-driven development has quite a few examples of successful deployment and no major failures to report. This combination of many successes and few failures puts feature-driven development somewhat ahead of conventional Agile development, where failures are increasing over time as Agile is pushed past its natural limits.*

The idea of feature-driven development (FDD) originated in Singapore in 1997. A software expert named Jeff DeLuca developed the concept for a major banking application with a team of 50 personnel. FDD is widely discussed in the literature and has been recommended by the software researcher Peter Coad.

FDD starts with an overall model of the entire system. Then, the full model is decomposed into a set of specific features based in part on the business value the features provide to customers. Once the feature list is complete and has been formally reviewed, the feature sets are assigned to various development groups. For small projects, there may be only one development group, but for large systems, there may be 10 or more.

Project planning and estimating are done by feature instead of by the full system. This adds rigor to planning and estimating. It is much more accurate to size and estimate pieces of a system and then assemble an overall estimate than it is to try to slice a big project into segments like slicing a stick of bologna.

FDD also uses frequent inspections, which gives FDD applications unusually high quality, since inspections have the highest defect removal efficiency (DRE) of any removal method: about 85%.

There are many commercial software tools that support FDD, and a few open-source tools as well. For example, CAST spec is a design tool aimed at FDD. An open-source package called FDD Tools also supports the method.

There are also many images on the web of FDD, including the one by Scott Ambler shown in Figure 22.1.

Note the frequent use of walk-throughs and formal inspections. Both of these have empirical data showing significant quality improvements.

The only notable gap in FDD is the lack of explicit recognition of reusable components. Ideally, FDD projects would seek out reusable materials early on and nominate some features as candidates for a reusable library.

The feature lists should also be screened and evaluated for inclusion in a global taxonomy of software application features: that would be extremely valuable to the industry as a whole.

Table 22.1 shows the FDD results for an application of 1000 function points and the Java programming language.

FDD is not as well known as the Agile family of methodologies, but it seems to be successful for larger applications that are stressful for Agile. FDD is also capable of being used on systems and embedded software, which are outside the normal range of the Agile family. Overall, FDD is a useful methodology, which might lead to even better results if the features could be cataloged and eventually turned into certified reusable features.

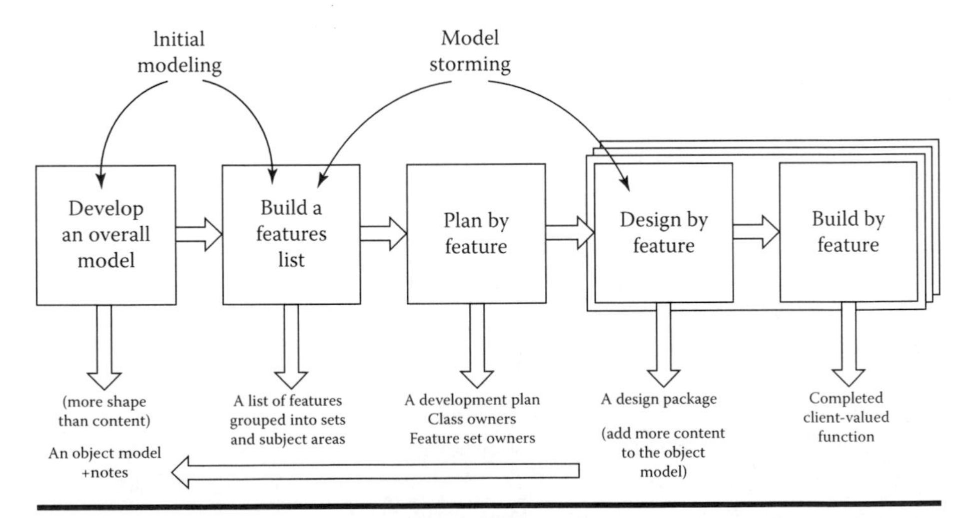

Figure 22.1 Illustration of feature-driven development (FDD). (Copyright 2002–2005 Scott W. Ambler. Original copyright S.R. Palmer and J.M. Felsing.)

Table 22.1 Example of 1000 Function Point FDD Project with Java Language

Project Size and Nature Scores		**Feature Driven**
1	Small projects <100 function points	10
2	New projects	10
3	IT projects	10
4	Systems/embedded projects	10
5	Web/cloud projects	10
6	Medium projects	9
7	Large projects >10,000 function points	8
8	High-reliability, high-security projects	6
9	Enhancement projects	5
10	Maintenance projects	2
	Subtotal	80
Project Activity Scores		**Feature Driven**
1	Pattern matching	10
2	Reusable components	10
3	Models	10
4	Design	10
5	Test case design/generation	10
6	Testing defect removal	9
7	Architecture	8
8	Change control	8
9	Integration	8
10	Defect prevention	8
11	Requirements	7
12	Requirements changes	7
13	Coding	7

(*Continued*)

Table 22.1 (Continued) Example of 1000 Function Point FDD Project with Java Language

14	Pre-test defect removal	7
15	Planning, estimating, measurement	5
	Subtotal	124
	Total	204
Regional Usage (Projects in 2016)		**Feature Driven**
1	North America	5,000
2	Pacific	4,000
3	Europe	5,000
4	South America/Central America	2,000
5	Africa/Middle East	500
	Total	16,500
Defect Potentials per Function Point (1000 function points; Java language)		**Feature Driven**
1	Requirements	0.2
2	Architecture	0.1
3	Design	0.3
4	Code	1.25
5	Documents	0.4
6	Bad fixes	0.2
	Total	2.45
	Total Defect Potential	2,450
Defect Removal Efficiency (DRE)		
1	Pre-test defect removal (%)	85.00
2	Testing defect removal (%)	90.00
	Total DRE (%)	98.50
	Delivered defects per function point	0.037
	Delivered defects	37

(Continued)

Table 22.1 (Continued) Example of 1000 Function Point FDD Project with Java Language

	High-severity defects	6
	Security defects	1
Productivity for 1000 Function Points		**Feature Driven**
	Work hours per function point 2016	11.20
	Function points per month 2016	11.80
	Schedule in calendar months	13.50

Data Source: Namcook analysis of 600 companies.

Feature-driven scoring
Best = +10
Worst = −10

Chapter 23

Git Software Development

Executive summary: *The name "Git" is British slang for an annoying person. Linus Torvalds developed Git in 2005 and gave it the name. (He said he was egotistical and named the project after himself.) Strange names are fairly common in the software industry, as can be seen from Yahoo and Google. Git is not a full software development methodology, as is Agile or team software process (TSP). Rather, Git is a collection of useful tools for keeping track of updates to code and documentation and also to assist in bug tracking. Git is somewhat agnostic, and it can be used in conjunction with most of the 60 methodologies discussed in this book. Git is an open-source product that is available for free, so its use is rapidly expanding among the software development communities, especially for Linux and Unix developers.*

The history of Git is interesting. Linus Torvalds and other developers of the Linux kernel up until 2005 had been using a free version of a source control system (SCM) called BitKeeper. The copyright holder of BitKeeper decided to withdraw the free version, claiming that it was being reverse engineered. That left Linus and colleagues without an effective version control package.

Linus wanted a combination of fast performance and a full feature set. Git has evolved rapidly since 2005, and it has had many other contributors. Although created for Linux, it now supports other operating systems such as Solaris, OS X, Windows, and others. Microsoft itself created a Git port for Windows.

This chapter is not a tutorial on Git features, but a Google search on the term "Git" will turn up quite a few. Rather, this chapter is a quantified comparison of Git development with 59 other development methodologies and platforms.

Git also supports a variety of common programming languages, including C, Haskell, Java, Ruby, and Python, and more are being added as Git evolves.

The expansion of Git has been so rapid that as of 2016, a survey found that about 43% of all programmers now use Git as their primary source control system. That number is likely to top 50% by 2017. Any method or tool set used by half of the world's programming population is a topic that needs to be understood by the software community, including C-level executives.

Security has been an issue with Git, and in 2014, a serious vulnerability was found in the Windows and Mac versions of Git. However, various patches and repairs were made by 2015. In any case, Git also needs external security measures in addition to those in Git itself.

Git has become a useful tool for software development teams, since it allows changes to be made individually and experimented with and then carefully promotes them into the official version of the application. Git is also a preferred tool among the open-source software community, since Git features are congruent with the open-source philosophy.

Table 23.1 shows the Git scores and the productivity and quality results of building a 1000 function point application using Git source control tools.

It can be seen that Git is the most widely used methodology or methodology control system in the world. Because Git speeds up software development, productivity has benefited. However, Git quality and Git security are not quite as impressive, although improvements in both topics have been made recently.

There is so much competition in the software tools world that it is something of a surprise to see a tool set such as Git achieve a market penetration of almost half the programmers in the United States and Europe between 2005 and 2016. Other older tools and software that had a similar rapid growth among the software community include DOS, Windows, Basic, and Android. However, as of 2016, it appears that Git use is expanding even faster. When use grows as fast as it has with Git, it usually means that a major weakness is being remedied. That seems to be what Git offers in terms of version controls.

Table 23.1 Example of 1000 Function Point Git Project with Java Language

Project Size and Nature Scores		Git
1	Small projects <100 function points	10
2	Medium projects	8
3	Large projects >10,000 function points	8
4	New projects	10
5	IT projects	10
6	Systems/embedded projects	8
7	Web/cloud projects	10

(*Continued*)

Table 23.1 (Continued) Example of 1000 Function Point Git Project with Java Language

8	High-reliability, high-security projects	6
9	Enhancement projects	3
10	Maintenance projects	1
	Subtotal	74
Project Activity Scores		**Git**
1	Pattern matching	10
2	Reusable components	10
3	Planning and estimating	6
4	Requirements	7
5	Architecture	6
6	Models	6
7	Design	6
8	Coding	10
9	Defect prevention	7
10	Test case design/generation	5
11	Testing defect removal	7
12	Requirements changes	10
13	Change control	10
14	Integration	9
15	Pre-test defect removal	6
	Subtotal	115
	Total	189
Regional Usage (Git projects 2016)		**Git**
1	North America	550,000
2	Pacific	450,000
3	Europe	600,000

(Continued)

Table 23.1 (Continued) Example of 1000 Function Point Git Project with Java Language

4	South America/Central America	400,000
5	Africa/Middle East	200,000
	Total	2,200,000
Defect Potentials per Function Point (1000 function points; Java language)		**Git**
1	Requirements	0.40
2	Architecture	0.30
3	Design	0.45
4	Code	0.95
5	Documents	0.25
6	Bad fixes	0.40
	Total	2.75
	Total defect potential	2,750
Defect Removal Efficiency (DRE)		
1	Pre-test defect removal (%)	65.00
2	Testing defect removal (%)	85.00
	Total DRE (%)	94.70
	Delivered defects per function point	0.145
	Delivered defects	142
	High-severity defects	22
	Security defects	2
Productivity for 1000 Function Points		
	Work hours per function point 2016	13.07
	Function points per month 2016	10.10
	Schedule in calendar months	13.80

Data Source: Namcook analysis of 600 companies.

Git scoring
Best = +10
Worst = –10

Chapter 24

Global 24 Hour Development

Executive summary: This method uses at least three physical locations that are located 8 hours apart in three different countries. At the end of each prime work shift, all the project materials are sent at 5 p.m. to the next location, moving from east to west around the world. Development never stops. This method is more difficult and more labor-intensive than other methods, but can shorten development schedules by 66% in theory and about 50% in practice. It is used primarily for large systems > 10,000 function points with extremely short deadlines. Surprisingly, 24 hour development frequently occurs with open-source projects. It is then called "internet speed development."

A similar idea would be to develop applications around the clock in one location using three shifts of software engineering personnel. This is more expensive due to overtime and also difficult to staff, since few people want to do software on second or third shifts.

The concept of continuous 24 hour development using geographically dispersed development locations can only work for software, because instantaneous delivery of work products to the next location via the Internet is needed, and only software can do this.

Continuous 24 hour development is technically more difficult than single-location development. It is also more expensive and more labor-intensive. It requires much better requirements and design than normal, and excellent broadband communications among all locations. This method raises costs compared with single-site development but can compress schedules by 66% in theory and over 50% in practice. Thus, it is a method best used for large systems with short deadlines. For example, it would be very useful for financial applications that are trying to bring out new products.

The idea of 24 hour development is not appropriate for small projects or for those with significant cost constraints. It is appropriate for large systems with short deadlines, including, but not limited to,

1. Large operating systems
2. Large telecommunication systems
3. Large government systems (outsourced to global service companies)
4. Large defense systems
5. Large financial systems introducing new concepts
6. Enterprise resource planning (ERP) packages
7. Big data repositories
8. Large defense software applications

Because this method is trading shorter schedules for higher costs, it is appropriate to look at typical results for a large 10,000 function point system built in a single location compared with building it using 24 hour, three-location development (see Table 24.1).

It can be seen that there is a substantial increase in project effort but a significant reduction in project schedule.

Companies that can use this method are large global companies, including IBM, Microsoft, Intel, Facebook, Google, SAP, Oracle, Siemens, Huawei, and quite a few others.

Smaller companies might use the approach if they use global outsource vendors, such as Accenture, Computer Aid, Inc., IBM, Infosys, and quite a few others.

(An exception to the rule that 24 hour development is aimed at large systems is the open-source community. In this case, independent developers all around the world may be working on the same application at the same time. See the chapter on "Open-Source Development" for additional information.)

To use three-location development, the last 30 minutes of each work day must be spent in contact with the next location to be sure that the next developers in line are up to speed on today's progress and technical changes. This will probably be done using intranets or webinar tools or services such as Skype or Webex or something similar. At each shift change, a kind of "mini scrum" will occur as the work products are moved from one time zone to the next.

The three-location method requires very solid architecture and design and also very good defect prevention and pre-test defect removal, including static analysis of each day's code changes. Since test cases and test scripts will either be transferred with other materials or perhaps run in one of the three locations, formal testing by certified test personnel is mandatory for success.

Reusable materials are congruent with global development and can lower costs as well as making schedules even shorter. Refer to the "Reuse-Based Development" chapter of this book for additional information.

Table 24.1 Impact of Multiple Locations on Productivity

Staff Months for One Location vs. Three Locations (Work hours per function point for 10,000 function points)			
	One Location	*Three Locations*	*Effort Difference*
Requirements	1.00	2.00	1.00
Architecture	2.00	3.00	1.00
Design	3.00	4.00	1.00
Coding	5.00	5.00	0.00
Testing	4.00	5.00	1.00
Integration	2.00	3.00	1.00
Documentation	1.00	1.00	0.00
Project management	2.00	3.00	1.00
Total	20.00	26.00	6.00
Calendar Months for One Location vs Three Locations			
Requirements	2.00	1.00	–1.00
Architecture	2.00	1.00	–1.00
Design	3.00	2.00	–1.00
Coding	7.00	3.50	–3.50
Testing	6.00	2.50	–3.50
Integration	2.00	1.00	–1.00
Documentation	2.00	1.00	–1.00
Project management	24.00	12.00	–12.00
Total	24.00	12.00	–12.00

Table 24.2 shows the scores and quantitative results for global three-location development. To be consistent with other examples, 1000 function points and the Java language are assumed, although in real life, global development applications are almost always over 10,000 function points in size, and some top 100,000 function points.

Because of the rather high costs of this methodology and the need for complex logistics involving at least three separate development centers located eight time

Table 24.2 Example of 1000 Function Point 24-Hour Global Project with Java Language

Project Size and Nature Scores		**Global 24 Hour**
1	Large projects >10,000 function points	10
2	New projects	10
3	IT projects	10
4	Systems/embedded projects	10
5	Medium projects	7
6	Web/cloud projects	6
7	Small projects <100 function points	5
8	Enhancement projects	4
9	High-reliability, high-security projects	4
10	Maintenance projects	3
	Subtotal	69
Project Activity Scores		**Global 24 Hour**
1	Pattern matching	10
2	Reusable components	10
3	Planning and estimating	10
4	Architecture	10
5	Models	10
6	Coding	10
7	Change control	10
8	Integration	10
9	Requirements	8
10	Pre-test defect removal	8
11	Test case design/generation	8
12	Testing defect removal	8
13	Defect prevention	7
14	Design	5

(*Continued*)

Table 24.2 (Continued) Example of 1000 Function Point 24-Hour Global Project with Java Language

15	Requirements changes	4
	Subtotal	128
	Total	197
Regional Usage (Projects in 2016)		**Global 24 Hour**
1	North America	25
2	Pacific	35
3	Europe	20
4	South America/Central America	10
5	Africa/Middle East	1
	Total	91
Defect Potentials per Function Point (1000 function points; Java language)		**Global 24 Hour**
1	Requirements	0.30
2	Architecture	0.20
3	Design	0.75
4	Code	1.15
5	Documents	0.20
6	Bad fixes	0.20
	Total	2.80
	Total defect potential	2,800
Defect Removal Efficiency (DRE)		
1	Pre-test defect removal (%)	75.00
2	Testing defect removal (%)	87.00
	Total DRE (%)	96.50
	Delivered defects per function point	0.98
	Delivered defects	98
	High-severity defects	20

(*Continued*)

Table 24.2 (Continued) Example of 1000 Function Point 24-Hour Global Project with Java Language

	Security defects	2
Productivity for 1000 Function Points		**Global 24 Hour**
	Work hours per function point 2016	15.60
	Function points per month 2016	8.46
	Schedule in calendar months	7.00

Data Source: Namcook analysis of 600 companies.

Global 24 hour scoring 2016
Best = +10
Worst = –10

zones apart, global 24 hour development is normally used only by global Fortune 500 companies that build large systems and have tight deadlines. This methodology trades about a 33% increase in development costs for about a 60% reduction in development schedules. Some application types that use this method include major ERP packages, operating systems, telephone switching systems, global banking software, and some defense applications.

In theory, global outsource vendors might offer global 24 hour development to clients for a premium fee, but that does not seem to have occurred as of 2016.

Chapter 25

Hybrid Agile/Waterfall Software Development

Executive summary: *Although Agile and waterfall seem to be at opposite ends of the spectrum, in fact, hybrid forms that combine the best features of Agile and waterfall have been surprisingly successful. In recent years, Agile has been hybridized with several methods and concepts, including, in alphabetical order, capability maturity model integrated (CMMI), Rational Unified Process (RUP), team software process (TSP), and waterfall. In general, the hybrid forms are slightly superior in terms of quality and productivity to the "pure" methods. This is due to condensing best practices and removing a few weak practices at the same time.*

Waterfall is a traditional method dating back to the late 1960s. Agile is a new method dating to the Agile Manifesto of February, 2001. Surprisingly, Agile and waterfall mix together quite well and offer some advantages compared with the "pure" methods.

Agile Concepts Used in Hybrid Agile/Waterfall Combinations

1. Individuals and interactions are better than formal processes and tools.
2. Working software is better than comprehensive documentation.
3. Customer collaboration is better than comprehensive contracts.
4. Responding to change is better than following a rigid plan.

5. Embedded users provide requirements.
6. User stories are used for requirements analysis.
7. Larger projects are divided into "sprints" that last about 2 weeks.
8. Daily "Scrum" sessions are conducted for status reports.
9. Pair programming is used, or two programmers taking turns coding or "navigating."
10. Test-driven development is used, or writing test cases before the code is written.
11. Specialized Agile metrics are used, such as velocity, burn down, burn up, and so on.
12. Agile "coaches" help introduce Agile concepts.

Some of the more useful waterfall concepts that are used in the hybrid forms are listed in the following section.

Waterfall Concepts Used in Hybrid Agile/Waterfall Combinations

1. Size the entire applications before starting, using function points and lines of code as well as story points.
2. Perform early risk analysis using data from similar projects.
3. Perform careful cost and schedule estimates using data from similar projects.
4. Use pre-test defect removal such as inspections and static analysis.
5. Use formal test methods as well as informal developer testing.
6. Use formal cost and progress tracking in addition to Scrum sessions.

This combination preserves the flexibility of Agile but adds some much-needed rigor to planning, estimating, and quality control. Many clients are unhappy with the fact that some Agile projects seem to want to jump in and get started without having a full budget or even a full development schedule.

In its pure form, Agile is intended for projects that probably have fewer than 100 users and are small enough to require fewer than 10 developers, which usually means projects <500 function points in size.

The hybrid Agile/waterfall method may not be the optimal choice for large systems with over 10,000 users and 100 development personnel, but it comes closer than pure Agile. However, readers should also consider team software process (TSP) and Rational Unified Process (RUP), which have proven success records in the 10,000 function point size range.

The hybrid approach is a better choice for projects that need government certification, such as medical devices and avionics packages. The casual Agile attitude about documentation is not suitable for the mandatory documentation forced on projects by government policies, but the hybrid method is able to handle these documents.

The Agile/waterfall hybrid is also better for projects that need certification by the Food and Drug Administration (FDA) or the Federal Aviation Authority (FAA). The hybrid form is also a better choice for financial applications that require Sarbanes–Oxley governance.

The numerous sprints and the Agile readiness to absorb frequent requirements changes have made pure Agile planning and estimating somewhat more difficult than for other methods such as RUP and TSP.

Pure Agile is also weak in measuring results, so quality and productivity data is sparse. The unique structure of Agile projects in terms of multiple sprints makes productivity comparisons with other methods difficult.

Several methods of planning, estimating, and measuring Agile have been developed, and a few support the hybrid concept too. One of these, which supports the hybrid Agile/waterfall combination, was developed by Namcook Analytics. It consists of converting Agile story points into function points and converting Agile sprints into a standard chart of accounts. In other words, Namcook consolidates all the Agile sprints into one set of activities, such as requirements, design, coding, testing, documentation, and so on.

These two Namcook techniques allow pure Agile and Agile hybrid methods to be compared with other methods using a side-by-side format. The Namcook Software Risk Master (SRM) tool can predict Agile sprints but convert overall Agile results into a standard chart of accounts.

Agile principles can be combined with other methods, and a number of "hybrid" methods exist in 2016, including, but not limited to, Agile/capability maturity model integrated (CMMI); Agile/waterfall; Agile/RUP; and Agile/TSP. In general, hybrids seem slightly superior to pure methods.

There is a significant amount of literature about Agile hybrids of various flavors. One of the recent studies of Agile/waterfall hybrids, published by CAST Software, is based on examination of more than 1300 projects from around the world.

The overall demographics of Agile/waterfall hybrids in 2016 and the results of a typical Agile hybrid project of 1000 function points using the Java programming language are shown in Table 25.1.

It is a bit surprising that hybrid software methodologies such as this amalgamation of Agile and waterfall turn out to be successful, but in fact, they sometimes seem to be more productive than either Agile or waterfall in its "pure" form. This is perhaps because the hybrid method selects the best features of each of the pure methodologies: flexibility from Agile and proper sizing, estimating, and risk analysis from waterfall.

Table 25.1 Example of 1000 Function Point Hybrid Project with Java Language

Project Size and Class Scores		Hybrid Agile/Waterfall
1	Small projects <100 function points	10
2	New projects	10
3	IT projects	10
4	Medium projects	8
5	High-reliability, high-security projects	7
6	Systems/embedded projects	6
7	Web/cloud projects	6
8	Large projects >10,000 function points	5
9	Enhancement projects	5
10	Maintenance projects	5
	Subtotal	72
Project Activity Scores		**Hybrid Agile/Waterfall**
1	Coding	10
2	Requirements changes	9
3	Change control	9
4	Requirements	8
5	Integration	8
6	Testing defect removal	8
7	Defect prevention	7
8	Design	6
9	Test case design/generation	6
10	Architecture	5
11	Pre-test defect removal	5
12	Reusable components	4
13	Models	4
14	Pattern matching	3

(Continued)

Table 25.1 (Continued) Example of 1000 Function Point Hybrid Project with Java Language

15	Planning, estimating, measurement	2
	Subtotal	94
	Total	166
Regional Usage (Projects in 2016)		**Hybrid Agile/Waterfall**
1	North America	2500
2	Pacific	1500
3	Europe	3000
4	South America/Central America	750
5	Africa/Middle East	500
	Total	8250
Defect Potentials per Function Point (1000 function points; Java language)		**Hybrid Agile/Waterfall**
1	Requirements	0.50
2	Architecture	0.15
3	Design	0.60
4	Code	1.10
5	Documents	0.40
6	Bad fixes	0.30
	Total	3.05
	Total defect potential	3050
Defect Removal Efficiency (DRE)		
1	Pre-test defect removal (%)	65.00
2	Testing defect removal (%)	85.00
	Total DRE (%)	94.70
	Delivered defects per function point	0.162
	Delivered defects	162
	High-severity defects	25
	Security defects	2

(Continued)

Table 25.1 (Continued) Example of 1000 Function Point Hybrid Project with Java Language

	Productivity for 1000 Function Points	Hybrid Agile/Waterfall
	Work hours per function point 2016	12.63
	Function points per month 2016	10.45
	Schedule in calendar months	13.60

Data Source: Namcook analysis of 600 companies.

Hybrid Agile/waterfall scoring
Best = +10
Worst = −10

Chapter 26

Information Engineering (IE) Software Development

Executive summary: *As the name "information engineering" (IE) implies, this method is aimed primarily at large information systems, often using major enterprise-level databases, repositories, or even "big data" in today's world. IE applications are not aimed at systems or embedded software. IE originated in Australia circa 1975 and then moved to other countries. IE has a broader set of requirements and design issues than other methods because it deals with database and information topics over and above pure software engineering.*

A researcher named Clive Finkelstein published a series of articles in 1976 that dealt with his ideas about what was needed to successfully build and deploy database applications. Another famous software researcher, James Martin, also adopted information engineering (IE) concepts and published a paper entitled "Information Engineering" in 1981. The Finkelstein and Martin approaches were somewhat different but congruent.

IE is heavily related to database concepts, which were also evolving in the 1970s and 1980s. Originally, data was stored on tape or punched cards and then moved to disk. Later, database design itself expanded under the influence of the relational database model.

Texas Instruments was a pioneer in the IE field, and developed one of the first IE tool suites, called the Information Engineering Facility (IEF). This tool still exists in 2016, although it is now owned by Computer Associates (CA). The tool is marketed under the name ALL Fusion.

Because IE concerns data, it is also concerned with business models, with database formats, and with overall business topics in addition to the specifics of building software itself. For example, IE development teams usually employ database analysts and business analysts, who may or may not be part of web or systems software projects. In fact, business models and software models often occur in the same IE projects. IE is one of very few development methods that have a dual-track development cycle.

IE is an amalgamation of external information planning and the specifics of how to best store and manipulate information with computers and software. In other words, IE is a multi-disciplinary approach that usually requires more kinds of domain experts than most other forms of software engineering.

To illustrate the additional kinds of personnel used with IE projects, Table 26.1 compares IE with Agile in terms of occupation groups.

Both have the same number of programmers, but the IE project employs specialists such as database analysts and business analysts, who are not part of Agile. Of course, the IE project has much more extensive data connected with it than the Agile project.

An example IE project of 1000 function points in the Java language is shown in Table 26.2. This is somewhat smaller than typical IE projects, which often run between 5,000 and 20,000 function points in size. However, the 1000 function point size was selected for this report, since it is among the most common size plateaus and is large enough to require serious engineering skills.

IE started as a methodology for applications with large volumes of data to process. This means it is a viable contender in today's "big data" world, where data volumes are even larger than they were 10 years ago.

A related but unmeasured topic of some importance is the fact that companies now own much more data than they do software. However, there is no "data point" metric to measure data size and no "data defect removal efficiency" (DDRE) to measure data quality. As a result, probably not even one company in the entire world actually has reliable and accurate information on data acquisition costs, data maintenance costs, and data quality costs.

High data costs and high cyber-security costs are two important fields that need new metrics based on function point logic and new quality metrics based on defect removal efficiency (DRE).

For example, "security flaw defect removal efficiency" (SFDRE) would be another useful quality metric, but it does not exist in 2016.

Worse, as far as the author knows, none of the function point organizations or any other associations or companies or government agencies are even looking at the problems and trying to gather quantified data.

You would think that perhaps the Department of Defense, IBM, Microsoft, Google, or Amazon would have active research in the topics of data costs, data

Table 26.1 IE versus Agile Staff Profiles for 1000 function points

		IE Staff	*Agile Staff*	*Difference*
1	Programmers	6.00	6.00	0.00
2	Testers	6.00	4.00	2.00
3	Business analysts	4.00	0.00	4.00
4	Database administration	3.00	0.00	3.00
5	Designers	2.00	1.00	1.00
6	First line managers	2.00	2.00	0.00
7	Technical writers	1.00	1.00	0.00
8	Project office staff	1.00	0.00	1.00
9	Configuration control	1.00	1.00	0.00
10	Project librarians	1.00	1.00	0.00
11	Estimating specialists	1.00	0.00	1.00
12	Architects	1.00	0.00	1.00
13	Security specialists	1.00	1.00	0.00
14	Function point counters	1.00	0.00	1.00
15	Human factors specialists	1.00	0.00	1.00
16	Quality assurance	1.00	1.00	0.00
17	Administrative support	1.00	1.00	0.00
18	Second line managers	0.00	0.00	0.00
19	Third line managers	0.00	0.00	0.00
20	Performance specialists	0.00	0.00	0.00
	Total	**34.00**	**19.00**	**15.00**

quality, cyber-security costs, and security flaw defect removal. If they do, they certainly do not seem to be publishing anything of interest.

Software bugs are like bacterial infections that can be diagnosed fairly easily and in many cases prevented from occurring or cured easily. Data bugs are more like immune system conditions with unknown etiologies and no effective prevention or control methods as of 2016. Cyber-security bugs are somewhat akin to

Table 26.2 Example of 1000 Function Point IE Project with Java Language

Project Size and Nature Scores		IE
1	Medium projects	10
2	Large projects >10,000 function points	10
3	New projects	10
4	IT projects	10
5	Small projects <100 function points	9
6	High-reliability, high-security projects	7
7	Enhancement projects	5
8	Maintenance projects	5
9	Systems/embedded projects	5
10	Web/cloud projects	5
	Subtotal	76
Project Activity Scores		**IE**
1	Requirements	10
2	Architecture	10
3	Models	10
4	Design	10
5	Pattern matching	8
6	Reusable components	8
7	Requirements changes	8
8	Coding	8
9	Change control	8
10	Integration	8
11	Defect prevention	8
12	Testing defect removal	8
13	Planning, estimating, measurement	7
14	Pre-test defect removal	7

(Continued)

Table 26.2 (Continued) Example of 1000 Function Point IE Project with Java Language

15	Test case design/generation	7
	Subtotal	125
	Total	201
Regional Usage (Projects in 2016)		**IE**
1	North America	4,000
2	Pacific	5,000
3	Europe	3,000
4	South America/Central America	2,000
5	Africa/Middle East	1,000
	Total	15,000
Defect Potentials per Function Point (1000 function points; Java language)		**IE**
1	Requirements	0.25
2	Architecture	0.20
3	Design	0.40
4	Code	1.15
5	Documents	0.40
6	Bad fixes	0.25
	Total	2.65
	Total defect potential	2,650
Defect Removal Efficiency (DRE)		
1	Pre-test defect removal (%)	70.00
2	Testing defect removal (%)	85.00
	Total DRE (%)	95.50
	Delivered defects per function point	0.120
	Delivered defects	119
	High-severity defects	18
	Security defects	2

(Continued)

Table 26.2 (Continued) Example of 1000 Function Point IE Project with Java Language

Productivity for 1000 Function Points		
	Work hours per function point 2016	14.20
	Function points per month 2016	9.29
	Schedule in calendar months	15.60

Data Source: Namcook analysis of 600 companies.

IE scoring
Best = +10
Worst = −10

viruses, which are not affected by antibiotics and are more difficult to diagnose and treat than bacterial infections.

Medicine has a wide-ranging scope of research topics that study essentially all known medical problems. Software, on the other hand, has a distressingly narrow scope of research that spends perhaps 85% of time and energy on code development and code defects, while ignoring more serious problems that originate in requirements, design, bad data, and security flaws.

There is quite a bit of money being spent on security problems, but much of the money is going toward firewalls and anti-virus techniques. There is little or no research money addressing the more fundamental topic of raising the immunity levels of software operating systems and applications so that they become intrinsically resistant to cyber-attacks.

These problems are relevant to IE, since applications with large volumes of data will obviously have more data errors, and they are also much more likely to be subject to cyber-attacks aiming to steal valuable data such as social security numbers, credit card numbers, bank account data, client lists, trade secrets, military secrets, military defense plans, and other high-value data.

Chapter 27

IntegraNova Development

Executive summary: *IntegraNova is a Spanish software research and product company. Its main product is a model-driven application generator based on the Object Management Group (OMG) meta-model. IntegraNova is used primarily in Europe, but it is being demonstrated to the U.S. Department of Defense. IntegraNova is aimed at information technology (IT) applications, because that is the primary focus of the OMG meta-model. IntegraNova officers and marketing personnel say that the IntegraNova application generator is not aimed at systems or embedded software. That being said, the generator seems to have satisfied customers, some of whom were interviewed for this report.*

The idea of software development from abstract models is a fairly old concept. Model-driven development and computer-aided software engineering (CASE) had a burst of popularity in the 1980s, almost as Agile is having today. Some of the companies involved include Bachman Systems, Higher-Order Software, Cadre Technologies, and Logic Works. More recently, companies such as IntegraNova in Spain are bringing out operational application generators based on modeling.

The modeling idea also influenced the Unified Modeling Language (UML) and was part of the concept of the universal systems language.

There are many images on the web of various kinds of modeling approaches; Figure 27.1 shows the IntegraNova model concepts.

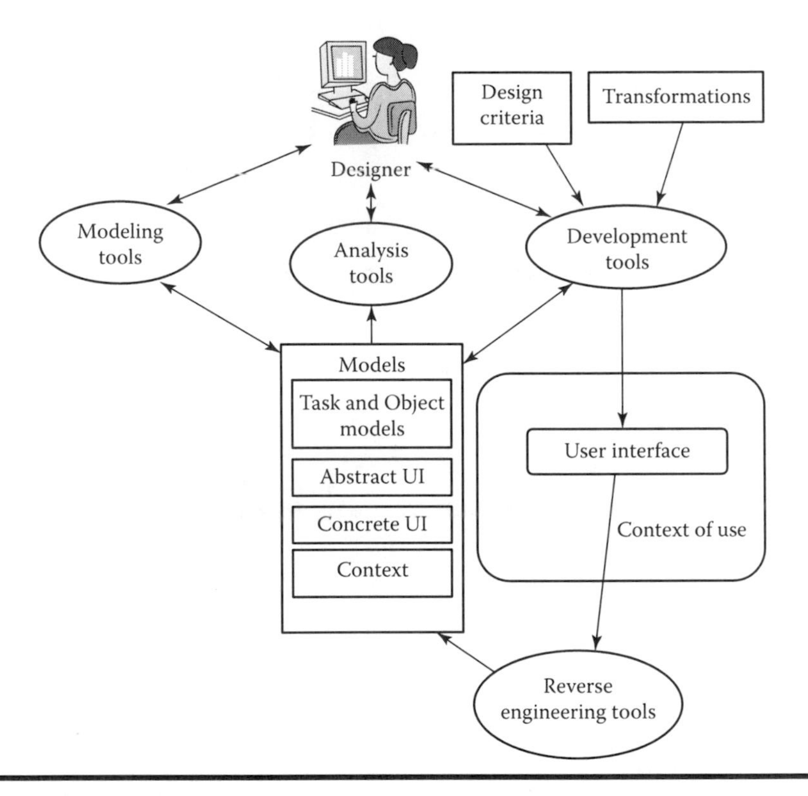

Figure 27.1 IntegraNova workflow illustration.

The IntegraNova product is based on the Object Management Group (OMG) meta-model. This is aimed mainly at information technology (IT) products and not at systems or embedded software. IntegraNova officers have told the author that IT projects are the current main focus of IntegraNova. See also the T-VEC chapter, which applies models to requirements, design, and code validation and verification.

In model-based development, requirements changes are handled by going back to the model and regenerating the application: actual manual code changes are discouraged. This is because manual code changes deviate from the underlying model and quickly lead to errors that are hard to find and hard to correct.

Once the code is generated and goes into production, it will fairly soon need to be updated, both to repair bugs and to add various enhancements or modifications based on external changes to business situations. If you go back and recreate the model, everything stays synchronized. But if you change the code itself, the code and the models are no longer in synch, and all future changes will have to be made to the code, with probably incomplete documentation.

In today's world, many modeling tools exist. Many are open source, and many more are commercial tools. An interesting survey of open-source modeling tools was published for Ericsson by Zeligsoft. The authors are Kenn Hussey, Brian Selec, and Toby McClean. Images of sample model tools are common on the web. Model-based development is fairly popular in the telecom industry due to its reasonably stable requirements and a large number of qualified domain experts.

One of the gaps in the model-driven literature is actual productivity and quality results. This was a shortcoming of CASE in the 1980s and is a chronic shortcoming of Agile today in 2016.

IntegraNova is probably unique among modeling tools in that it includes a built-in function point generator. Therefore, every IntegraNova application comes with size in function points. This makes it quite easy to show productivity in work hours per function point and quality in terms of defects per function point.

This feature of function point sizing should be part of all meta-models and application generators, but unfortunately, most ignore size completely.

Function point metrics are the best choice for both model-driven productivity and model-driven quality. Sample results for model-driven development for an IntegraNova application of 1000 function points and generated Java code are shown in Table 27.1.

IntegraNova is moving into an interesting niche among software development methodologies. It generates applications from a repository of reusable features. IntegraNova also has a built-in function point sizing tool, which makes it unique among all of the methodologies in this book. IntegraNova is also the only methodology that generates applications directly instead of just providing guidelines on how to build them.

Currently, IntegraNova only generates IT applications, but if these are successful, new features might be included to generate real-time, embedded, and systems software. IntegraNova offers an impressive combination of high quality and high productivity due in part of the use of reusable features.

Table 27.1 Sample of 1000 function point Integra Nova project with Java Language

Project Size and Nature Scores		**IntegraNova**
1	Small projects <100 function points	10
2	Medium projects	10
3	New projects	10
4	Systems/embedded projects	10
5	Large projects >10,000 function points	9
6	IT projects	9
7	Web/cloud projects	4
8	Enhancement projects	2
9	Maintenance projects	2
10	High-reliability, high-security projects	0
	Subtotal	66
Project Activity Scores		**IntegraNova**
1	Pattern matching	10
2	Reusable components	10
3	Planning, estimating, measurement	10
4	Requirements	10
5	Requirements changes	10
6	Design	10
7	Defect prevention	10
8	Pre-test defect removal	10
9	Architecture	9
10	Models	9
11	Coding	8
12	Change control	8
13	Test case design/generation	8

(Continued)

Table 27.1 (Continued) Sample of 1000 function point Integra Nova project with Java Language

14	Testing defect removal	8
15	Integration	7
	Subtotal	137
	Total	203
Regional Usage (Projects in 2016)		**IntegraNova**
1	North America	15
2	Pacific	–
3	Europe	200
4	South America/Central America	–
5	Africa/Middle East	–
	Total	215
Defect Potentials per Function Point (1000 function points; Java language)		**IntegraNova**
1	Requirements	0.15
2	Architecture	0.15
3	Design	0.30
4	Code	0.50
5	Documents	0.40
6	Bad fixes	0.10
	Total	1.60
	Total defect potential	1600
Defect Removal Efficiency (DRE)		
1	Pre-test defect removal (%)	85.00
2	Testing defect removal (%)	90.00
	Total DRE (%)	98.50
	Delivered defects per function point	0.024
	Delivered defects	24

(*Continued*)

Table 27.1 (Continued) Sample of 1000 function point Integra Nova project with Java Language

	High-severity defects	4
	Security defects	0
Productivity for 1000 Function Points		**IntegraNova**
	Work hours per function point 2016	6.00
	Function points per month 2016	22.00
	Schedule in calendar months	5.70

Data Source: Namcook analysis of 600 companies.

IntegraNova scoring
Best = +10
Worst = −10

Chapter 28

Iterative Software Development

Executive summary: *The word "iterative" means doing one thing after another in sequence. In the term "iterative software development," it means building software incrementally in a sequence of separate usable features rather than as an entire package. This is an old concept, which was used on the NASA Mercury project in the 1960s. The essential idea of iterative development was adopted by Agile, extreme programming (XP), feature-based development, specifications by example, and many other more recent methodologies. Although the concept has many proofs of success, it cannot always be applied. For example, you cannot build embedded medical device software such as a cochlear implant in pieces: the device won't work at all and cannot be implanted in a patient unless every feature is working. Of course, small segments can be developed individually, but they have no intrinsic user value. Only the full package allows patients to hear with a cochlear implant.*

The idea of iterative development was among the first to move away from waterfall development and try a new direction. The method starts with an overall depiction of the full application. Then, it is divided into separate components for individual construction. For example, an application of 1000 function points may have five components of 200 function points each. For architecture, the overall application is the basis. For detail design, the separate components are the basis. Each component is built and tested separately. But of course, the full package is system tested and acceptance tested after all components are complete.

Iterative development was successfully used on the NASA space shuttle between 1977 and 1980. A total of 17 iterative increments were constructed over a period of 31 months. Iterative development is useful for projects in which requirements are likely to change during development (as is Agile). Iterative is still widely deployed for defense and some avionics software packages, with fairly good success.

Unlike modern Agile, iterative development does not assume that all developers are co-located. Iterative is much older than Scrum, too, and status checking is in the form of conference calls and written reports.

Iterative projects, on average, are somewhat larger than Agile projects, and they often have development teams of more than 100 personnel, who may work in different locations and even in different cities. In other words, iterative is a good fit for space and defense projects, where multiple sub-contractors are all working toward a common application.

Average Agile projects among the author's clients are <300 function points in size with fewer than eight developers co-located in the same building. Average iterative projects among the author's clients are >2500 function points in size with about 15 developers located in three separate geographic locations.

Because iterative has been used since the 1960s, it is a normal methodology estimated by commercial parametric cost estimating tools. The available parametric tools in 2016 include, in alphabetical order,

1. COCOMO III
2. CostXpert
3. ExcelerPlan
4. KnowledgePlan
5. R2Estimator
6. RASS Estimate (United Kingdom)
7. SEER
8. SLIM
9. Software Risk Master (SRM) by the author of this report
10. True Price

Agile projects, DevOps, and container development are not supported by all these estimation tools, but they are supported by the author's SRM estimation tool.

There are many images of iterative development on the web, including the one shown in Figure 28.1. Another web image shows a comparison between waterfall and iterative development and provides a kind of pseudo meta-model of how it works (Figure 28.2).

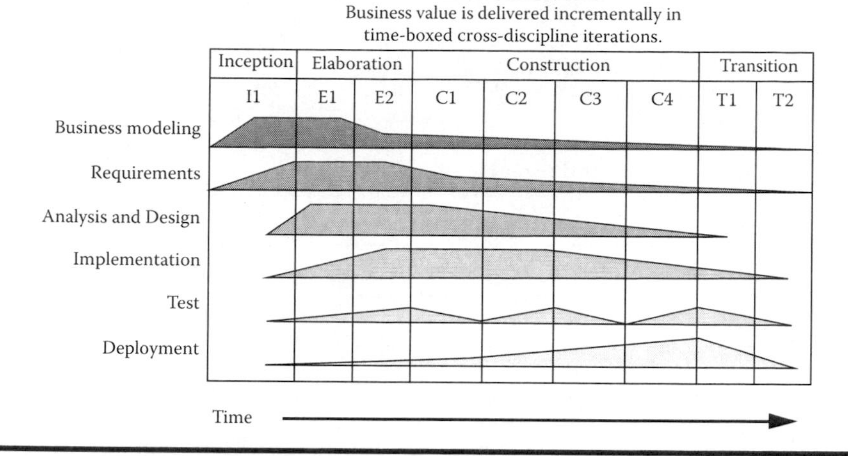

Figure 28.1 Illustration of iterative development.

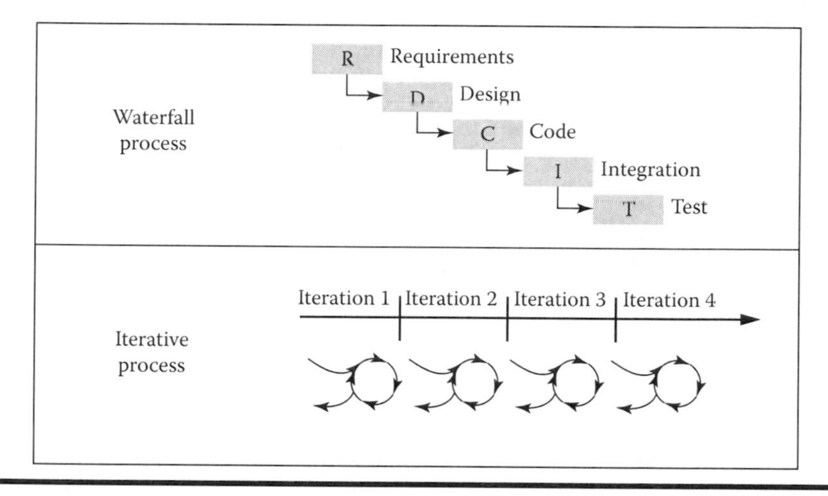

Figure 28.2 Comparison of waterfall and iterative development.

Table 28.1 shows the Namcook results for iterative development based on a project of 1000 function points and the Java programming language. The results assume five iterations, each of 200 function points.

Iterative development is one of the older development methodologies in this book, but also one of the stronger methods, with substantial proofs of success over many years. The typical project that uses iterative development is probably close to 10 times larger than a typical Agile project.

Table 28.1 Example of 1000 Function Point Iterative Project with Java Language

Project Size and Nature Scores		Iterative
1	New projects	10
2	Systems/embedded projects	10
3	Large projects >10,000 function points	9
4	IT projects	9
5	Medium projects	8
6	Web/cloud projects	8
7	High-reliability, high-security projects	8
8	Small projects <100 function points	7
9	Enhancement projects	6
10	Maintenance projects	5
	Subtotal	80
		Iterative
1	Coding	9
2	Change control	9
3	Integration	9
4	Defect prevention	9
5	Pre-test defect removal	8
6	Testing defect removal	8
7	Planning, estimating, measurement	7
8	Requirements	7
9	Requirements changes	7
10	Design	7
11	Test case design/generation	7
12	Reusable components	6
13	Architecture	6
14	Models	6

(Continued)

Table 28.1 (Continued) Example of 1000 Function Point Iterative Project with Java Language

15	Pattern matching	4
	Subtotal	109
	Total	189
Regional Usage (Projects in 2016)		**Iterative**
1	North America	10,000
2	Pacific	12,000
3	Europe	9,000
4	South America/Central America	8,000
5	Africa/Middle East	4,000
	Total	43,000
Defect Potentials per Function Point (1000 function points; Java language)		**Iterative**
1	Requirements	0.35
2	Architecture	0.30
3	Design	0.35
4	Code	1.50
5	Documents	0.40
6	Bad fixes	0.40
	Total	3.30
	Total defect potential	3,300
Defect Removal Efficiency (DRE)		
1	Pre-test defect removal (%)	70.00
2	Testing defect removal (%)	85.00
	Total DRE (%)	95.50
	Delivered defects per function point	0.149
	Delivered defects	149
	High-severity defects	23
	Security defects	2

(Continued)

Table 28.1 (Continued) Example of 1000 Function Point Iterative Project with Java Language

Productivity for 1000 Function Points		Iterative
	Work hours per function point 2016	16.10
	Function points per month 2016	8.20
	Schedule in calendar months	16.50

Data Source: Namcook analysis of 600 companies.

Iterative scoring
Best = +10
Worst = −10

Chapter 29

Kaizen

Executive summary: *Kaizen is one of a set of inter-related "tools" included in the overall concept of Lean, developed primarily by Toyota in Japan. While Lean and Kaizen are not software methodologies per se, they have nonetheless been quite influential in the field of software engineering—Lean thinking is the foundation for all Agile approaches. Kaizen is an approach to continuous improvement that emphasizes small incremental improvements rather than large step changes. To grasp the full power of Kaizen, it is useful to understand some of the related concepts, including Poke Yoke, quality circles, Kanban, 3-Ms, demand management, and standardized work. All these ideas are potentially relevant to the optimization of any selected software development methodology.*

Terms and Definitions

Kaizen means "continuous improvement" in Japanese. As illustrated in Figure 29.1, there are several different types of Kaizen events, which may vary in duration from a few minutes to perhaps several months.

- System-level (also called "flow") Kaizen addresses overall value stream improvements and is generally the province of mid- to senior-level management. These improvements generally deal with the elimination of "Muri" (overburden), one of the 3-Ms, and with the elimination of variability (i.e., demand management).
- Process-level Kaizen is normally the province of a work group or quality circle and is concerned with the elimination of "Muda" (waste). Daily Kaizen is

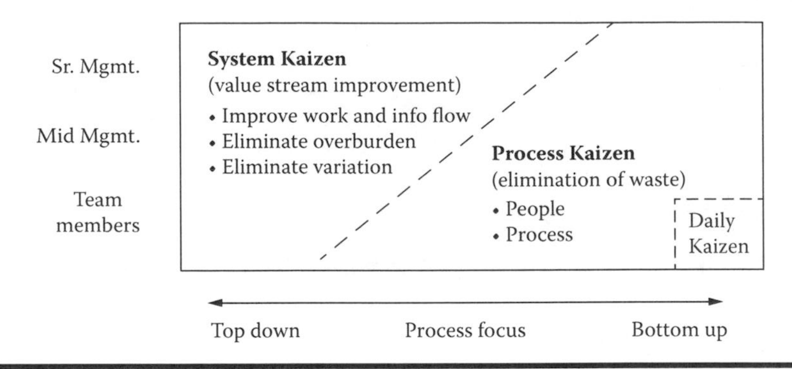

Figure 29.1 Illustration of Kaizen concepts.

a special case of the process type, which is intended to address an out-of-the-ordinary event that merits immediate attention but is small and local in scope.

Poka Yoke means "mistake proofing" in Japanese. The idea here is to design processes so that human mistakes are prevented or at least immediately recognized. This is relevant to software, and instances include inspections and static analysis.

A *quality circle* is a group of workers who do the same or similar work and who meet regularly to identify, analyze, and solve work-related problems. Agile methods generally advocate a period of reflection and adjustment at the end of each iteration—essentially, the quality circle concept in action. Historically, quality circles have done well in Japan but not as well in the United States. However, Agile and team software process (TSP) include some quality circle concepts, and they both seem to work well in the United States.

Kanban refers to a signaling device—that is, "I'm ready for more work." This is one of the fundamental ideas that enable a "pull" approach to production. As it applies to software development, this means avoiding an up-stream build-up of work in process. For example, extensive requirements are not defined till there is the capacity to act on the requirement—only a small buffer of "stories" or use cases is created at any given time. This is the fundamental notion of "just in time:" rather than "building" to create inventory (e.g., requirements documents), build only on demand—that is, when the capacity exists to act on the requirement. A related aspect of Kanban is the notion of visible workflow. Individual work units (e.g., a story) are represented by cards displayed on a board that is structured to reflect the flow of work (e.g., story—design—test case—code). Kanban is a tool that can be used to manage and minimize work in process and to ensure single-piece uninterrupted flow. Kanban has been implemented in conjunction with Agile methods.

The *3-Ms* include

- *Mura*, which refers to variability in flow. Lean has shown that productivity and quality are optimized when variability in demand is minimized. A steady cadence is preferred. Some of the Agile methods advocate a sustainable cadence of activity to avoid burnout and uneven flow.
- *Muri*, which refers to "overburden," in which demand exceeds capacity—e.g., unrealistic workload, a common concern in software work. Muri leads to stress, low morale, defects, and rework. As manifested in software, this means managing to a realistic schedule with quality given more emphasis than schedule, avoiding excessive schedule pressures from senior managers, clients, or both.
- *Muda*, which refers to the "seven wastes": overproduction (e.g., volumes of requirements sitting on the shelf), inventory (e.g., a backlog of unresolved defects), waiting (for approvals), transportation (moving information or persons), over-processing (gold-plating), motion (workers not co-located), and correction (fixing defects that could have been prevented or detected earlier).

Demand management addresses balancing supply and demand. Imbalances lead to shortcuts that compromise quality and ultimately damage productivity.

Standardized work defines and documents the most effective method to accomplish a task—variation in methods leads to variation in time, quality, and cost. These definitions are established and owned by those doing the work, not by management, but they apply to all levels—for example, managers decide how best to do their jobs. Standardized work is designed to promote agility—that is, a sound tradeoff between efficiency and flexibility.

- An efficiency perspective values stable processes, standard methods, predictable outcomes, minimal disruption, and rigidity; create value by minimizing cost and time. The focus is on waste reduction, visual management à la Kanban, and ongoing improvement using Kaizen.
- A flexibility perspective values fluid processes, responsive procedures, less predictable outcomes, frequent changes, and adaptability; create value by maximizing responsiveness. The focus is on the voice of the customer, demand management, just-in-time thinking, level scheduling, small lot sizes, cross-training, and flow.

Since many of these concepts originated in automotive manufacturing, yet another aspect of that field is relevant to software: construction from standard reusable components rather than custom designs and manual coding of every feature.

When considering how to address standardized work, it is useful to distinguish between process work and practice work.

- Processes are series of actions or operations supported by structured information—generally repetitive, well defined, routine, controllable, and standardized. Data center operations and perhaps help desks fit this category.
- Practices are non-routine, highly variable, and loosely defined, and require judgment and experience. Frameworks are useful, but there is no one best way. Law, medicine, and software engineering are found in this category.

Many of the Lean tools and ideas are applicable to projects both large and small and can generally be combined with any chosen development methodology.

See also the chapters on pattern-based development; reuse-based development; and Lean, Agile, and model-based development.

As of 2016, Japanese software has the highest quality of any country yet examined. Table 29.1 shows the top 25 countries in terms of overall software quality results. These results include applications from 100 to 100,000 function points and a mix of information technology (IT), web, systems, embedded, and other types of projects. Table 29.1 is based on small samples and extrapolation, but there is no other equivalent source.

Although China is not in the top 25 countries in 2016, the rate of improvement and the study of software technologies may well bring China into the top 10 countries by 2020. Better measures of quality and more emphasis on reuse, defect prevention, pre-test defect removal, and formal testing are all needed to achieve top 10 status.

It is interesting that the South Korean government is not only a leader in software quality but has joined Brazil in requiring the use of function point metrics for all government software projects. In fact, the author at Namcook was visited by a delegation from South Korea specifically to explore function point measures for quality.

In addition to being #1 in quality, Japan has the second largest function point users' group in the world after the United States (Brazil is #3). In fact, the 2016 world conference on function points was held in Tokyo in October. China should increase function point use for both quality studies and economic studies.

A sample project is shown in Table 29.2, in which Kaizen and Kanban were joined to the TSP methodology. The sample is for 1000 function points and the Java language.

As of 2016, Kaizen is used primarily in Japan, with considerable success for both hardware and software projects. Japan was influenced by several of the American quality experts, including Joseph Juran and W. Edwards Deming. But their influence would not have been effective if Japan had not already had a long national tradition of achieving high quality in many products, such as swords, pottery, silk, and woodwork, dating back over 1000 years.

Table 29.1 2016 Global Defect Potentials and Defect Removal Efficiency (DRE)

		Defect Potential in 2016	*DRE in 2016 (%)*	*Delivered Defects in 2016*
1	Japan	4.50	96.31	0.17
2	India	4.90	95.79	0.21
3	Denmark	4.80	94.76	0.25
4	South Korea	4.90	94.76	0.26
5	Canada	4.75	94.50	0.26
6	Switzerland	5.00	94.76	0.26
7	Israel	5.10	94.76	0.27
8	United Kingdom	4.75	94.25	0.27
9	Sweden	4.75	93.73	0.30
10	Norway	4.75	93.73	0.30
11	Netherlands	4.80	93.73	0.30
12	Hungary	4.60	93.22	0.31
13	Ireland	4.85	93.22	0.33
14	United States	4.82	92.85	0.34
15	Brazil	4.75	92.70	0.35
16	France	4.85	92.70	0.35
17	Australia	4.85	92.70	0.35
18	Austria	4.75	92.19	0.37
19	Belgium	4.70	91.82	0.38
20	Finland	4.70	91.67	0.39
21	Hong Kong	4.75	91.67	0.40
22	Mexico	4.85	90.64	0.45
23	Germany	4.95	90.64	0.46
24	Philippines	5.00	90.64	0.47
25	New Zealand	4.85	90.13	0.48

Table 29.2 Example of 1000 Function Point Kaizen Project with Java Language

Project Size and Nature Scores		**Kaizen**
1	Large projects >10,000 function points	10
2	New projects	10
3	Systems/embedded projects	10
4	High-reliability, high-security projects	10
5	Medium projects	9
6	IT projects	9
7	Small projects <100 function points	8
8	Web/cloud projects	8
9	Enhancement projects	5
10	Maintenance projects	3
	Subtotal	82
	Project Activity Scores	**Kaizen**
1	Pattern matching	10
2	Planning, estimating, measurement	10
3	Architecture	10
4	Coding	10
5	Change control	10
6	Integration	10
7	Defect prevention	10
8	Pre-test defect removal	10
9	Test case design/generation	10
10	Testing defect removal	10
11	Reusable components	9
12	Requirements changes	9
13	Models	9
14	Design	9

(Continued)

Table 29.2 (Continued) Example of 1000 Function Point Kaizen Project with Java Language

15	Requirements	8
	Subtotal	144
	Total	226
Regional Usage (Projects in 2016)		**Kaizen**
1	North America	1,000
2	Pacific	15,000
3	Europe	3,000
4	South America/Central America	3,500
5	Africa/Middle East	500
	Total	23,000
Defect Potentials per Function Point (1000 function points; Java language)		**Kaizen**
1	Requirements	0.1
2	Architecture	0.15
3	Design	0.2
4	Code	0.9
5	Documents	0.3
6	Bad fixes	0.1
	Total	1.75
	Total defect potential	1,750
Defect Removal Efficiency (DRE)		
1	Pre-test defect removal (%)	85.00
2	Testing defect removal (%)	90.00
	Total DRE (%)	98.50
	Delivered defects per function point	0.026
	Delivered defects	26
	High-severity defects	4
	Security defects	0

(*Continued*)

Table 29.2 (Continued) Example of 1000 Function Point Kaizen Project with Java Language

Productivity for 1000 Function Points		Kaizen
	Work hours per function point 2016	12.20
	Function points per month 2016	10.82
	Schedule in calendar months	13.20

Data Source: Namcook analysis of 600 companies.

Kaizen scoring
Best = +10
Worst = −10

Chapter 30

Lean Software Development

Executive summary: *Lean software development is an attempt to move the very successful Toyota Lean manufacturing approach to a software context. Lean is included in the Agile family of development methods, but of course, it has some technical differences. However, Lean shares the general weakness of Agile in planning, estimating, measurement, architecture, and design. Like Agile, Lean does not stress reusable components, and therefore, it is more labor intensive and slower than methodologies that support significant volumes of reuse, such as product-line engineering, 85% reuse, pattern matching, and several others.*

Toyota started as a small automobile company in Japan. It has grown steadily to become the world's largest automobile company. Along the way, Toyota gained an enviable reputation for generally good product quality and for extremely good manufacturing economics.

Part of the Toyota manufacturing success was due to the "Lean" concept, which was originally called the *Toyota Production System* (TPS) before the name "Lean" was created. The specific term "Lean" was first applied to Toyota's manufacturing process in a 1988 article by John Crafcik entitled "The Triumph of the Lean Production System."

One interesting aspect of Lean manufacturing is "just in time" assembly line operations and the stocking of parts for an assembly line. This approach was actually derived from studies of supermarket stocking, where shelves are refilled as customers buy products but are not over-filled and are not allowed to become empty.

Both Lean manufacturing and Lean software development emphasize seven key tenets:

1. Eliminate waste
2. Amplify learning
3. Decide as late as possible
4. Deliver as fast as possible
5. Empower the team
6. Build integrity in
7. See the whole

The web has substantial citations of Lean software, and there are also many books available. The web also has numerous images of Lean software development. As would be expected, Lean must deal with the normal topics of a software methodology, such as roles, deliverables, policies, and so on.

Although Lean nominally includes measures, it seems to share the endemic Agile gap of rather sparse measurements and rather poor planning and estimating techniques. Neither Lean nor Agile emphasizes architecture or design. For requirements, embedded users are normal, assuming small projects in which one user can speak for the others.

Also, some of the common agile metrics, such as story points, velocity, burn up, burn down, and so on, have no International Organization for Standardization (ISO) standards and no certification exams. They are hopelessly inconsistent and worthless for large-scale statistical studies.

The author's approach in this book and in both estimating and benchmark data collection is to convert the inaccurate Agile metrics into International Function Point Users Group (IFPUG) function points and consolidate all the Agile sprints into a standard chart of accounts. This method allows side-by-side comparisons of Agile and Lean with other methods such as team software process (TSP) and Rational Unified Process (RUP). Comparisons are impossible with the quirky and unstable Agile metrics.

A major gap in both Lean and Agile is that they both more or less ignore the economic advantages of reusable materials and assume custom designs and manual coding. These are intrinsically expensive and error prone and cannot possibly achieve the same quality and productivity rates as methods where reusability is a central concept.

Ignoring reuse is surprising, since a major economic advantage of automobiles in general and Toyota in particular is that there are between 30% to more than 50% standard reusable parts in the full Toyota and Lexus automobile line-up.

For some reason, the Toyota concept of reusability did not seem to be part of the Lean software concept. In fact, the "Product line engineering" chapter of this book (Chapter 46) shows the benefits of reusable materials among families of related software projects. The specific chapters on reuse-centered software development show the large economic and quality value associated with using certified reusable components instead of custom designs and manual coding.

Table 30.1 Example of 1000 Function Point Lean Project with Java Language

Project Size and Nature Scores		**Lean**
1	Small projects <100 function points	10
2	New projects	10
3	IT projects	10
4	Web/cloud projects	9
5	Medium projects	7
6	Systems/embedded projects	7
7	High-reliability, high-security projects	7
8	Large projects >10,000 function points	5
9	Maintenance projects	5
10	Enhancement projects	4
	Subtotal	74
Project Activity Scores		**Lean**
1	Requirements	9
2	Integration	9
3	Coding	8
4	Change control	8
5	Testing defect removal	8
6	Defect prevention	7
7	Requirements changes	6
8	Design	6
9	Test case design/generation	5
10	Architecture	4
11	Pre-test defect removal	4
12	Pattern matching	3
13	Planning, estimating, measurement	2
14	Models	2

(Continued)

Table 30.1 (Continued) Example of 1000 Function Point Lean Project with Java Language

15	Reusable components	–1
	Subtotal	80
	Total	154
Regional Usage (Projects in 2016)		**Lean**
1	North America	5,000
2	Pacific	30,000
3	Europe	5,000
4	South America/Central America	4,000
5	Africa/Middle East	2,500
	Total	46,500
Defect Potentials per Function Point (1000 function points; Java language)		**Lean**
1	Requirements	0.30
2	Architecture	0.15
3	Design	0.40
4	Code	1.20
5	Documents	0.40
6	Bad fixes	0.30
	Total	2.75
	Total defect potential	2,750
Defect Removal Efficiency (DRE)		
1	Pre-test defect removal (%)	70.00
2	Testing defect removal (%)	85.00
	Total DRE (%)	95.14
	Delivered defects per function point	0.134
	Delivered defects	134
	High-severity defects	21
	Security defects	2

(Continued)

Table 30.1 (Continued) Example of 1000 Function Point Lean Project with Java Language

Productivity for 1000 Function Points		
	Work hours per function point 2014	12.40
	Function points per month 2016	10.65
	Schedule in calendar months	13.70

Data Source: Namcook analysis of 600 companies.

Lean scoring
Best = +10
Worst = −10

A sample Lean software project of 1000 function points coded in the Java language is shown in Table 30.1.

Lean manufacturing as practiced by Toyota for automobiles and trucks has been very successful and has become a pattern for many other manufacturing companies. Lean applied to software has been popular but not as effective as for hardware. This is primarily due to the fact that the use of standard reusable components became lost in the translation between hardware Lean and software Lean. Also, software Lean did not pick up either function point metrics or the defect removal efficiency (DRE) metric, and hence it has been running blind without any solid quantitative data.

Chapter 31

Legacy Data Mining

Executive summary: *Success in software development as a business relies on a combination of processes (methodologies), technologies employed (toolsets), and domain knowledge related to the nature and uses of the software being developed. Legacy data mining is partially technology and partially application domain specific—it refers to "knowledge discovery" derived from databases and data warehouses. Data mining, in general, is about finding information by identifying patterns, associations, and relationships within large sets of data. Some of the raw data exists in readily accessible form within relational databases and data warehouses, while other data may exist in a less readily accessible form within legacy application systems. Data mining has grown rapidly and continues to do so. An awareness of data mining approaches and potential uses may be relevant when developing many different kinds of software systems.*

Data mining software analyzes relationships and patterns in stored transaction data based on open-ended queries. The types of software used to do this include statistical analysis packages, machine learning, and neural networks. In general, the tools involve six common classes of tasks:

- Anomaly detection: Finding outliers, changes or deviations from past patterns—finding unusual records that may be interesting or may be bad data
- Association rule learning (dependency modeling): Also known as *market basket analysis*—identifying products commonly purchased together
- Clustering: Discovering groups and structures in the data that are in some way similar—tends to find unexpected patterns

- Classification: Generalizing from what is known to enable screening of new data—for example, differentiating between spam and legitimate e-mail messages
- Regression: Applying the statistical method of multi-regression analysis to build predictive models
- Summarization: Reducing the volume of data by means such as graphical representation

By using pattern recognition technologies and statistical methods, data mining helps analysts recognize significant facts, relationships, trends, patterns, exceptions, and anomalies that might otherwise be unrecognized. Specific uses might include

- Market segmentation: Identifying the common characteristics of persons who purchase certain products or services.
- Customer churn: Predicting which customers are likely to buy from competitors.
- Fraud detection: Identifying transactions that are likely fraudulent. Credit card companies have become quite adept at early detection of potentially fraudulent transactions.
- Direct marketing: Determining which customers or prospects to include in mail or tele-marketing campaigns to maximize response rates.
- Interactive marketing: Predicting what an individual visiting a website might be interested in seeing.
- Market basket analysis: What are certain groups of customers likely to purchase together?
- Trend analysis: Determining changing purchase patterns from month to month or season to season. Many retailers change what they stock based on historical patterns.

Tools commonly used for data mining include

- Artificial neural networks: A family of non-linear predictive models that learn through training with a test data set and are then validated against a larger set (similar to the approach typically used to develop regression models). These networks resemble biological neural networks in structure.
- Decision trees: Examples include classification and regression trees (CART) and chi square automatic interaction detection (CHAID). They provide a set of rules that can be applied to a new data asset not used to develop these models.
- Rule induction: The extraction of if-then rules from a data set.
- Genetic algorithms: Optimization techniques that use processes such as genetic combination, mutation, and natural selection to devise a design based on natural selection concepts.

- Nearest neighbor method: A technique that classifies each record in a data set in terms of similarity to other records.
- Data visualization: The use of graphical tools to illustrate data relationships.

Two companies have interesting tools for legacy data mining. One of these is Relativity Technologies, and the other is CAST software.

These tools extract hidden business rules and requirements from legacy source code and create a kind of synthetic specification that can be used to assist in making necessary changes to legacy applications. In addition, these tools calculate the function point totals of the legacy applications and provide source code counts. This data is useful for long-term total cost of ownership (TCO) studies.

Table 31.1 shows the scores and results of building a new feature of 1,000 function points in the Java language for a legacy application of 10,000 function points, also in the Java language. The example shows the value of the synthetic specification. The development approach for the new work is spiral.

Legacy data mining is becoming more important as the software industry ages. Today in 2016, probably the average age of operational software in U.S. Fortune 500 companies is more than 10 years, and some applications may be approaching 25 years old.

The main software activity in the United States is no longer new development, but rather, maintenance, repairs, and occasional renovation of aging legacy software. Software does not age gracefully, so the legacy data mining tools are becoming more important.

Table 31.1 Example of 1000 Function Point Legacy Project with Java Language

Project Size and Nature Scores		Legacy Data Mining
1	Enhancement projects	10
2	Maintenance projects	10
3	Large projects >10,000 function points	9
4	New projects	9
5	Medium projects	8
6	IT projects	8
7	Systems/embedded projects	8
8	Web/cloud projects	8
9	High-reliability, high-security projects	8

(*Continued*)

Table 31.1 (Continued) Example of 1000 Function Point Legacy Project with Java Language

10	Small projects <100 function points	7
	Subtotal	85
Project Activity Scores		**Legacy Data Mining**
1	Requirements	10
2	Coding	9
3	Change control	9
4	Integration	9
5	Reusable components	8
6	Planning, estimating, measurement	8
7	Defect prevention	8
8	Pre-test defect removal	8
9	Testing defect removal	8
10	Requirements changes	7
11	Design	7
12	Test case design/generation	7
13	Pattern matching	6
14	Architecture	6
15	Models	6
	Subtotal	116
	Total	201
Regional Usage (Projects in 2016)		**Legacy Data Mining**
1	North America	2500
2	Pacific	2000
3	Europe	1500
4	South America/Central America	1000

(*Continued*)

Table 31.1 (Continued) Example of 1000 Function Point Legacy Project with Java Language

5	Africa/Middle East	1000
	Total	8000
Defect Potentials per Function Point (1000 function points; Java language)		**Legacy Data Mining**
1	Requirements	0.20
2	Architecture	0.30
3	Design	0.45
4	Code	1.45
5	Documents	0.40
6	Bad fixes	0.45
	Total	3.25
	Total defect potential	3250
Defect Removal Efficiency (DRE)		
1	Pre-test defect removal (%)	55.00
2	Testing defect removal (%)	85.00
	Total DRE (%)	93.30
	Delivered defects per function point	0.228
	Delivered defects	228
	High-severity defects	35
	Security defects	3
Productivity for 1000 Function Points		**Legacy Data Mining**
	Work hours per function point 2014	13.90
	Function points per month 2014	9.42
	Schedule in calendar months	14.10

Data Source: Namcook analysis of 600 companies.

Legacy data mining scoring 2016
Best = +10
worst = +10

Chapter 32

Legacy Renovation Development

Executive summary: *The term "legacy renovation" refers to keeping the essential features and some or all of the source code from aging software but removing bugs and possibly improving the structure and documentation. This term should be compared to "reverse engineering," "reengineering," "legacy data mining," and "legacy redevelopment," all of which center on aging legacy applications.*

The first step in legacy renovation may be to review and update existing documentation. Renovation may also involve restructuring of the system architecture and design, and potentially moving to a more modern programming, language, operating system, and so on. Renovation may also involve changing the data structures associated with the system. Typically, renovation does not entail changes to the functionality of the system.

It has been estimated that there are hundreds of billions of lines of legacy code in existence and hundreds of millions of function points, much of it in old languages such as COBOL, MUMPS, Chill, CORAL, and assembly, which are no longer used and are unknown to most of the current generation of programmers.

(The author has suggested an international museum of programming languages together with working compilers and other tools. This is needed as aging software in obsolete languages becomes more and more common. Some of these aging applications have national importance, such as air-traffic control, power generation, and telecommunications switching systems.)

Table 32.1 shows U.S. software history by decade starting in 1950, when 100% of software applications were brand new and 0% needed maintenance or renovation.

Table 32.1 Distribution of New, Enhanced, and Replaced Projects

Year	*New Software*	*Enhanced Software*	*Replaced Software*	*Total Projects*	*Average Age (Years)*
1950	50	0	0	50	1.00
1960	1,200	350	25	1,575	2.50
1970	12,500	5,000	1,000	18,500	5.00
1980	65,000	30,000	4,000	99,000	7.50
1990	245,000	115,000	30,000	390,000	9.00
2000	900,000	800,000	125,000	1,825,000	12.00
2010	1,000,000	1,100,000	450,000	2,550,000	15.00
2020	1,200,000	1,300,000	800,000	3,300,000	20.00

The United States in 2016 has an aging human population, as we all know. But the average age of operating software projects is also increasing with each year that passes. Because renovation is difficult, and replacement is expensive, many important mission-critical applications are still in daily use even though they are slow, error prone, and vulnerable to cyber-attacks.

In the early days of software, new projects were the norm, and maintenance and renovation projects were few in number. But sometime between 1990 and 2000, the number of maintenance and renovation projects pulled ahead of new software development in the United States. By 2020, maintenance and renovation projects will be the dominant work for software engineering in every country. (For additional information, refer to the author's book *The Technical and Social History of Software Engineering*, Addison Wesley 2014.)

From the fact that the majority of the 60 software methodologies in this book focus on development and ignore maintenance and renovation, it can be seen that the software engineering fields does not really understand the steady migration to maintenance and renovation and away from new custom development.

Given the huge quantity of existing legacy software, it is obvious that every effective software development methodology should start by examining similar legacy applications. But only pattern matching and reuse development do this. Most methodologies, such as Agile, Crystal, extreme programming (XP), iterative, spiral, and so on, assume, mistakenly, that every application is brand new and has never been done before.

For that matter, most of the existing meta-models, such as International Organization for Standardization (ISO), Object Management Group (OMG), and Software Engineering Method and Theory (SEMAT), also focus on unique custom development and ignore patterns derived from similar legacy software, and they also ignore reusable components taken from legacy software.

Although most software development methodologies ignore maintenance and renovation, there are a number of companies and tools aimed specifically at renovation projects. These include, but are not limited to,

1. Maintenance work benches
2. Code restructuring tools
3. Translators from old languages (i.e., COBOL) to modern languages (i.e., Java)
4. Data mining tools for extracting lost requirements and algorithms from old code
5. Automatic function point counting of legacy applications
6. Cyclomatic complexity analyzers
7. Static analysis tools for locating hidden bugs in legacy code
8. Test coverage tools
9. Security analysis tools to guard against cyber-attacks
10. Re-documentation tools to refresh obsolete requirements

Recreating all this legacy code would entail immense cost and substantial risk—hence, renovation is often an appealing strategy. Renovation is often cost-effective when the system in question has both high value and high maintenance costs—in addition, it is generally a less risky approach than complete redevelopment.

The principal factors that influence the cost of renovation include the following:

- The scope of the system: How big it is
- The quality of the existing software: The extent to which it is or is not structured, the degree of cohesion and coupling found in the code
- Tool support for reengineering: Availability tools to ease the process
- The extent of data conversion that may be required – for example, moving the underlying data structure from sequential to relational
- The availability of staff who are knowledgeable about the application and its technologies

The renovation process will typically entail the following steps:

- Source code translation: The existing code is converted, ideally by automated methods, from an old to a new version of the same language, or possibly to a different language. In many cases, a degree of manual effort may be unavoidable. There are a number of reasons why code translation may be necessary, including
  - Hardware platform update: The new platform compilers may not support the older language or version.

  - It may be difficult to find staff fluent in the old language (e.g., COBOL).
  - Organizational policy: The licensing cost for older operating systems, compilers, and so on may be prohibitive.
- Reverse engineering: This process is described elsewhere and may or may not be necessary.
- Program structure improvement: Many legacy systems exhibit high coupling and low cohesion, the opposite of what is desirable for maintainability. It may be desirable, for example, to apply the concepts of object-oriented development to improve maintainability. Short of that level of change, it may be sufficient to simply improve structure and documentation—much will depend on the skills and experience of the reengineering team.
- Program modularization: This step, similar to structure improvement, may be desirable to eliminate redundancy. This may include
  - Creating data abstraction hierarchies consistent with object-oriented principles.
  - Centralizing hardware control facilities such as device drivers.
  - Grouping similar functionality together.
  - Grouping business process functionality together.
- Data reengineering: For example, moving from sequential to relational data sets. Other data-related considerations that may be addressed include
  - Adding edits to improve data quality, potentially including scrubbing and conversion of existing data.
  - Removing limits included in the original code that are no longer appropriate—for example, field size limitations.
  - Architectural considerations—for example, moving from centralized to distributed.
  - Elimination of hard-coded literals.

Today in 2016, about 90% of all software projects are not "new" in the sense that they have never been done before. They are either enhancements to legacy applications or attempts to build new and better versions of legacy applications.

In general, attempting to re-engineer an aging legacy system will take longer and cost more than the original version, even if modern languages are used. The problem is that the original requirements, architecture, and design cannot be ignored or left out, but these are essentially invisible until some kind of data mining or reengineering takes place.

Table 32.2 shows the results of renovating an application of 1000 function points in the Java programming language, using a code converter that translated legacy COBOL into modern Java.

In 2016, legacy renovation has become an important activity of the U.S. software industry and also the global software industry. It is unfortunate that most of the methodologies discussed in this book ignore the maintenance of legacy software and assume brand new applications that have never been built before. Worse,

Table 32.2 Example of 1000 Function Point Renovation Project with Java Language

Project Size and Nature Scores		**Renovate**
1	High-reliability, high-security projects	10
2	Small projects <100 function points	9
3	Medium projects	8
4	Maintenance projects	8
5	IT projects	8
6	Systems/embedded projects	8
7	Large projects >10,000 function points	7
8	Enhancement projects	7
9	Web/cloud projects	4
10	New projects	0
	Subtotal	69
Project Activity Scores		**Legacy Renovate**
1	Coding	9
2	Integration	9
3	Change control	8
4	Defect prevention	8
5	Requirements changes	7
6	Architecture	7
7	Models	7
8	Design	7
9	Testing defect removal	7
10	Requirements	6
11	Test case design/generation	6
12	Pattern matching	5
13	Reusable components	5
14	Planning, estimating, measurement	5

(Continued)

Table 32.2 (Continued) Example of 1000 Function Point Renovation Project with Java Language

15	Pre-test defect removal	5
	Subtotal	101
	Total	170
Regional Usage (Projects in 2016)		**Legacy Renovate**
1	North America	15,000
2	Pacific	15,000
3	Europe	15,000
4	South America/Central America	8,000
5	Africa/Middle East	8,000
	Total	61,000
Defect Potentials per Function Point (1000 function points; Java language)		**Legacy Renovate**
1	Requirements	0.25
2	Architecture	0.35
3	Design	0.33
4	Code	1.25
5	Documents	0.30
6	Bad fixes	0.20
	Total	2.68
	Total defect potential	2,680
Defect Removal Efficiency (DRE)		
1	Pre-test defect removal (%)	70.00
2	Testing defect removal (%)	85.00
	Total DRE (%)	95.50
	Delivered defects per function point	0.121
	Delivered defects	121
	High-severity defects	19
	Security flaws	2

(Continued)

Table 32.2 (Continued) Example of 1000 Function Point Renovation Project with Java Language

Productivity for 1000 Function Points		Legacy Renovate
	Work hours per function point 2016	13.60
	Function points per month 2016	9.71
	Schedule in calendar months	15.20

Data Source: Namcook analysis of 600 companies.

Legacy renovation scores
Best = +10
Worst = −10

they also assume custom designs and manual coding instead of construction from standard reusable components.

Fortunately, the software industry does have some companies and some tool suites that are aimed at aging legacy applications, but the huge volume of legacy software does not get the attention that it needs in the software literature. In fact, the outsource industry is now over 70% involved with maintaining legacy software rather than new development for clients.

Chapter 33

Legacy Repair Development

Executive summary: *Legacy repairs are a major software activity in 2016 due to the huge number of aging legacy applications. Legacy repairs are often difficult due to the high cyclomatic complexity of many legacy applications. Also, quite a few legacy applications are written in antique programming languages such as MUMPS or CORAL, which are not well known by modern programmers. As a result, bad-fix injections (new bugs in bug repairs) remain an endemic problem of the software industry and a continuing problem for legacy maintenance. See also legacy renovation, legacy data mining, and legacy redevelopment. Commercial off-the-shelf (COTS) modifications and enterprise resource planning (ERP) modifications are also relevant, since a prime reason for acquiring such packages is to replace aging legacy software.*

In the calendar year 2016, there are probably about 7,000,000 aging legacy applications running in the United States. These would total perhaps 5,250,000,000 function points and 750,000,000 Software Non-Functional Assessment Process (SNAP) points. The estimated number of software engineers working on maintaining these legacy applications is perhaps 1,750,000. Legacy applications contain large numbers of latent bugs: probably about 0.5 per function point or 375 latent bugs per application.

Assuming latent bugs are discovered at a rate of about 10% per year, each legacy application probably generates roughly 35 bug reports per year, of which about 6 would be high-severity defects that interfere with using the applications. Legacy applications also contain security flaws, which might total as many as 59,000,000 security flaws for all U.S. legacy applications combined or perhaps 8 security flaws per legacy software package.

Finding and fixing bugs in legacy software applications is now a major component of total U.S. software budgets for all companies and all government agencies, without exception.

Because legacy defect repairs are not new development, essentially none of the methodologies discussed in this book even discusses the issues except for the other legacy chapters.

Also, maintenance and bug repairs are very difficult to measure, because most bug repairs do not add features or new function points, and some only require a single line of code to be changed.

Assume that you have a legacy application of 1000 function points that was developed for 15 work hours per function point or 15,000 total hours. This is equal to 8.8 function points per month. Assume that in Year 1 of deployment, clients reported 10 post-release bugs that need to be fixed.

Assume that nine of these bugs involved sections of the application that totaled one function point each or nine function points in total. Assume that the 10th bug required 1 line of code to be changed for zero function points. Assume that each bug required 5 h for repairs and testing, including the 10th bug at zero function points. How do you measure these bug repairs?

What should be done is to keep a running total of all hours expended on the application year by year and also the annual change in size for function points. Each calendar or fiscal year, the data should be re-normalized to show current accumulated values. Table 33.1 illustrates the example just discussed.

It is interesting that while total cumulative work hours increased for the year, normalized productivity in terms of work hours per function point went down, due to the fact that the bug repairs were faster than the original development.

The really tricky topic is Bug 10, which had zero function points but required 5 h of work to fix the problem. Obviously, normal function point counting will not work, since there are no new features. A possible surrogate is to use "backfiring" or mathematical conversion from lines of code.

Assume that Bug 10 required one line of Java code in the application to be changed. Since Java contains about 53 logical code statements per function point, it can be assumed that Bug 10 is roughly 1/50th of a function point. In this case, work hours per function point would be about 250, and function points per month would be only about 1.8.

In real life, the situation is made more complex because of other factors not shown in Table 33.1:

1. About 7% of post-release bug repairs are "bad fixes" and contain new bugs.
2. Software keeps growing at about 8% per calendar year.

Table 33.1 Post-Release Bug Repairs for Year 1

Size	*Function Points*	*Work Hours*	*Work Hours per Function Point*
Application	1,000	15,000.00	15.00
Bug 1	1	5.00	5.00
Bug 2	1	5.00	5.00
Bug 3	1	5.00	5.00
Bug 4	1	5.00	5.00
Bug 5	1	5.00	5.00
Bug 6	1	5.00	5.00
Bug 7	1	5.00	5.00
Bug 8	1	5.00	5.00
Bug 9	1	5.00	5.00
Bug 10	0	5.00	0
Year 1 total	**1,009.0**	**15,050.00**	**14.92**

3. Cyclomatic complexity goes up each year due to bug repairs.
4. Bugs tend to clump in a small number of "error-prone modules."
5. Duplicate and invalid bug reports require effort but add zero functionality.
6. There are some 23 different kinds of work all called "maintenance."
7. Customer support also needs to be measured and included.

The bottom line is that estimating and measuring post-release defect repairs and maintenance costs are a lot trickier than just measuring software development projects.

Most benchmarks assume a constant size value once function points are originally counted. This is a mistake, because software projects have been measured to grow and change at up to 3% per calendar month during development and up to 8% per calendar year after release. Table 33.2 shows a 10 year pattern of applications growth for a large system of 10,000 function points as predicted by the author's Software Risk Master (SRM) tool.

The bottom line is that software maintenance estimating and software maintenance measures are complex and difficult. The 23 kinds of work all called "maintenance" are listed in the following section.

Table 33.2 SRM Multi-Year Sizing Example

	Nominal application size in IFPUG function points	10,000			
	SNAP points	1,389			
	Language	C			
	Language level	2.50			
	Logical code statements	1,280,000			
		Function Points	**SNAP Points**	**Logical Code**	
1	Size at end of requirements	10,000	1,389	1,280,000	
2	Size of requirement creep	2,000	278	256,000	
3	Size of planned delivery	12,000	1,667	1,536,000	
4	Size of deferred features	−4,800	(667)	(614,400)	
5	Size of actual delivery	7,200	1,000	921,600	
6	Year 1 usage	12,000	1,667	1,536,000	Kicker
7	Year 1 usage	13,000	1,806	1,664,000	
8	Year 1 usage	14,000	1,945	1,792,000	
9	Year 1 usage	17,000	2,361	2,176,000	Kicker
10	Year 1 usage	18,000	2,500	2,304,000	
11	Year 1 usage	19,000	2,639	2,432,000	
12	Year 1 usage	20,000	2,778	2,560,000	
13	Year 1 usage	23,000	3,195	2,944,000	Kicker
14	Year 1 usage	24,000	3,334	3,072,000	
15	Year 1 usage	25,000	3,473	3,200,000	

Notes: Simplified example with whole numbers for clarity.

Deferred features usually due to schedule deadlines.
IFPUG = International function point users group.
Kicker = Extra features added to defeat competitors.

Patent application 61434091. February 2012.

Major Kinds of Work Performed under the Generic Term "Maintenance"

1. Major enhancements (new features of >50 function points)
2. Minor enhancements (new features of <5 function points)
3. Maintenance (repairing customer defects for good will and pro bono)
4. Warranty repairs (repairing defects under formal contract or for a fee)
5. Customer support (responding to client phone calls or problem reports)
6. Error-prone module removal (eliminating very troublesome code segments)
7. Mandatory changes (required or statutory changes such as new tax laws)
8. Complexity or structural analysis (charting control flow plus complexity metrics)
9. Code restructuring (reducing cyclomatic and essential complexity)
10. Optimization (increasing performance or throughput)
11. Migration (moving software from one platform to another)
12. Conversion (changing the interface or file structure)
13. Reverse engineering (extracting latent design information from code)
14. Reengineering (transforming legacy application to modern forms)
15. Dead code removal (removing segments no longer used)
16. Dormant application elimination (archiving unused software)
17. Nationalization (modifying software for international use)
18. Mass updates such as Euro or Year 2000 repairs
19. Refactoring, or reprogramming applications to improve clarity
20. Retirement (withdrawing an application from active service)
21. Field service (sending maintenance members to client locations)
22. Reporting bugs or defects to software vendors
23. Installing updates received from software vendors

Because of the intrinsic difficulty of quantifying software maintenance, Table 33.3, showing the scores and quantified results, needs to be considered carefully. Although the application being maintained is 1000 function points and coded in Java, the sum total of the annual maintenance for the application was only 25 bug repairs 25 function points, but still in Java.

This table is something of an anomaly, because it shows the results of defect repairs of 25 function points in an application sized at 1000 function points. One of the repairs had zero function points but still required code changes and testing.

Legacy software repairs are a major effort for the entire global software industry in 2016. Unfortunately, most software development methodologies ignore software legacy maintenance, and the software literature overall has only sparse coverage. Software measures are difficult, too, due to the very small or even zero function point sizes of many repairs.

Table 33.3 Example of 1000 Function Point Legacy Repair Project with Java Language

Project Size and Nature Scores		Legacy Repairs
1	Enhancement projects	6
2	Maintenance projects	10
3	Large projects >10,000 function points	6
4	New projects	6
5	Medium projects	8
6	IT projects	8
7	Systems/embedded projects	8
8	Web/cloud projects	8
9	High-reliability, high-security projects	8
10	Small projects <100 function points	7
	Subtotal	75
Project Activity Scores		**Legacy Repairs**
1	Requirements	10
2	Coding	6
3	Change control	9
4	Integration	7
5	Reusable components	0
6	Planning, estimating, measurement	–4
7	Defect prevention	4
8	Pre-test defect removal	3
9	Testing defect removal	8
10	Requirements changes	7
11	Design	3
12	Test case design/generation	9
13	Pattern matching	4
14	Architecture	0

(Continued)

Table 33.3 (Continued) Example of 1000 Function Point Legacy Repair Project with Java Language

15	Models	0
	Subtotal	66
	Total	124
	Regional Usage (Projects in 2016)	**Legacy Repairs**
1	North America	200,000
2	Pacific	185,000
3	Europe	190,000
4	South America/Central America	90,000
5	Africa/Middle East	60,000
	Total	725,000
Defect Potentials per Function Point (25 function points; Java language)		**Legacy Repairs**
1	Requirements	0.10
2	Architecture	0.10
3	Design	0.10
4	Code	1.00
5	Documents	0.00
6	Bad fixes	0.70
	Total	2.00
	Total defect potential	50.00
Defect Removal Efficiency (DRE)		
1	Pre-test defect removal (%)	55.00
2	Testing defect removal (%)	82.00
	Total DRE (%)	91.90
	Delivered defects per function point	0.292
	Delivered defects	4
	High-severity defects	1
	Security defects	0

(Continued)

Table 33.3 (Continued) Example of 1000 Function Point Legacy Repair Project with Java Language

Productivity for 1000 Function Points		Legacy Repairs
	Work hours per function point 2014	75.00
	Function points per month 2014	1.76
	Schedule in calendar months	1.00

Data Source: Namcook analysis of 600 companies.

Legacy repair scoring 2016
Best = +10
Worst = −10

Chapter 34

Legacy Replacement Development

Executive summary: *In 2016, the average business application is more than 10 years old and often unstable. Some are over 25 years old and very unstable. The term "legacy replacement" refers to keeping the essential features of an existing legacy application but replacing it with a new version using improved technologies and modern programming languages. The first step in legacy replacement may be to review and update existing documentation. If documentation is incorrect or missing, data mining can create synthetic documents.*

In 2016, the average age of business applications among the author's clients is more than 10 years. Some business and also technical applications such as air-traffic control and switching systems are more than 25 years of age.

Software does not age well. Over time, the numerous small changes made to legacy applications increase entropy or disorder. The practical effect of increasing software entropy is a slow increase in average cyclomatic complexity. High cyclomatic complexity makes bug repairs difficult and raises "bad-fix" probabilities.

Table 34.1 shows a typical situation that can lead to either renovation of legacy software or replacement of legacy software. The most troublesome problems are in bold for easy visibility.

As illustrated in Table 34.1, a company owns an aging application that is high in cyclomatic complexity, unstable, unreliable, and filled with both bugs and security flaws. It also suffers from low customer satisfaction rates.

It is obvious that renovation of this application would not be a success. The language is obsolete; the file structure is not suitable; cyclomatic complexity is

Table 34.1 Overview of Reasons for Legacy Software Redevelopment

	Legacy Application	*Replacement Application*	*Difference*
Age (years)	12.50	0.00	12.50
Platform of application	Mainframe	Cloud	
Size (function points)	1,000	1,000	0
Language	Assembly	Java	
Size (lines of code)	320,000	53,000	267,000
Application file structure	Flat file	Relational	
Average cyclomatic complexity	**35**	8	27
Number of bugs per year	**65**	3	62
High-severity bugs per year	**10**	1	9
EPM	**16**	0	16
Number of known security flaws	**15**	0	15
Bad-fix injection (%)	**17.50**	2.00	15.50
Reliability: mean time to failure (hours)	23.00	655.00	−632.00
Maintenance effort (months per) year	32.00	3.00	29.00
Mean time to repair bugs (days)	5.50	0.50	5.00
Maintenance ($ per year)	320,000	30,000	290,000
Security recovery (effort per year)	6.00	0.00	6.00
Annual security recovery ($)	60,000	0.00	60,000
Annual $ per year	380,000	$30,000	350,000
Customer satisfaction	**Poor**	Good	

dangerously high; and 17.5% of bug repairs contain new bugs called *bad fixes*. Clearly, replacement of this aging application is the only choice that is economically viable.

Also, the legacy application has an alarming number of "error-prone modules" (EPM). These were discovered by IBM in the 1970s. An analysis of the IMS database application showed that out of 425 modules, 300 had zero-defect reports from customers. About 57% of all customer-reported bugs were against 32 modules developed in one department.

EPM are an endemic problem for legacy software, and they have been found in essentially all legacy applications coded in low-level languages. The main reasons for EPM include, but are not limited to,

1. Low-level programming languages
2. High cyclomatic complexity and excessive branching
3. Poor training of developers
4. Poor training of test personnel
5. Failure to use code inspections
6. Failure to use static analysis
7. Poor quality measures
8. Lack of quality measures that show errors for specific modules
9. Poor test coverage, often due to high cyclomatic complexity
10. Mediocre test case and test script designs

EPM are fully preventable before they occur but difficult to treat afterwards. Usually, EPM require surgical removal and complete replacement.

Prevention of EPM is fairly easy and extremely cost-effective. The steps that can prevent EPM include

1. High-level programming languages
2. Adherence to structured programming principles
3. Low cyclomatic complexity
4. Daily use of static analysis
5. Code inspections of all critical modules
6. Defect tracking that maps bugs to specific modules
7. Using test coverage tools for all modules
8. Better training of developers and testers

Modules are normally based on source code rather than function points. If you assume that an average module will contain about 150 source code statements, the legacy application in assembly language will have about 2100 modules. Of these, about 5%, or 105 modules out of 2100 modules, will probably be error prone.

The new version coded in Java will probably have about 350 modules in total. Due to hopefully better coding practices in 2016, there will be zero EPM.

EPM are extremely expensive, since they contain the bulk of all software errors. Worse, they often are so poorly structured and have such high cyclomatic complexity that bad-fix injections can exceed 25%, as opposed to the U.S. average of about 7% bad-fix injections.

Make no mistake, the legacy application illustrated in this chapter is not going to stabilize or run out of bugs. Aging applications are difficult to enhance or modify, and a high percentage of changes introduce regressions or "bad fixes," which are new bugs in bug repairs themselves.

Of course, in today's world, businesses have several options to consider before jumping into replacement of legacy software:

1. Buy a commercial package that has most of the needed features.
2. Find an open-source package that has most of the needed features.
3. Try to renovate the legacy application—*hazardous in this case*.
4. Use data mining to extract legacy requirements and design factors.
5. Build the replacement application internally.
6. Outsource the replacement application domestically.
7. Outsource the application overseas.
8. Keep running the legacy application as long as possible.

These choices are listed in approximate rank order of economic success. Buying is usually better than building. Building yourself is usually better than outsourcing.

However, a hybrid approach is worth considering. A colleague, Dr. Tom Love, has suggested that the renovation of legacy applications can stretch their useful working lives. Also, the savings in annual maintenance costs from the renovations can at least partly fund redevelopment.

(The author has suggested an international museum of programming languages together with working compilers and other tools. This is needed, as aging software in obsolete languages becomes more and more common. Some of these aging applications have national importance, such as air-traffic control, power generation, and telecommunications switching systems.)

Table 34.2 shows U.S. software history by decade starting in 1950, when 100% of software applications were brand new and 0% needed maintenance or renovation.

The United States in 2016 has an aging human population, as we all know. But the average age of operating software projects is also increasing with each year that passes. Because renovation is difficult and replacement may be expensive, many important mission-critical applications are still in daily use, even though they are slow, error prone, and vulnerable to cyber-attacks.

(For additional information, refer to the author's book *The Technical and Social History of Software Engineering*, Addison Wesley 2014.)

From the fact that a majority of methodologies in this report focus on development and ignore maintenance and renovation, it can be seen that the software engineering field does not really understand the steady migration to maintenance and renovation and away from new custom development.

Given the huge quantity of existing legacy software, it is obvious that every effective software development methodology should start by examining similar

Table 34.2 Distribution of New, Enhance, and Replace Projects

Year	*New Software*	*Enhanced Software*	*Replaced Software*	*Total Projects*	*Average Age (Years)*
1950	50	0	0	50	1.00
1960	1,200	350	25	1,575	2.50
1970	12,500	5,000	1,000	18,500	5.00
1980	65,000	30,000	4,000	99,000	7.50
1990	245,000	115,000	30,000	390,000	9.00
2000	900,000	800,000	125,000	1,825,000	12.00
2010	1,000,000	1,100,000	450,000	2,550,000	15.00
2020	1,200,000	1,300,000	800,000	3,300,000	20.00

legacy applications. But only pattern matching and reuse development do this. Most methodologies, such as Agile, Crystal, extreme programming (XP), iterative, spiral, and so on, assume, mistakenly, that every application is brand new and has never been done before.

For that matter, most of the existing meta-models such as International Organization for Standardization (ISO), Object Management Group (OMG), and Software Engineering Method and Theory (SEMAT), also focus on unique custom development and ignore patterns derived from similar legacy software, and they also ignore reusable components taken from legacy software.

Although most software development methodologies ignore maintenance, renovation, and replacement, there are a number of companies and tools aimed specifically at renovation or replacement projects. These include, but are not limited to,

1. Maintenance work benches
2. Code restructuring tools
3. Translators from old languages (i.e., COBOL) to modern languages (i.e., Java)
4. Data mining tools for extracting lost requirements and algorithms from old code
5. Automatic function point counting of legacy applications
6. Cyclomatic complexity analyzers
7. Static analysis tools for locating hidden bugs in legacy code
8. Test coverage tools
9. Security analysis tools to guard against cyber-attacks
10. Re-documentation tools to refresh obsolete requirements and design

For legacy repairs, there are serious consequences from increased entropy and high levels of cyclomatic complexity. Even applications that are well structured at release will gradually become unstructured due to the accumulation of small changes over more than 10 years.

The principal factors that influence the cost of replacement are similar to those for renovation and include the following:

- The scope of the system: How big it is
- The quality of the existing software: The extent to which it is or is not structured, the degree of cohesion and coupling found in the code
- Tool support for reengineering: Availability tools to ease the process
- The extent of data conversion that may be required: For example, moving the underlying data structure from sequential to relational
- The availability of staff knowledgeable about the application and its technologies
- Data reengineering: For example, moving from sequential to relational data sets.

Today in 2016, about 90% of all software projects are not "new" in the sense that they have never been done before. They are either enhancements to legacy applications or attempts to build new and better versions of legacy applications.

In general, attempting to re-engineer and replace an aging legacy system will take longer and cost more than the original version, even if modern languages are used. The problem is that the original requirements, architecture, and design cannot be ignored or left out, but these are essentially invisible until some kind of data mining or reengineering takes place.

Table 34.3 shows the results of redeveloping an application of 1000 function points in the Java programming language. The replacement version used the Rational Unified Process (RUP) because Agile is not a good choice for replacing legacy applications.

In 2016, legacy software applications outnumber new development projects in the United States and around the world by probably a ratio of 10:1. Unfortunately, most methodologies focus on new development and ignore aging legacy applications.

Legacy redevelopment is an expensive but necessary activity in 2016. It suffers from a shortage of effective procedures, tools, and training of software engineering personnel. Most academic training of software engineering students deals with new development, while maintenance of legacy software is either ignored or taught in passing.

Table 34.3 Example of 1000 Function Point Legacy Replacement Project with Java Language

Project Size and Nature Scores		**Legacy Replace**
1	Web/cloud projects	10
2	High-reliability, high-security projects	10
3	Small projects < 100 function points	9
4	Medium projects	8
5	IT projects	8
6	Systems/embedded projects	7
7	Large projects > 10,000 function points	6
8	Enhancement projects	6
9	Maintenance projects	6
10	New projects	0
	Subtotal	70
Project Activity Scores		**Legacy Replace**
1	Coding	9
2	Integration	9
3	Change control	8
4	Defect prevention	8
5	Pattern matching	7
6	Planning, estimating, measurement	7
7	Requirements	7
8	Requirements changes	7
9	Architecture	7
10	Models	7
11	Design	7
12	Pre-test defect removal	7
13	Test case design/generation	7
14	Testing defect removal	7

(Continued)

Table 34.3 (Continued) Example of 1000 Function Point Legacy Replacement Project with Java Language

15	Reusable components	6
	Subtotal	110
	Total	180
Regional Usage (Projects in 2016)		**Legacy Replace**
1	North America	12,000
2	Pacific	10,000
3	Europe	13,000
4	South America/Central America	7,000
5	Africa/Middle East	5,000
	Total	47,000
Defect Potentials per Function Point (1000 function points; Java language)		**Legacy Replace**
1	Requirements	0.35
2	Architecture	0.30
3	Design	0.55
4	Code	1.20
5	Documents	0.30
6	Bad fixes	0.15
	Total	2.85
	Total defect potential	2,850
Defect Removal Efficiency (DRE)		
1	Pre-test defect removal (%)	65.00
2	Testing defect removal (%)	85.00
	Total DRE (%)	94.80
	Delivered defects per function point	0.148
	Delivered defects	148
	High-severity defects	23
	Security defects	2

(*Continued*)

Table 34.3 (Continued) Example of 1000 Function Point Legacy Replacement Project with Java Language

Productivity for 1000 Function Points		Legacy Replace
	Work hours per function point 2016	13.35
	Function points per month 2016	9.89
	Schedule in calendar months	14.05

Data Source: Namcook analysis of 600 companies.

Legacy replacement scoring
Best = +10
Worst = −10

Chapter 35

Mashup Development

Executive summary: *The term "mashup development" is fairly new and refers to Web 2.0 applications that are created using pieces of existing web applications. Although a new methodology, mashups are very popular and have rapidly increasing tools sets. Their popularity has led to dozens of conferences all over the world. IBM has become a major tool vendor to the mashup community. While mashups are a form of reuse, the term tends to be used mainly for Web 2.0 applications rather than general software products.*

The idea of building new software applications from pieces of older software has been around since the 1960s, but has been somewhat difficult to do well or safely. However, with the advent of the Web and especially Web 2.0, constructing applications from existing web projects is getting easier and more reliable.

The term *mashup* is curious in that it comes from the British West Indies and is much older than software. The original meaning of mashup was to be intoxicated and hence, not functioning normally. The term mashup is not only restricted to modern software but it is also used in music, to refer to creating new musical compositions spliced together from segments of existing musical compositions.

A minor example of a mashup software project might be for a state government to combine the addresses and photographs of all state office buildings with a Google map to create a mashup that shows citizens how to find and drive to any desired state office. Most of us have seen hundreds of similar mashups when doing web searches to look for restaurants, real estate, or other tangible objects that need a good map to provide directions to consumers. Mashups have become common features of social networks and Web 2.0.

Mashups depend on fairly clear and standardized application program interfaces (API) in order to work successfully. Today, in 2016, these are fairly common and are available for many web applications.

Mashup development may run into copyright and ownership issues if the components used for the new application are owned by some other company. If only reusable components are acquired from the company building the new mashup there will be no problem, but if the components come from other companies then legal issues and copyright violations need to be considered.

Mashup development does not usually require actual programming in the sense of writing computer code in languages such as Java or C#. Instead, various mashup composition tools are used and these are simple enough for casual end users, assuming fairly good computer literacy, which in 2016 is the norm among young people and not uncommon among mature people.

Mashup development is based on mashup composition tools that allow visual splicing of web services from multiple sources into a new application. Today, in 2016, there are dozens of mashup composition tools from dozens of vendors. There are also many web images of mashup composition and a Google search will show many.

The results of mashup development for a small application of 1000 function points with the code itself in Java are shown in Table 35.1. However, the mashup model does not imply actual hand coding of Java, merely that some Java code will be included in the mashup application.

For applications where mashup technology is feasible, the results are a useful combination of very low effort, very short schedules, and acceptable quality assuming that the selected components were of good quality before being utilized. Since mashups are much quicker and cheaper than conventional development, it is easy to see why they have become popular.

Mashup software development is a fairly new phenomenon that is growing rapidly. It works very well for Web 2.0 and cloud applications that are similar to existing applications so that pieces can be extracted and spliced together to create a new application.

Obviously, mashup development is not a good choice for systems and embedded applications that run on custom hardware devices. It is probably not a good choice for applications that need Food and Drug Administration (FDA) or Federal Aviation Administration (FAA) certification, unless the older applications used for the reusable features are already certified.

Manual function point sizing is difficult with mashups since the applications are constructed from pieces and chunks of existing software. Software non-functional assessment process (SNAP) sizing is also difficult for the same reason. However, sizing via pattern matching as done in the author's Software Risk Master (SRM) tool allows mashup applications to be sized rapidly and early.

Table 35.1 Example of 1000 Function Point Mashup Project with Java Language

Project Size and Nature Scores		**Mashup**
1	Small projects <100 FP	10
2	Medium projects	10
3	New projects	10
4	IT projects	10
5	Web/cloud projects	10
6	Large projects >10,000 FP	6
7	Systems/embedded projects	4
8	High-reliability, high-security projects	4
9	Enhancement projects	2
10	Maintenance projects	2
	Subtotal	68
Project Activity Scores Mashup		
1	Pattern matching	10
2	Reusable components	10
3	Models	10
4	Coding	10
5	Requirements	8
6	Requirements changes	8
7	Change control	8
8	Integration	8
9	Testing defect removal	8
10	Design	7
11	Defect prevention	7
12	Pre-test defect removal	7
13	Planning, estimating, measurement	5
14	Architecture	4

(Continued)

Table 35.1 (Continued) Example of 1000 Function Point Mashup Project with Java Language

15	Test case design/generation	4
	Subtotal	114
	Total	182
Regional Usage (Projects in 2016)		**Mashup**
1	North America	17,000
2	Pacific	16,000
3	Europe	15,000
4	South America/Central America	10,000
5	Africa/Middle East	5,000
	Total	63,000
Defect Potentials per Function Point (1000 function points; Java language)		**Mashup**
1	Requirements	0.10
2	Architecture	0.20
3	Design	0.50
4	Code	1.00
5	Documents	0.40
6	Bad fixes	0.20
	Total	2.40
	Total defect potential	2400
Defect Removal Efficiency (DRE)		
1	Pre-test defect removal	70.00%
2	Testing defect removal	85.00%
	Total DRE	95.16%
	Delivered defects per function point	0.117
	Delivered defects	117
	High-severity defects	18
	Security defects	2

(Continued)

Table 35.1 (Continued) Example of 1000 Function Point Mashup Project with Java Language

Productivity for 1000 Function Points		Mashup
	Work Hours perFunction Point 2016	6.50
	Function Points per month 2016	20.31
	Schedule in calendar months	5.00

Data Source: Namcook analysis of 600 companies.

Mashup Scoring
Best = +10
Worst = –10

Chapter 36

Merise

Executive summary: *Merise is an older methodology that was developed and used primarily in France for mainframe applications during the 1970s and 1980s. It is rarely seen today, and is almost never seen outside of France or French territories. Merise was heavily influenced by data-oriented approaches associated with the advent of re databases. Merise relies heavily on up-front planning and modeling. As such, Merise is a "heavy" methodology and by no means a "light" methodology.*

Merise was an attempt to integrate data-oriented (such as Information Engineering) and process-oriented modeling approaches. In practice, these ideas tended to lead to "analysis paralysis." The Merise approach dealt with three different levels of abstraction:

- A conceptual view: What
- An organizational view: Who does what, when, where
- An operational view: How are things done

Merise consisted of three different cycles:

- An abstraction cycle: The levels of conceptual modeling of both data and process (six models in total).
  - Abstract-level data modeling using entity-relationship (ER) diagrams
  - Organizational-level data model consisting of tables for the relational model
  - Operational-level physical data model expressed in a data definition language

  - Conceptual-level process models consisting of events, event synchronization, and operations
  - Organizational-level process models detailed to a task level and associated with actors
  - Operational-level process models dealing with computer actors only modeled as transactions (e.g., temporal events)
- A development life cycle: Four steps; when reverse engineering a legacy system, three levels of abstraction again appear
  - Strategic planning at the corporate level
  - Preliminary study for domain of interest
  - Detailed study for a particular project
  - Schedule for development, implementation, and maintenance
- A decision cycle: A joint process between senior managers, users, and developers

Most of the documentation of Merise is in French and is not considered relevant for current generation projects. Merise usage centers around France with some North American usage in Quebec in Canada. Use elsewhere is sparse.

An example of Merise for a project of 1000 function points using the Java language is shown in Table 36.1.

Merise is no longer an active methodology, but it served a useful purpose at the beginning of the era when large databases began to arrive in the software world. Merise was effective for database and repository software where relational databases were utilized. Some Merise concepts are still valid and have spread to other methodologies such as information engineering.

Table 36.1 Example of 1000 Function Point Merise Project with Java Language

Project Size and Nature Meta-Model		Merise	
1	Small projects <100 FP	9	
2	Medium projects	8	
3	IT projects	8	
4	Large projects >10,000 FP	7	
5	New projects	7	
6	High-reliability, high-security projects	6	
7	Systems/embedded projects	5	
8	Web/cloud projects	5	
9	Enhancement projects	4	
10	Maintenance projects	3	
	Subtotal	62	
Project Activity Strength Scores Merise			
1	Coding	8	
2	Change control	8	
3	Integration	8	
4	Reusable components	7	
5	Planning, estimating, measurement	7	
6	Requirements	7	
7	Requirements changes	7	
8	Architecture	7	
9	Models	7	
10	Design	7	
11	Defect prevention	7	
12	Pre-test defect removal	7	
13	Test case design/generation	7	
14	Testing defect removal	7	

(Continued)

Table 36.1 (Continued) Example of 1000 Function Point Merise Project with Java Language

15	Pattern matching	6	
	Subtotal	107	
	Total	169	
Regional Usage (Projects in 2016)		**Merise**	
1	North America	500	Quebec, Montreal
2	Pacific	200	
3	Europe	5000	France
4	South America/Central America	100	
5	Africa/Middle East	100	
	Total	5900	
Defect Potentials per Function Point (1000 function points; Java language)		**Merise**	
1	Requirements	0.3	
2	Architecture	0.4	
3	Design	0.3	
4	Code	1.4	
5	Documents	0.4	
6	Bad fixes	0.3	
	Total	3.1	
	Total defect potential	3100	
Defect Removal Efficiency (DRE)			
1	Pre-test defect removal	50.00%	
2	Testing defect removal	85.00%	
	Total DRE	92.50%	
	Delivered defects per function point	0.233	
	Delivered defects per function point		

(Continued)

Table 36.1 (Continued) Example of 1000 Function Point Merise Project with Java Language

	High-severity defects		
	Security defects		
Productivity for 1000 Function Points		**Merise**	
	Work hours per function point 2016	16.00	
	Function points per month 2016	8.25	
	Schedule in calendar months	17.30	

Data Source: Namcook analysis of 600 companies.

Merise Scoring
Best = +10
Worst = –10

Chapter 37

Micro Service Software Development

Executive summary: *This is one of the newer and more interesting software development methodologies. Micro service development is a kind of derivative extension of the older service-oriented architecture (SAR). Both involve the construction of software in large part from collections of standard reusable components or "services." SAR deals with fairly large chunks or components that usually top 100 function points in size and may top 1000 function points. Micro services, on the other hand, deal with very small services often below 5 function points in size and seldom topping 10 function points.*

The term *micro service* was introduced by Dr. Peter Rodgers in a speech at the Cloud Computing Expo conference in 2005. Micro service development is often used in conjunction with DevOps, and involves the construction of software using large numbers of fairly small reusable components many of which are perhaps 5 function points or smaller in size.

Some examples of micro service components would include currency conversions; inflation rate calculations; compound interest calculations; temperature conversions from Celsius to Fahrenheit; calculating the median, mean, and mode from lists of values; software size conversion from story points to International Function Point Users Group (IFPUG) function points; calculating automobile fuel consumption in terms of miles per gallon or liters per mile. These have in common a compact size below 5 function points and constrained fairly simple inputs and outputs.

Micro services, of course, need common interface conventions and use generic pipelines and common message formats. This discussion is not a tutorial on SAR or micro service architecture as such, but rather a comparison of the quantified results of using micro services in real-world applications.

To date, micro services are used primarily for web and cloud applications, and they have proved to be both robust and successful but with some caveats. The individual micro services can be coded in different programming languages and were originally developed via an array of development methodologies ranging from Agile to waterfall. Micro service applications can also use modern container development approaches.

Among the caveats and cautions is the fact that micro services are not a panacea and may not always be the best choice. There are also warnings of slower performance speeds and increased cyclomatic complexity levels, which may make testing of full applications more expensive and difficult. Of course, testing small individual micro service components is fairly straightforward.

An extension of the micro service concept down to even smaller chunks called *nano services* (below 1 function point in size) sometimes occurs but not successfully. Nano service development is even claimed to be anti-pattern or harmful because of its very high cyclomatic complexity and possibly very sluggish performance. An example of a nano service might be the rounding of numeric decimal places in mathematical calculations. Another nano service would be backfiring or mathematical conversion from lines of code to IFPUG function points. (Java averages about 53 code statements per function point; C averages about 128 code statements per function point. Tables of conversion values exist for over 600 programming languages in 2016.)

Overall, applications built with the micro service concept may not only use 100% micro services but may also include perhaps 10%–20% custom designs and custom code for features where no micro service components actually exist in 2016.

While micro service applications themselves can be constructed at rates in excess of less than 2 hours per function points, the custom designs and custom code portions would drop back to industry averages in the range of 10–15 work hours per function point. The net productivity would be the combination of custom work and reusable micro service features in the final application; probably in the 5–10 work hour per function point range.

As of 2016, the majority of micro service components come from third parties or from local collections developed over several years. As a result, most micro service components have not been sized using the new SNAP metric for non-functional requirements. In fact, most micro service components probably don't include many non-functional requirements.

There are expanding numbers of tools that support the micro service concept including frameworks for construction, test tools, and hosting. Some of the many tool sets include Docker, Microsoft's Microservice4Net, NetKernal, and Red Hat open-source tools. There are dozens of others, far too many to list here. A Google search on micro service software development will turn up hundreds of citations, as of 2016.

The overall scores for the micro service approach for 1000 function points coded in Java are shown in Table 37.1, as are schedules, quality, and work hours

Table 37.1 Example of 1000 Function Point Micro Service Project with Java Language

Project Size and Nature Scores		**Micro Service**
1	Small projects <100 FP	10
2	Medium projects	8
3	Large projects >10,000 FP	6
4	New projects	10
5	IT projects	10
6	Systems/embedded projects	7
7	Web/cloud projects	10
8	High-reliability, high-security projects	5
9	Enhancement projects	5
10	Maintenance projects	3
	Subtotal	74
Project Activity Scores		**Micro Service**
1	Pattern matching	10
2	Reusable components	10
3	Planning and estimating	7
4	Requirements	7
5	Architecture	5
6	Models	8
7	Design	9
8	Coding	9
9	Defect prevention	6
10	Test case design/generation	6
11	Testing defect removal	7
12	Requirements changes	10
13	Change control	10

(Continued)

Table 37.1 (Continued) Example of 1000 Function Point Micro Service Project with Java Language

14	Integration	9
15	Pre-test defect removal	5
	Subtotal	118
	Total	192
Regional Usage (Micro Service 2017) Service		**Micro Service**
1	North America	5,000
2	Pacific	2,500
3	Europe	3,600
4	South America/Central America	1,800
5	Africa/Middle East	500
	Total	13,400
Defect Potentials per Function Point (1000 function points; Java language)		**Micro Service**
1	Requirements	0.30
2	Architecture	0.10
3	Design	0.35
4	Code	0.40
5	Documents	0.10
6	Bad fixes	0.10
	Total	1.35
	Total defect potential	1350
Defect Removal Efficiency (DRE)		
1	Pre-test defect removal	70.00%
2	Testing defect removal	85.00%
	Total DRE	95.50%
	Delivered defects per function point	0.061

(Continued)

Table 37.1 (Continued) Example of 1000 Function Point Micro Service Project with Java Language

	Delivered defects	61
	High-severity defects	9
	Security defects	1
Productivity for 1000 Function Points		**Micro Service**
	Work hours per function point 2016	11.80
	Function points per month 2016	11.18
	Schedule in calendar months	9.50

Data Source: Namcook analysis of 600 companies.

Micro Service Scoring
Best = +10
Worst = –10

per function point. The work hours per function point results in this table and the quality results assume 95% reusable micro service construction and only 5% custom design and custom code.

In 2016, micro service development is new and still evolving. As of 2016, the total volume of micro service components is probably in the range of 2500. No doubt, many additional micro service functions will become available in future years.

However, micro services, like all other software development methods, would benefit from forensic analysis of existing applications in order to provide a full taxonomy of software features that might be utilized in the form of standard reusable components, either as micro services below 5 function points in size or as larger chunks of up to 1000 function points in size.

Micro service defect potentials are low because of the large volume of reusable components that have already gone through pre-test and test stages plus one or more years of customer use. They may not be zero-defect features but they are definitely better than custom designs and manual coding.

The author's Software Risk Master (SRM) estimating tool can predict micro service productivity and quality by selecting micro service development and then providing a user input for the distribution of custom code versus reusable micro service features, such as 90% reusable micro services and 10% custom designs and custom code. The SRM default value for micro service development is 95% reuse and 5% custom designs and custom code.

Chapter 38

Microsoft Solutions Framework (MSF) Development

Executive summary: *Microsoft is the world's largest software company and has thousands of developers. The Microsoft Solutions Framework (MSF) is a set of practices recommended and endorsed by Microsoft. The set combines a meta-model with a collection of sub-processes and tool suites. The Microsoft approach claims to be methodology neutral and can be used with Agile, waterfall, rational unified process (RUP), or other methodologies.*

Microsoft was incorporated in 1975 and its history is well known. As Microsoft grew into one of the largest software companies in the world, they found frequent problems with internal development approaches. By using a combination of external consultants and their own research personnel, Microsoft gradually smoothed out their internal methods.

Starting in 1993, Microsoft began to offer software process suggestions supported by tool suites to customers. The first release of the MSF was in 1993. This was followed by Release 2 in 1997; Release 3 in 2002; Release 4 in 2005; and Release 5 in 2013.

The MSF is not a "pure" development methodology such as Agile, RUP, or team software process (TSP). It is an amalgamation of a meta-model and a suite of sub-models, tools, and suggestions.

Microsoft states that the solutions framework is methodology independent and can be used with other approaches such as Agile, capability maturity model integrated (CMMI), RUP, TSP, and waterfall in alphabetical order. Further, Microsoft states that the MSF approach is scalable from small projects up to large systems. This is an unusual feature because without explicitly saying so, most methodologies are aimed at specific size ranges. Agile is aimed at small projects below 1000 function points. TSP is aimed at large systems in the 10,000 function point size range. Scalability is both rare and hard to accomplish.

Another fairly unique feature of the MSF is the inclusion of governance, or keeping management in control of software projects. As many readers know, large software projects are often out of control and keep slipping their schedules and overrunning their budgets. An image of the Microsoft governance track is shown in Figure 38.1

Yet another interesting feature of MSF is that it includes a specific risk identification and abatement model. Risk analysis is not part of many methodologies, but it should be part of all of them. Moreover, risk identification should be the first step in software development and performed as soon as a project can be placed on a formal taxonomy.

The risks of similar projects should be examined and risk solutions should be applied to the current project even before full requirements are known. Early risk management tools such as Software Risk Master (SRM) are available, which can identify risks via pattern matching from similar projects, and therefore, can be used very early prior to full requirements.

There is an extensive literature on MSF, with dozens of papers and some books published by Microsoft itself. There are also many images on the web of MSF concepts and tool suites that can be found via Google searches.

Because Microsoft has its own publishing house, there is no shortage of books and journal articles on the MSF.

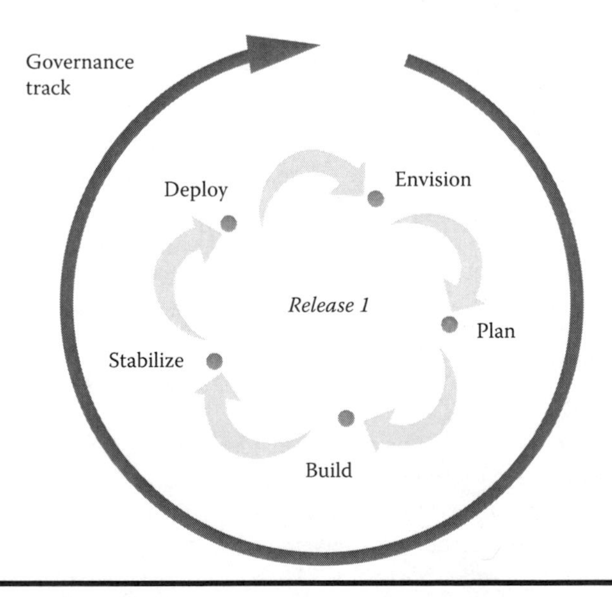

Figure 38.1 Illustration of Microsoft Solutions Framework.

While MSF is widely deployed and popular with users, there is a shortage of quantified empirical data on quality, productivity, schedules, and costs. This is not a unique Microsoft problem but rather an endemic problem of the software industry.

Most methodologies, except TSP, do not even include measurements as part of their fundamental activities. Thus, there is little solid empirical data on Agile, extreme programming, model-based development, pair programming, and almost every other method.

Fortunately, there are third-party benchmark organizations that do provide quantitative data, such as the International Software Benchmark Standards Group (ISBSG), the Quality/Productivity Management Group (Q/P), Namcook Analytics, and a number of others.

An MSF example is shown in Table 38.1 for 1000 function points and the Java programming language.

The MSF is backed by Microsoft and has been fairly successful for a wide range of applications and also for a wide variety of sizes ranging from less than 100 function points to over 10,000 function points. Some Microsoft projects, such as Windows 10 and the Office suite, top 100,000 function points when total features are installed. Average users only utilize a small fraction of the total features.

Table 38.1 Example of 1000 Function Point Microsoft Solutions Project with Java Language

Project Size and Nature Scores		Microsoft Solutions
1	Small projects < 100 FP	10
2	New projects	10
3	IT projects	10
4	Web/cloud projects	10
5	Medium projects	9
6	Large projects > 10,000 FP	9
7	Systems/embedded projects	8
8	Enhancement projects	6
9	High-reliability, high-security projects	6
10	Maintenance projects	4
	Subtotal	82
Project Activity Scores		**Microsoft Solutions**
1	Reusable components	10
2	Coding	10

(*Continued*)

Table 38.1 (Continued) Example of 1000 Function Point Microsoft Solutions Project with Java Language

3	Change control	10
4	Integration	10
5	Design	9
6	Requirements	8
7	Models	8
8	Test case design/generation	8
9	Testing defect removal	8
10	Requirements changes	7
11	Defect prevention	7
12	Architecture	6
13	Pattern matching	5
14	Planning, estimating, measurement	5
15	Pre-test defect removal	5
	Subtotal	116
	Total	198
Regional Usage (Projects in 2016)		**Microsoft Solutions**
1	North America	25,000
2	Pacific	17,000
3	Europe	16,000
4	South America/Central America	8,000
5	Africa/Middle East	7,000
	Total	73,000
Defect Potentials per Function Point (1000 function points; Java language)		**Microsoft Solutions**
1	Requirements	0.30
2	Architecture	0.25
3	Design	0.40

(Continued)

Table 38.1 (Continued) Example of 1000 Function Point Microsoft Solutions Project with Java Language

4	Code	1.30
5	Documents	0.40
6	Bad fixes	0.35
	Total	3.00
	Total defect potential	3000
Defect Removal Efficiency (DRE)		
1	Pre-test defect removal	70.00%
2	Testing defect removal	85.00%
	Total DRE	95.50%
	Delivered defects per function point	0.135
	Delivered defects	135
	High-severity defects	21
	Security defects	2
Productivity for 1000 Function Points		**Microsoft Solutions**
	Work hours per function point 2016	12.85
	Function points per month 2016	10.27
	Schedule in calendar months	14.00

Data Source: Namcook analysis of 600 companies.

Microsoft Solutions Scoring
Best = +10
Worst = −10

Chapter 39

Model-Based Development

Executive summary: *The term "model-based development" refers to creating an abstract model of a software application that can be used to generate executable code. Model-driven development has a fairly long history dating back to the 1980s. Although the ideas are powerful and appealing, some practical problems have kept model-based development from achieving a high market share. Changing the code generated from a model without changing the model itself causes serious incompatibilities that are hard to overcome. Model-based development is also a form of reuse development.*

The idea of software development from abstract models is a fairly old concept. Model-driven development and computer-aided software engineering (CASE) had a burst of popularity in the 1980s, almost like Agile is having today. Some of the companies involved include Bachman Systems, Higher-Order Software, Cadre Technologies, and Logic Works.

The modeling idea also influenced Unified Modeling Language (UML) and was part of the concept of the Universal Systems Language.

However, the early popularity of CASE and models did not last. It was too easy for the model and the generated code to get out of synchronization, which spoiled the advantages of model-based development.

There are many images on the web of various kinds of modeling approaches, with the V-Model diagram being the most common.

Model-based development is best suited for projects that have comparatively stable requirements that don't change very much, such as sorts, compilers, and perhaps private branch exchange (PBX) switching systems. It may also be a good choice for embedded applications that won't change much after installation. It is not an optimal choice for applications with rapidly changing requirements and rapidly evolving sets of diverse user needs.

Some specific model-based successes have been in the space program; the T-Vec model–based software was used for the Mars Rover. However, models tend to be somewhat rigid and are often limited to specific domains.

IBM supports model-based development with its Rational Rhapsody tool set, as shown in Figure 39.1.

Another interesting and more general illustration of model-based development shows the close relationship between models and software reuse, as shown in Figure 39.2. This is an important benefit of models.

Once the code is generated and goes into production, it will need to be updated fairly soon both to repair bugs and to add various enhancements or modifications based on external changes to business situations. If you go back and recreate the model, then everything stays synchronized. But if you change the code itself, then the code and the model are no longer in synch, and all future changes will have to be done to the code with probably incomplete documentation.

In today's world, many modeling tools exist. Many are open source and many more are commercial tools. An interesting survey of open-source modeling tools was published for Ericsson by Zeligsoft. The authors are Kenn Hussey, Brian Selec, and Toby McClean. Images of sample model tools are common on the web. Model-based development is fairly popular in the telecom industry due to reasonably stable requirements and a large number of qualified domain experts.

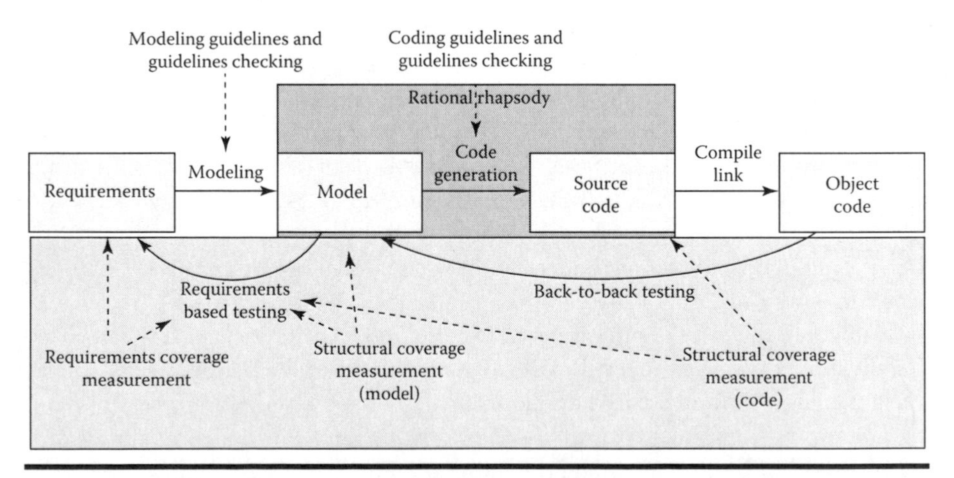

Figure 39.1 Illustration of IBM rhapsody model-based development.

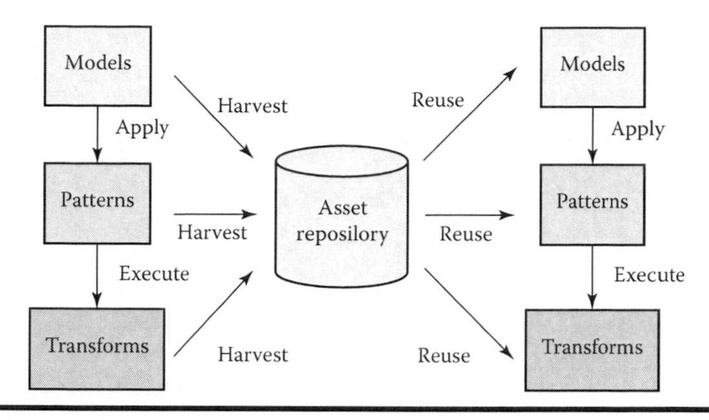

Figure 39.2 Illustration of models and reuse.

One of the gaps in the model-driven literature is that of actual productivity and quality results. This was a shortcoming of CASE in the 1980s and a chronic shortcoming of Agile today in 2016.

In general, software models and meta-models are static diagrams. But software is the most dynamic, frequently changing manufactured product in human history. It is obvious that software engineering models and meta-models should be animated, full color, and in three dimensions to capture the realities of software.

Animated dynamic models could handle issues such as requirements growth, bugs entering and being found, and cyber-attack defense strategies, which are not possible with today's primitive static models. The software industry is at least 10 years behind other engineering fields such as aeronautical and automotive engineering in the sophistication of design and modeling tools.

Function point metrics are the best choice for both model-driven productivity and model-driven quality. Sample results for model-driven development for an application of 1000 function points and generated Java code are shown in Table 39.1.

Model-based development is technically challenging and requires above average skill sets on the part of software engineers. This is probably why model-based development has mainly been deployed in academic environments or for complex real-time and systems software applications. Model-based development has not yet been widely used for ordinary information technology projects in banks and insurance companies or for web applications.

Table 39.1 Example of 1000 Function Point Model-Based Project with Java Language

Project Size and Nature Scores		**Models**
1	Small projects <100 FP	10
2	Medium projects	10
3	New projects	10
4	Systems/embedded projects	10
5	Large projects >10,000 FP	9
6	IT projects	9
7	High-reliability, high-security projects	9
8	Enhancement projects	6
9	Web/cloud projects	6
10	Maintenance projects	2
	Subtotal	81
Project Activity Scores Models		
1	Pattern matching	10
2	Reusable components	10
3	Requirements	10
4	Design	10
5	Defect prevention	10
6	Pre-test defect removal	10
7	Test case design/generation	10
8	Testing defect removal	10
9	Architecture	9
10	Models	9
11	Coding	8
12	Change control	8
13	Integration	8
14	Planning, estimating, measurement	7

(Continued)

Table 39.1 (Continued) Example of 1000 Function Point Model-Based Project with Java Language

15	Requirements changes	7
	Subtotal	136
	Total	217
Regional Usage (Projects in 2016)		**Models**
1	North America	5,000
2	Pacific	4,000
3	Europe	5,000
4	South America/Central America	3,000
5	Africa/Middle East	1,000
	Total	18,000
Defect Potentials per Function Point (1000 function points; Java language)		**Models**
1	Requirements	0.15
2	Architecture	0.15
3	Design	0.30
4	Code	1.00
5	Documents	0.40
6	Bad fixes	0.15
	Total	2.15
	Total defect potential	2015
Defect Removal Efficiency (DRE)		
1	Pre-test defect removal	85.00%
2	Testing defect removal	90.00%
	Total DRE	98.50%
	Delivered defects per function point	0.03
	Delivered defects	32
	High-severity defects	5
	Security defects	0

(*Continued*)

Table 39.1 (Continued) Example of 1000 Function Point Model-Based Project with Java Language

Productivity for 1000 Function Points		Models
	Work hours per function point 2016	10.80
	Function points per month 2016	12.22
	Schedule in calendar months	11.50

Data Source: Namcook analysis of 600 companies.

Model-based Development Scores
Best = +10
Worst = −10

Chapter 40

Object-Oriented (OO) Methods

Executive summary: *The object-oriented (OO) approach may be considered as an evolution in thinking about how to relate data structures and programs to one another—it is a "post-relational" approach that intends to facilitate modularity, flexibility, and hence, maintainability. OO is not a full life cycle methodology—rather, it is a way to design and implement software programs. Typically, OO development uses languages (such as Simula, Smalltalk, Objective C, and C++) specifically designed to support the OO framework. OO introduces a set of terminology, described in this chapter, which must be carefully understood. The following description is technically correct but intentionally written from a management perspective.*

OO introduces 10 basic terms that capture the gist of the approach—*object*, *method*, *message*, *class*, *sub-class*, *instance*, *inheritance*, *encapsulation*, *abstraction*, and *polymorphism*. Each of these ideas is discussed next.

Objects: A software "package" that contains related procedures and data. Procedures are known as *methods* and the data elements are referred to as *variables*. By way of example, a certain type of robot, called an automated guided vehicle (AGV), is found in many factories—it can be represented as an *object* in software. The AGV can exhibit a variety of behaviors (*methods*)—it can load, unload, move to a location, and so on. It requires certain information about itself—load capacity, turning radius, and so on—which are *variables* that may in some combination be relevant to its behaviors. Everything an object can do is represented in its methods, and everything it knows is represented in its variables—in all, a nice tight piece of modular software that is fully self-contained.

Messages: Real-world objects interact and are in many cases interdependent—*messages* are the means that software objects use to communicate with one another. A *message* is the name of an *object* (the "addressee") followed by the name of the *method* to be invoked and any *variables* that may be needed—for example, AGV101 (the *object*) move to (the *method*) bin#5 (a *variable*). *Messages* support *all* interactions among objects.

Classes: Define "templates" of methods and variables associated with a particular type of object. Objects that belong to a class are *instances* of a class that contains only those variables peculiar to the specific *instance.* Continuing the previous example, AGV may be considered a *class* while the individual vehicles would be *instances*—AGV1, AGV2, and so on—each instance might have different variables (carrying capacity, speed, and so on) but all may have the same set of *methods.* An *object* is a specific *instance* of a *class*.

Inheritance: A mechanism that allows one class of objects to be defined as a special case of a more general class (a super class). These special cases "inherit" the methods and variables of the super class(es) and are called *subclasses.* For example, one subclass of AGV may be designed to carry pallets, another to carry barrels. Subclasses may define their own methods and/or variables, and may override inherited characteristics. Any number of levels of super classes and subclasses may be defined to create a class hierarchy.

Encapsulation: Means packaging data and procedures together. This is an "information hiding" strategy. Data inside an object is accessed *only* by that object's methods—hence, there is no possibility of unintended data contamination resulting in modules that are self-protecting. This approach greatly facilitates changes as a change to one object has no impact or dependency on any other object.

Abstraction: Rests on the notion that it is possible to define "virtual" classes that are not in themselves meaningful, but are created solely to facilitate inheritance of particular variables and/or methods by other classes lower in the class hierarchy. This is a very powerful notion that can greatly reduce redundant code within a larger application while simplifying maintenance. Going one step further, it is possible for an object to inherit from more than one super class.

Polymorphism: "many forms" in Greek, refers to hiding alternative procedures behind a common interface—for example, the same method name can be used in more than one class. This notion promotes safe reuse and makes objects more independent of one another.

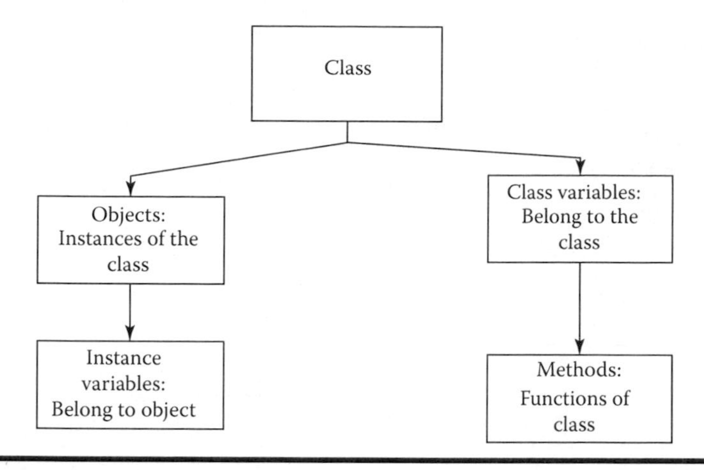

Figure 40.1 Object-oriented programming.

Associated with object methods are a newer generation of database managers known as object database management systems (ODBMSs). These systems, for example, are ideal for complex data structures such as bills of materials—imagine, for example, a complete bill of materials for the Boeing Dreamliner.

OO concepts can be and are used within many different methodologies and are applicable to both very small and very large systems. They represent a significant paradigm shift in software design and development.

There are quite a few images of OO programming and development on the web including the one shown in Figure 40.1.

As can be seen, the terminology of OO development is somewhat unique and requires study and learning to make the transition from conventional procedural programming.

Shown in Table 40.1 are the scores and results for an OO project of 1000 function points coded in the Java language.

OO programming languages and OO development entered the mainstream of software methodologies in the 1980s and continue to be used successfully on a variety of applications. As many readers know, Apple selected the Objective C language as its primary language for all Apple products in the 1980s including operating systems, smartphones, and tablets. Novices to the OO world need to undergo a learning experience since the OO terminology is somewhat different from that of the older procedural world.

Table 40.1 Example of 1000 Function Point Object-Oriented Project with Java Language

Project Size and Nature Scores		**OO**
1	Small projects < 100 FP	10
2	Medium projects	10
3	New projects	10
4	IT projects	10
5	Large projects > 10,000 FP	9
6	Systems/embedded projects	9
7	Web/cloud projects	9
8	High-reliability, high-security projects	9
9	Enhancement projects	4
10	Maintenance projects	3
	Subtotal	83
Project Activity Scores OO		
1	Pattern matching	10
2	Reusable components	10
3	Coding	10
4	Change control	10
5	Requirements changes	8
6	Models	8
7	Integration	8
8	Requirements	7
9	Design	7
10	Defect prevention	7
11	Pre-test defect removal	7
12	Testing defect removal	7
13	Architecture	6
14	Test case design/generation	6

(*Continued*)

Table 40.1 (Continued) Example of 1000 Function Point Object-Oriented Project with Java Language

15	Planning, estimating, measurement	5
	Subtotal	116
	Total	199
Regional Usage (Projects in 2016)		**OO**
1	North America	15,000
2	Pacific	15,000
3	Europe	10,000
4	South America/Central America	12,000
5	Africa/Middle East	5,000
	Total	57,000
Defect Potentials per Function Point (1000 function points; Java language)		**OO**
1	Requirements	0.35
2	Architecture	0.20
3	Design	0.35
4	Code	1.15
5	Documents	0.40
6	Bad fixes	0.25
	Total	2.70
	Total defect potential	2700
Defect Removal Efficiency (DRE)		
1	Pre-test defect removal	75.00%
2	Testing defect removal	90.00%
	Total DRE	97.04%
	Delivered defects per function point	0.08
	Delivered defects	80
	High-severity defects	12
	Security defects	4

(Continued)

Table 40.1 (Continued) Example of 1000 Function Point Object-Oriented Project with Java Language

Productivity for 1000 Function Points		OO
	Work hours per function point 2016	13.4
	Function points per month 2016	9.85
	Schedule in calendar months	14.1

Data Source: Namcook analysis of 600 companies.

Object-Oriented Scores
Best = +10
Worst = –10

Chapter 41

Open-Source Software Development

Executive summary: *The phrase "open-source software" refers to applications where the software can be examined and also modified by almost anyone. The software usually requires licensing such as the GNU open-source license, but the software itself may be available for free, for a voluntary donation, or for a small support fee. The open-source business model is almost unique and there are few other products of any kind that are developed by such a random collection of individuals, who may even live on different continents. The open-source model has led to some very successful software packages including Google, Firefox, Android, Libre Office, Apache Open Office, and even the hypertext markup language (HTML), which was contributed by Tim Berners-Lee, the famous Internet pioneer.*

Open-source software is a remarkable business that is quite different from conventional for-profit business ventures that charge for products and services. Open-source software is also different from non-profit organizations, which may or may not have tax exempt status.

The most surprising thing about open-source software is that not only does it exist, but it is doing well and has led to some very successful applications such as Google, Firefox, Android, WordPress, Facebook, and many others. It has also led to some successful companies such as Red Hat.

Open-source organizations generate revenues in a variety of ways that include:

Open-Source Software Funding Models 2016

1. Selling services such as training and maintenance
2. Fremium software (entry-level free versions and more complete fee versions)
3. Dual licensing (open-source plus commercial licensing)
4. Donations from users of the software
5. Grants from large companies such as IBM and Microsoft
6. Crowd funding
7. Ads attached to or displayed by open-source packages

Several open-source companies generate well over $1 billion in annual revenues, including Red Hat, Google, and Facebook. Many open-source companies are in the $10 million revenue range. For a business that originated with random developers, some of whom never meet, putting together software packages and giving them away, it is a considerable surprise that so much money is to be had.

Open-source development is not like the structured methods used for closed applications inside companies or government agencies. The sequence with open-source development follows.

Open-Source Software Development Sequence

1. Someone with a good idea announces the idea or releases a working version to the public.
2. The original developer and others work on improvements to the software.
3. The original developer sets up an e-mail or communications network for all to use.
4. The original code and the improvements probably use static analysis tools.
5. The individual developers test their work prior to submittal.
6. The improvements are submitted to a central source, probably the first developer.
7. The product is kept current using version control software packages.
8. Various developers may review the work of other developers.
9. New developers may join and old developers may drop out.
10. A variety of support tools, bug trackers, and test tools will be used.

Note that formal phases such as requirements and design are not prominent in the open-source model. The requirements are probably one person's good idea. The design is probably embodied in prototypes or, at best, pseudo code with little or no paper documentation.

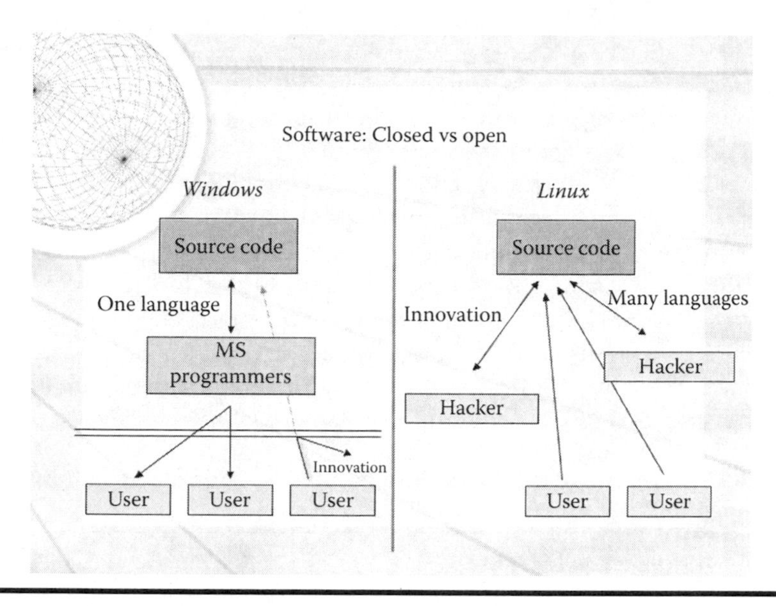

Figure 41.1 Illustration of open-source concepts.

There are a number of illustrations on the web of open-source development practices including the one shown in Figure 41.1.

Some corporate and government security experts are reluctant to use open-source packages on corporate or government computers, due to potential security flaws or even deliberate inclusion of spyware. However, the security issues of open source are outside the scope of this chapter. In any case, open-source security problems do not appear to be any worse than those in proprietary closed systems, surprisingly.

Open-source quality compares favorably with that of proprietary software applications. In part, this is due to very widespread usage of static analysis tools among the open-source community. In fact, some static analysis tool vendors such as Coverity (now owned by SAP) provide no-cost static analysis support for open-source developers.

A chronic gap in open-source data is that of productivity and quality data. There have been studies of open-source code quality by companies such as Coverity, CAST, and OptiMyth.

Productivity is harder to study due to the diverse community of software personnel who build open-source applications and the lack of any central organization that cares about productivity. Worse, there is no central repository of costs and effort for most open-source applications since their features are built independently, often by volunteers who do not receive payment.

Also difficult to measure is the productivity of open-source clients who may make private changes for their own private internal use, rather than providing changes back to the open-source community. The correct way to measure this is to treat the open-source package as "reused code" and measure *delivery* productivity rather than *development* productivity.

Assume your company acquires an open-source package of 1000 function points and then adds a proprietary new feature of 100 function points at a rather slow rate of 20 work hours per function point for a total of 2000 hours or only 6.6 function points per month for development productivity.

If you measure *delivery* productivity for the total of 1100 function points to internal users rather than *development* productivity for 100 function points, the net result would be 1.8 hours per function point or 73.3 function points per month.

Table 41.1 shows the results for both productivity and quality for a 1000 function point open-source package using Java as the programming languages. The productivity data is extrapolated from small samples.

Table 41.1 Example of 1000 Function Point Open-Source Project with Java Language

Project Size and Nature Scores		Open Source
1	New projects	10
2	Enhancement projects	10
3	Maintenance projects	10
4	IT projects	10
5	Web/cloud projects	10
6	Small projects < 100 FP	9
7	Medium projects	7
8	Large projects > 10,000 FP	6
9	Systems/embedded projects	4
10	High-reliability, high-security projects	–5
	Subtotal	71
Project Activity Scores		**Open Source**
1	Reusable components	10
2	Requirements changes	10
3	Coding	10
4	Change control	10
5	Integration	10

(*Continued*)

Table 41.1 (Continued) Example of 1000 Function Point Open-Source Project with Java Language

6	Requirements	9
7	Pre-test defect removal	9
8	Testing defect removal	8
9	Design	7
10	Pattern matching	6
11	Defect prevention	6
12	Test case design/generation	6
13	Architecture	5
14	Models	4
15	Planning, estimating, measurement	–1
	Subtotal	109
	Total	180
Regional Usage (Projects in 2016)		**Open Source**
1	North America	4,000
2	Pacific	3,000
3	Europe	3,000
4	South America/Central America	2,000
5	Africa/Middle East	500
	Total	12,500
Defect Potentials per Function Point (1000 function points; Java language)		**Open Source**
1	Requirements	0.40
2	Architecture	0.25
3	Design	0.40
4	Code	1.30
5	Documents	0.40
6	Bad fixes	0.20

(*Continued*)

Table 41.1 (Continued) Example of 1000 Function Point Open-Source Project with Java Language

	Total	2.95
	Total defect potential	2950
Defect Removal Efficiency (DRE)		
1	Pre-test defect removal	82.00%
2	Testing defect removal	87.00%
	Total DRE	97.33%
	Delivered defects per function point	0.079
	Delivered defects	79
	High-severity defects	12
	Security defects	1
Productivity for 1000 Function Points		**Open Source**
	Work hours per function point 2016	12.60
	Function points per month 2016	10.47
	Schedule in calendar months	12.50

Data Source: Namcook analysis of 600 companies.

Open-Source Scoring
Best = +10
Worst = −10

Open-source software is a unique sub-industry of the overall software industry. So far as the author can tell, no other industries in history have used distributed cooperative development teams who build products that are often given away for free. It is surprising that open-source software exists at all and even more surprising that it has been successful in building a number of popular applications such as Firefox and Linux. The open-source sub-industry could not exist without the Internet and the ability to move software code and other artifacts rapidly from place to place.

The closest related business sectors to open-source software also depend on the Internet. One of these is the creation of "Wiki" sites where hundreds or thousands of individuals work on a common product such as the well-known Wikipedia, which is now the largest encyclopedia in human history.

Other topics that depend on the shared work of many volunteers who probably will never meet include consumer service websites such as Angie's List and some of the online travel services that show hotel accommodation and prices all over the world based in part on thousands of reviews.

Chapter 42

Pair Programming Software Development

Executive summary: *The term "pair programming" refers to having software coded by a two-person team instead of a single programmer. One programmer codes while the other "navigates" or watches each line and suggests changes as needed. The roles switch back and forth. Pair programming is one of the Agile family of methods, although it is only used by about 15% of Agile projects. Pair programming is of course very expensive and the results are ambiguous. The studies of pair programming are woefully bad, comparing only unaided teams against unaided individuals. No study to date has even considered the impact of static analysis or inspections. Pair programming only exists because the software industry does not measure well and does not know how to achieve good software quality. Individual programmers using static analysis or inspections or both are more cost-effective and have quality equal to or superior to pairs.*

Agile is a blanket methodology with many variations and sub-methods. One of the more curious sub-methods is that of pair programming. Under this approach two programmers take turns writing code while the other observes or navigates. The two members of the pair change roles from time to time.

The literature on pair programming is embarrassingly bad. Academic studies compared one unaided individual against unaided pairs, and assert that pairs take 15% more time but produce 15% fewer bugs, for example, the study by Laurie Williams of the University of Utah. That may be true for unaided comparisons, but, of course, individual programmers who use static analysis and inspections have 70% fewer bugs than unaided individuals and fewer bugs than pairs as well.

The pair comparisons in the literature seem to assume only small projects with individual cowboy programming and ignore quality-strong development methods such as the team software process (TSP). The pair literature is also silent on building larger applications where programming may be less than 30% of total effort and paper documents cost more than source code.

The literature also totally ignores the quality and cost impact of including certified reusable components, or building new applications from 85% reusable code that approaches zero defects. The pair literature also ignores the use of static analysis, formal inspections, and even automated testing. Basically, the pair literature seems to consider only one unaided cowboy programmer compared with two unaided cowboy programmers.

The sociology of pair programming is interesting. Some programmers love the pair concept and claim both social and learning advantages. But many programmers hate pair programming and the literature also includes many papers by programmers who changed jobs just to get away from pair programming.

Pair programming has attracted considerable interest from both enthusiasts and detractors. In a larger sense, pair programming illustrates chronic problems of software engineering that set it apart from more mature engineering fields:

Deficiencies of Software Engineering Research

1. Releasing new methods without any testing or validation.
2. Making claims of high quality with only partial and imperfect data.
3. Making claims of high productivity with only partial and imperfect data.
4. Lack of consistent and reliable measurements of results.
5. Failure to examine limits or projects where the method is ineffective.
6. Failure to consider alternate methods that might be superior.

Pair programming was released to the outside world without any validation of empirical results. This is not uncommon in software and has occurred with many other "silver bullet" concepts such as rapid application development (RAD), computer-aided software engineering (CASE), and now pair programming.

The quantitative results of pair programming are inconsistent in the literature. To evaluate the results of pair programming or any other software engineering practice, all of the possible costs and benefits need to be considered and measured. The potential costs and benefits of software methods, languages, tools, and pair programming include:

Tangible Results

- Learning curve to acquire skills in the method
- Variations in how the method is performed
- Certification or formal examination of skill in the method

- Size of work being studied (module, application, system, etc.)
- Costs of production of code or other deliverables
- Effort for production in terms of months, days, hours, and so on
- Schedule or calendar time for production
- Quality of delivered materials such as code
- Ranges and limitations of the method
- Failures and abandonment of the method
- Comparison to alternate methods

Intangible Results

- Team morale and job satisfaction
- Precursor activities the affect the method (i.e., requirements, design)
- Successor activities affected by the method (i.e., testing, quality assurance)
- Team fatigue or medical issues due to method
- Customer satisfaction with deliverables
- Voluntary attrition of personnel who do not like the method

Factors that Influence Results

- Experience of personnel
- Complexity of problems
- Executive willingness to fund costly methods
- Programming languages or combinations of languages
- Changes in products during development
- Distractions that interfere with work progress
- Work hours per day, week, month, and year
- Paid and unpaid overtime
- Possible medical problems for workers
- Natural problems such as floods or earthquakes

The literature and experiments with pairs are deficient because they do not consider all of the variables that can affect the outcome of pairs versus single programmers. Some of the topics that should have been included but were not are

1. Single programmers who use static analysis
2. Single programmers who use inspections
3. Single programmers who use both static analysis and inspections
4. Expert single programmers compared with average pairs
5. Expert pairs compared with average single programmers
6. Novice single programmers compared with average pairs
7. Novice pairs compared with expert single programmers

8. Novice single programmers compared with novice pairs
9. Average single programmers compared with average pairs
10. Expert single programmers compared with expert pairs

Many of the studies appear to take random pairs and compare the results with random single programmers without either side using static analysis or inspections. This is not a realistic comparison. Both sides should use at least static analysis.

The literature asserts that average pairs of programmers code about 15% more slowly than average single programmers. This reduction in speed would only have a positive value if defects were reduced to almost zero but that is not the case. Also, requirements and design defects outnumber code defects on large systems and pair programmers have little impact on non-code defects.

Table 42.1 is extracted from a pair programming calculator developed by the author. The actual calculator allows researchers to input a number of variables including staff compensation, application size in lines of code (LOC), and coding speeds for both pairs and individual programmers. Table 42.1 shows a typical pattern for average pairs and average individual programmers for 1000 code statements.

As can be seen from Table 42.1, the economics of pair programming are severely negative. Pair programming costs were $7576 versus $2841 for an increase of over

Table 42.1 Cost Comparison of Single versus Pair Programming

	Application size in LOC		1000
	Single coding speed in LOC per hour		20
	Pair coding speed in LOC per hour		15
	Monthly compensation		$7500
	Hourly compensation		$56.82
	Single Programmer	**Pair Programming**	**Difference**
Clock hours	50.00	66.67	–16.67
Staff hours	50.00	133.33	–83.33
Code cost	$2841	$7576	–$4735
Cost percentage			266.67%
Schedule days	8.33	11.11	–2.78
Schedule months	0.42	0.56	–0.14
Schedule percentage			133.33%

266%. Pair programming schedules were 11.11 days as opposed to 8.33 days for an increase of 133%.

Unless the quality results for the pair approached zero defects and the quality results for the individual programmer were very poor, there would seem to be no economic justification for pair programming.

Unfortunately, the quality data from the pair programming literature is not adequate because it omits the impacts of inspections and static analysis. Table 42.2 shows the quality results of a single programmer using both static analysis and inspections versus a pair that uses only testing.

If the individual programmer used a combination of static analysis and inspections, then the quality would be much better than the pair. However, this is not a fair comparison because the pair could also use static analysis.

The point of Table 42.2 is that the literature on pair programming fails to include the potential impacts of either static analysis or inspections on quality. If a single programmer uses static analysis and inspections, the quality results will be much better than the pair. If both the pair and the individual use static analysis, the quality results will be about the same but the pairs are still more than 250% as expensive.

In pair programming, both team members share a single work station. Over and above the issue of high costs, pair programming also tends to raise the incidence of airborne infection such as flu and the common cold.

There are a total of 126 software occupations groups involved with software. The pair literature is totally silent on the topics of possibly pairing business analysts, architects, certified test personnel, technical writers, or any of the other software knowledge workers.

Even worse, the pair programming literature assumes 100% manual coding of custom code. The pair programming literature totally ignores comparisons to

Table 42.2 Pair Programming vs. Single Programming Quality

	Single Programmer	*Pair Programming*	*Difference*
Defect potential	13	12	1
Static analysis (%)	50		
Defects remaining	7		7
Inspection (%)	65		
Defects remaining	2		2
Test (%)	75		
Defects remaining	1	3	–2
Defects delivered	1	3	–2

mashups or application development where certified reusable modules are utilized rather than labor-intensive and error-prone line-by-line coding.

Table 42.3 shows the results of a typical paired information technology (IT) project of 1000 function points using the Java programming language.

Pair programming is mainly proof that the software industry is incompetent in measuring either quality or productivity. It is also a sad demonstration that the software industry lacks any notion of certification or validation of results. New medicines are tested and validated before being released to patients in order to guarantee effectiveness and to identify possible harmful side effects. The software industry does not seem to understand the concepts of proving that methods are effective and not harmful before putting them into service.

Table 42.3 Example of 1000 Function Point Pair Programming Project with Java Language

Project Size and Nature Scores		Pairs
1	Small projects <100 FP	8
2	Enhancement projects	8
3	Maintenance projects	8
4	New projects	7
5	Web/cloud projects	7
6	IT projects	5
7	High-reliability, high-security projects	5
8	Medium projects	4
9	Systems/embedded projects	4
10	Large projects >10,000 FP	**–3**
	Subtotal	53
Project Activity Scores		**Pairs**
1	Reusable components	9
2	Coding	7
3	Defect prevention	7
4	Change control	6
5	Integration	5
6	Pre-test defect removal	5
7	Test case design/generation	5

(Continued)

Table 42.3 (Continued) Example of 1000 Function Point Pair Programming Project with Java Language

8	Testing defect removal	5
9	Requirements changes	4
10	Pattern matching	3
11	Requirements	3
12	Architecture	3
13	Models	3
14	Design	3
15	Planning, estimating, measurement	**–1**
	Subtotal	67
	Total	120
Regional Usage (Projects in 2016)		**Pairs**
1	North America	3,000
2	Pacific	2,000
3	Europe	2,000
4	South America/Central America	2,000
5	Africa/Middle East	1,000
	Total	10,000
Defect Potentials per Function Point (1000 function points; Java language)		**Pairs**
1	Requirements	0.30
2	Architecture	0.30
3	Design	0.60
4	Code	1.10
5	Documents	0.40
6	Bad fixes	0.20
	Total	2.90
	Total defect potential	2900

(*Continued*)

Table 42.3 (Continued) Example of 1000 Function Point Pair Programming Project with Java Language

Defect Removal Efficiency (DRE)		
1	Pre-test defect removal	70.00%
2	Testing defect removal	85.00%
	Total DRE	95.50%
	Delivered defects per function point	0.131
	Delivered defects	131
	High-severity defects	20
	Security defects	2
Productivity for 1000 Function Points		**Pairs**
	Work hours per function point 2016	18.50
	Function points per month 2016	7.13
	Schedule in calendar months	17.00

Data Source: Namcook analysis of 600 companies.

Pair Programming Scoring
Best = +10
Worst = –10

Chapter 43

Pattern-Based Development

Executive summary: *The term "pattern based development" is somewhat similar to "model-based development" but includes a much larger selection of patterns. Model-based development deals mainly with technical application development. Pattern-based development also includes risk patterns, cost patterns, schedule patterns, quality patterns, usage patterns, and technical documentation patterns as well as software construction patterns. See also the chapters on "model-based development" and "85% software reuse," since reuse is based on formal patterns. Mashups are also based on reuse, but are somewhat more casual than planned reuse. Patterns are selected because they work; see also the chapter on "anti-patterns" for a discussion of patterns that don't work and are hazardous.*

The idea of pattern-based development has been evolving since the 1970s, but the evolution has not been smooth or comprehensive. Some topics such as "design patterns" have an extensive literature and are well understood. Other topics such as "cost and schedule patterns" are hardly understood at all and sometimes are not even included in the concept.

The full set of patterns that are needed for major software projects include, but are not limited to, the following.

Elements of Pattern-Based Software Development

1. Taxonomy patterns
2. Risk patterns

3. Size patterns
4. Size change patterns
5. Schedule patterns
6. Staffing and occupation patterns
7. Departmental organization patterns (matrix, hierarchical)
8. Cost patterns
9. Value and return on investment (ROI) patterns
10. Intellectual property and patent patterns
11. Governance patterns
12. Security patterns
13. Cyber-attack patterns
14. Hardware platform patterns
15. Software platform patterns
16. Algorithm patterns
17. Requirements patterns
18. Requirements creep patterns
19. Architectural patterns
20. Design patterns
21. Code structure patterns
22. Defect removal patterns
23. Test patterns
24. Data patterns
25. Deployment patterns
26. Maintenance patterns
27. Enhancement patterns
28. Customer support patterns
29. Usage patterns and use cases
30. Litigation patterns (if any)

Unfortunately, the bulk of the software pattern literature, as of 2016, centers on design patterns and ignores most other pattern elements. Of course, design is a critical issue and quite important, but it is far from being the only pattern needed for successful software development.

Consider the various kinds of patterns needed to design an ordinary home of about 3000 square feet. First is the external appearance pattern and also the siting of the home within its property boundaries. But for actual construction, separate patterns and diagrams are needed for window placement, framing, electricity, Internet connection, plumbing, water supply, waste disposal, and interior features such as built-in shelves and cabinets.

Architects know all of these patterns, but software personnel are unfamiliar with many kinds of useful patterns. This is why there are so many cost and schedule overruns even for projects of a size and type that have been built hundreds of times. Nobody looks at the patterns before starting a new project!

The true starting point for every project should be examining the results of historical projects with the same patterns; not jumping into the dark without a clue as to how similar projects were built and what kind of quality and productivity they achieved.

The new Software Engineering Methods and Theory (SEMAT) concept is adding new patterns to software engineering. Some of these are new topics such as "essence," "alphas," and "kernels." The SEMAT concept is now an Object Management Group (OMG) standard but, as of 2016, is new enough so that there is little empirical data available.

However, it is likely to be a valuable addition to the collection of software engineering patterns that should be considered at the start of every significant software project. SEMAT itself is outside the scope of this chapter, so a Google search is recommended to find current papers, presentations, and books on this emerging topic.

SEMAT is likely to prove useful in creating families of reusable components that can be utilized in many applications. It is obvious that SEMAT patterns need to be included in future pattern libraries.

The SEMAT literature asserts that SEMAT is not a "methodology" itself, but rather a way of viewing fundamental software topics that can add strength and value to many software methodologies such as Agile, team software process (TSP), rational unified process (RUP), and many others. However, SEMAT is still too new to have accumulated much empirical data.

A useful property of patterns, in general, is that many of them are "reusable" in the sense that they occur on many projects within many industries. The reuse may not top the 85% shown in the chapter on "85% reuse" but will probably approach 45% for architecture, design, and code components. Test cases are also reusable and may approach 50%.

One gap in the literature that needs additional research is "what do patterns look like?" It is obvious that patterns need a graphical element so Agile user stories are not suitable. Graphical diagrams such as Unified Modeling Language (UML) diagrams are a step in the right direction but they are static, and the real need is for dynamic moving patterns.

Since the first "pattern" that needs to be used in pattern matching is a taxonomy, it is useful to show the taxonomy pattern used by Namcook Analytics in its Software Risk Master (SRM) tool for benchmarks and for estimating:

Software Risk Master Application Taxonomy

1. Project nature: New or enhancement or package modification, and so on
2. Hardware platform: Cloud, smartphone, tablet, personal computer, mainframe, and so on
3. Operating system: Android, Linux, Unix, Apple, Windows, and so on
4. Project scope: Program, departmental system, enterprise system, and so on

5. Project class: Internal, commercial, open source, military, government, and so on
6. Project type: Telecom, medical device, avionics, utility, tool, social, and so on
7. Problem complexity: Very low to very high
8. Code complexity: Very low to very high
9. Data complexity: Very low to very high
10. Country of origin: Any country or combination of countries (using telephone codes)
11. Geographic region: State or province (using telephone codes)
12. Industry: North American Industry Classification (NAIC) codes
13. Project dates: Start date; delivery date (if known)
14. Methodology: Agile, RUP, TSP, waterfall, Prince2, and so on
15. Languages: C, C#, Java, Objective C, combinations, and so on
16. Reuse volumes: 0%–100%
17. Capability maturity model integrated (CMMI) level: 0 for no use of CMMI; 1–5 for actual levels
18. SEMAT use: Using or not using software engineering methods and practices
19. Start date: Planned start date for application construction
20. Deployment date: Planned date of deployment of the application to intended clients

As can easily be seen from the Namcook taxonomy pattern, if you are looking for similar projects to compare against a planned project, this taxonomy will bring in identical or very similar projects to the one under construction. It is useless to compare unlike projects such as a defense application compared with a civilian web application. Comparisons need to be between the same size, class, type, and industry.

Note that the author has built a proprietary method of early sizing of software projects before requirements based on pattern matching. This method uses the aforementioned taxonomy and can size applications in a total of 23 metrics in about 1.8 minutes. The metrics include several varieties of function points, story points, use-case points, and both logical and physical source code.

Once the project is placed on a formal taxonomy, then it is possible to show clients and developers many useful patterns from projects with the same taxonomy.

One useful pattern, for example, is that of risks. Table 43.1 shows a typical risk pattern for an IT application of 10,000 function points in size, as produced by SRM.

As can easily be seen from Table 43.2, pattern-based development combined with a formal taxonomy brings a great deal of early visibility to software that is not readily available from other methods.

Another useful pattern is the pattern of defect origination and defect removal. Effective quality control requires more than just testing; it needs a

Table 43.1 Software Risk Master Application (SRM) Risk Assessment

Risk Analysis from Similar Projects	**Normal Odds (%)**
Optimistic cost estimates	35.00
Inadequate quality control using only testing	30.00
Excessive schedule pressure from clients, executives	29.50
Technical problems hidden from clients, executives	28.60
Executive dissatisfaction with progress	28.50
Client dissatisfaction with progress	28.50
Poor quality and defect measures (omits >10% of bugs)	28.00
Poor status tracking	27.80
Significant requirements creep (>10%)	26.30
Poor cost accounting (omits >10% of actual costs)	24.91
Schedule slip (>10% later than plan)	22.44
Feature bloat and useless features (>10% not used)	22.00
Unhappy customers (>10% dissatisfied)	20.00
Cost overrun (>10% of planned budget)	18.52
High warranty and maintenance costs	15.80
Cancellation of project due to poor performance	14.50
Low reliability after deployment	12.50
Negative ROI due to poor performance	11.00
Litigation (patent violation)	9.63
Security vulnerabilities in software	9.60
Theft of intellectual property	8.45
Litigation (breach of contract)	7.41
Toxic requirements that should be avoided	5.60
Low team morale	4.65
Average risks for this size and type of project	**18.44**
Financial risk: (cancel; cost overrun; negative ROI)	**44.02**

Table 43.2 Example of 1000 Function Point Pattern-Based Project with Java Language

Project Size and Nature Scores		Patterns
1	Small projects <100 FP	10
2	Medium projects	10
3	New projects	10
4	IT projects	10
5	Systems/embedded projects	10
6	Web/cloud projects	10
7	High-reliability, high-security projects	10
8	Large projects >10,000 FP	9
9	Enhancement projects	6
10	Maintenance projects	3
	Subtotal	88
Project Activity Scores		**Patterns**
1	Pattern matching	10
2	Reusable components	10
3	Planning, estimating, measurement	10
4	Requirements	10
5	Architecture	10
6	Models	10
7	Design	10
8	Coding	10
9	Integration	10
10	Defect prevention	10
11	Test case design/generation	10
12	Testing defect removal	10

(Continued)

Table 43.2 (Continued) Example of 1000 Function Point Pattern-Based Project with Java Language

13	Change control	8
14	Requirements changes	7
15	Pre-test defect removal	7
	Subtotal	142
	Total	230
Regional Usage (Projects in 2016)		**Patterns**
1	North America	1000
2	Pacific	500
3	Europe	400
4	South America/Central America	300
5	Africa/Middle East	100
	Total	2300
Defect Potentials per Function Point (1000 function points; Java language)		**Patterns**
1	Requirements	0.10
2	Architecture	0.15
3	Design	0.20
4	Code	0.80
5	Documents	0.30
6	Bad fixes	0.15
	Total	1.70
	Total defect potential	1700
Defect Removal Efficiency (DRE)		
1	Pre-test defect removal	95.00%
2	Testing defect removal	95.00%
	Total DRE	99.75%

(*Continued*)

Table 43.2 (Continued) Example of 1000 Function Point Pattern-Based Project with Java Language

	Delivered defects per function point	0.009
	Delivered defects	9
	High-severity defects	1
	Security defects	0
Productivity for 1000 Function Points		**Patterns**
	Work hours per function point 2016	9.50
	Function points per month 2016	13.89
	Schedule in calendar months	10.00

Data Source: Namcook analysis of 600 companies.

Pattern-based Scoring
Best = +10
Worst = –10

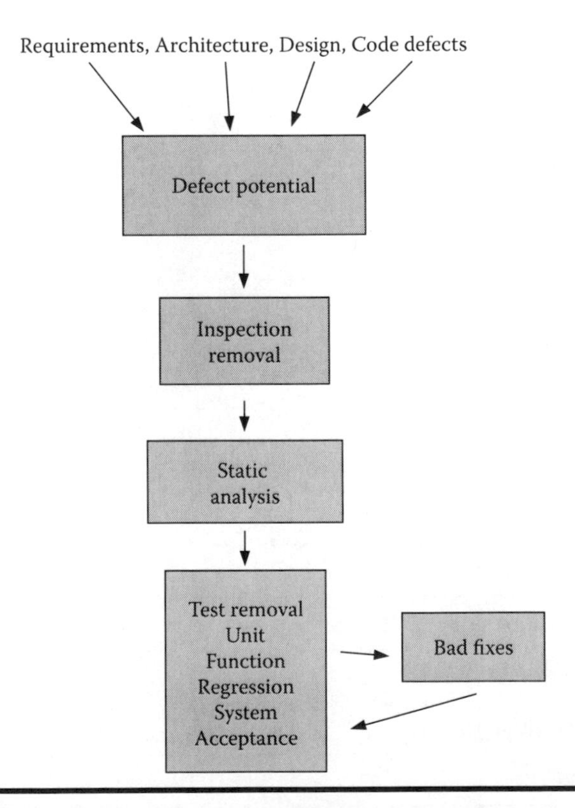

Figure 43.1 Illustration of defect potentials and defect removal.

synergistic combination of defect prevention, pre-test defect removal, and formal testing (Figure 43.1).

Note a minor point: About 7% of bug repairs contain new bugs. These are called *bad fixes* and are illustrated on the lower right-hand section of the defect pattern.

The new software engineering methods and theory approach (SEMAT) is attempting to both analyze software engineering patterns and also improve them, with interesting but sparse results as of 2014.

There are many images of patterns on the web, including the generic example shown in Figure 43.2.

Patterns can be used with any size, type, and form of software.

Custom designs and manual coding are intrinsically error prone, expensive, and insufficient for achieving high security and high quality. The combination of pattern-based development, models, and 85% software reuse can potentially transform software from among the most labor-intensive of modern industries to the most fully automated.

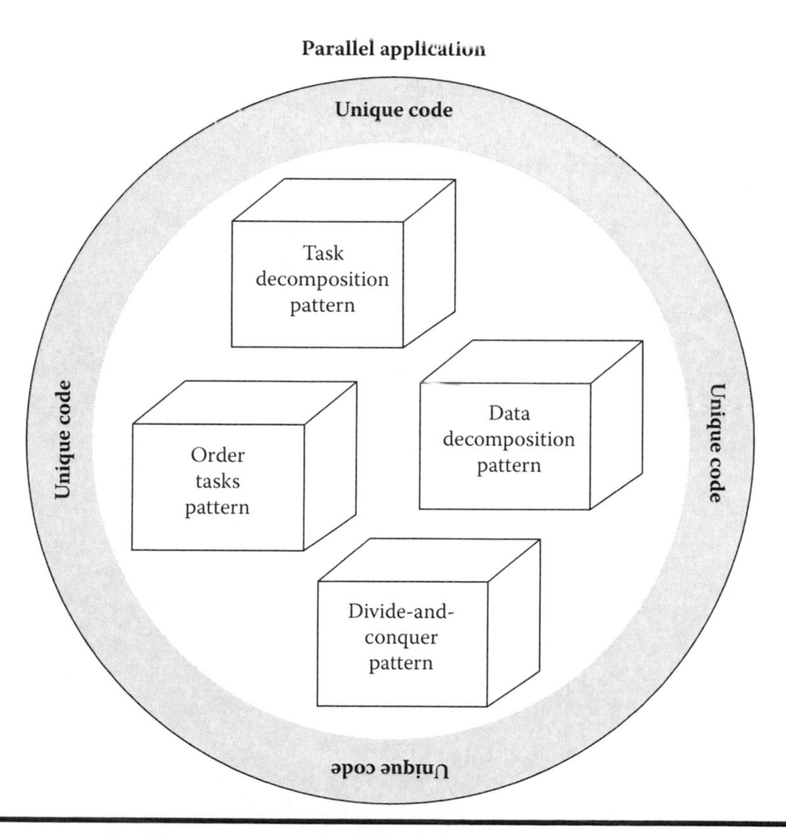

Figure 43.2 Illustration of software patterns.

Table 43.2 shows the probable results of pattern matching on an application of 1000 function points coded in Java with 45% reusable materials.

The pattern-matching software development methodology is still under development in 2016 and needs a great deal of additional work. The main tasks are to identify the full set of useful patterns and then make them available for software development purposes. Because pattern matching combines high quality and high productivity, there is a strong economic incentive to expand the available patterns from 2016 into a fuller and more useful set.

Chapter 44

Personal Software Process (PSP) Development

Executive summary: *The Personal Software Process (PSP) was created by Watts Humphrey and is aimed at individual software engineers. Although PSP is usually partnered with Watts' team software process (TSP) methodology, it can be used by itself for small individual projects. Because of its discipline and excellent measurement practices, PSP is a good background for many methods such as Agile, RUP, spiral, extreme programming (XP), and others.*

The TSP and PSP methodologies were developed by the late Watts Humphrey. The two methodologies are designed to be used together, with students learning PSP first, before learning TSP. These two methods are based in part on IBM's internal development methods for building systems software such as operating systems.

However, for smaller projects that require only one software engineer, the PSP program can be used by itself. PSP can also be used with other methodologies such as Agile, DevOps, or container development.

(Watts Humphrey was IBM's corporate director of programming technology before he left to join the Software Engineering Institute [SEI]. At SEI, Watts developed the essential concepts for the famous "Capability Maturity Model®" or CMM®.)

TSP and PSP are supported by rigorous training. PSP and TSP users of the methodologies are expected to complete formal courses prior to utilization. TSP and PSP are also supported by a growing library of books including several excellent books authored by Watts Humphrey himself. The web has an image collection of interesting PSP and TSP diagrams, including Figure 44.1 for PSP.

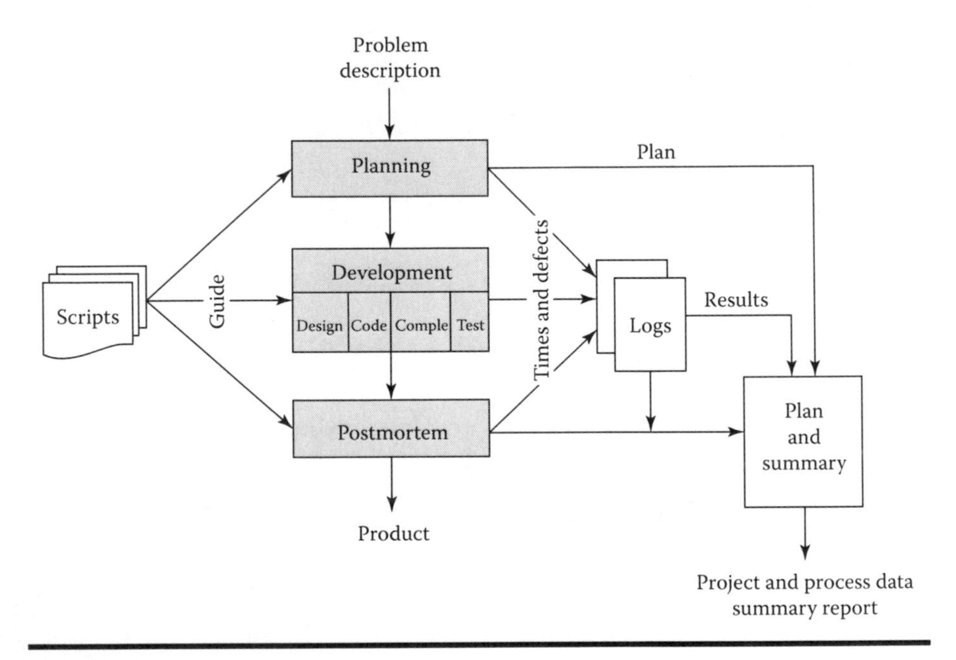

Figure 44.1 Illustration of personal software process (PSP).

Unlike Agile and many other methods, PSP developers keep very good records of the time they spend on various activities and also on the numbers of bugs or defects found. In general, PSP is an excellent background approach to accumulating highly accurate benchmarks for future analysis and use with other projects downstream. Figure 44.2 shows PSP in context with TSP.

Due to Watts heading up the SEI appraisal program and writing several books on TSP and PSP while employed at SEI, these methods are strongly endorsed and supported by the SEI. Among Namcook clients, the majority of PSP TSP users are in organizations that are ranked at capability maturity model integrated (CMMI) levels 3 or 5. In other words, TSP is strongly associated with fairly sophisticated companies. This may be one of the reasons why TSP quality is among the best of any software methodology.

Due to the background of these methods for large and complex systems software such as operating systems, TSP and PSP are among the best for systems software, embedded software, telecommunications software, medical devices, and defense software for weapons systems and other technical software projects.

TSP and PSP are "quality-strong" development methods that include rigorous inspections prior to testing and, more recently, static analysis prior to testing. TSP and PSP are also very strong on quality and productivity measurements, and keep

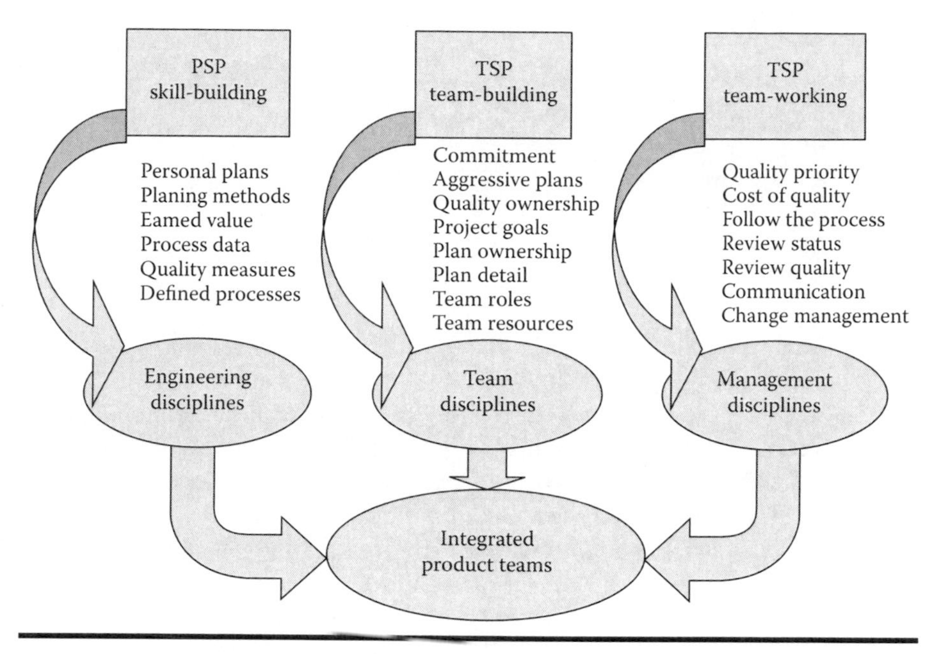

Figure 44.2 Illustration of synergy between TSP and PSP development methodologies.

much more data than many methodologies. In particular, TSP and PSP keep much more data than Agile, which is very sparse in keeping either quality or productivity data.

Measures of TSP and PSP quality show an interesting and unusual trend validated by Watts Humphrey personally. TSP and PSP projects show almost steady high-quality results even for large projects that approach 100,000 function points. Most software methodologies have elevated defect potentials and reduced defect removal efficiency (DRE) for large systems, but TSP seems to have overcome that endemic problem.

The excellent training provided by the PSP methodology for software engineers makes PSP an excellent choice for many "hybrid" approaches. Among the PSP hybrids noted among Namcook clients are

1. PSP/Agile
2. PSP/CMMI3
3. PSP/container
4. PSP/DevOps
5. PSP/evolutionary development (EVO)
6. PSP/iterative
7. PSP/model driven

8. PSP/object oriented (OO)
9. PSP/patterns
10. PSP/prototypes
11. PSP/spiral
12. PSP/structured
13. PSP/TSP
14. PSP/waterfall
15. PSP/XP

Although the combination of PSP/TSP is best for larger systems, PSP alone is effective for smaller sizes down to as low as a single function point.

Together, TSP and PSP show their strengths for larger and more complex software projects >5000 function points. They are also a good choice for projects that need Food and Drug Administration (FDA) or Federal Aviation Administration (FAA) certification, or which need Sarbanes–Oxley governance.

Because TSP and PSP originated inside IBM before function point metrics were the major software metric, some of the measures were originally based on lines of code (LOC). Function points are a much better choice than LOC for both quality and productivity measurements with TSP in 2016.

TSP and PSP were selected by the government of Mexico for internal software projects, and the method is also used elsewhere in Central and South America. Because of the strong endorsement of the SEI, TSP, and PSP are also used on many military and defense software projects.

TSP and PSP have annual conferences with hundreds of attendees. The method is not as popular as Agile, but then embedded and systems software projects are not as numerous as web and information technology (IT) projects.

Overall, Namcook scores and quantitative results for PSP in 2016 are shown in Table 44.1. The quantitative results assume 1000 function points and the Java programming languages. However, PSP alone is seldom used for projects >100 function points, since that is the upper limit of individual software engineering. Therefore, the example assumes a PSP/TSP combination and hence, has the same results as the TSP/PSP chapter, although the PSP process scores are different.

PSP has proved to be an effective and versatile software development methodology. It can be used by itself with small one-person projects, or used in conjunction with other methodologies for large team projects. The actual measured benefits of PSP go up with application size, which is not always the case. In other words, PSP is even more valuable at 10,000 function points than at 100 function points. This is the reverse of Agile, whose value declines as application size gets larger. PSP is often paired with TSP, and both methodologies were created by the late Watts Humphrey, who also developed the SEI capability maturity model. PSP can be used with any methodology including Agile, DevOps, Containers, Iterative, and so on.

Table 44.1 Example of 1000 Function Point PSP Project with Java Language

Project Size and Nature Scores		**PSP**
1	Small projects <100 FP	10
2	Medium projects	10
3	Large projects >10,000 FP	10
4	New projects	10
5	Systems/embedded projects	10
6	Web/cloud projects	10
7	High-reliability, high-security projects	10
8	IT projects	8
9	Enhancement projects	7
10	Maintenance projects	6
	Subtotal	91
Project Activity Scores		**PSP**
1	Coding	10
2	Pre-test defect removal	10
3	Planning, estimating, measurement	9
4	Defect prevention	8
5	Testing defect removal	8
6	Reusable components	6
7	Change control	6
8	Integration	6
9	Test case design/generation	6
10	Pattern matching	5
11	Requirements	5
12	Requirements changes	5
13	Design	5

(Continued)

Table 44.1 (Continued) Example of 1000 Function Point PSP Project with Java Language

14	Architecture	3
15	Models	3
	Subtotal	95
	Total	186
Regional Usage (Projects in 2016)		**PSP**
1	North America	6,000
2	Pacific	1,000
3	Europe	1,000
4	South America/Central America	2,500
5	Africa/Middle East	1,000
	Total	11,500
Defect Potentials per Function Point (1000 function points; Java language)		**PSP**
1	Requirements	0.25
2	Architecture	0.20
3	Design	0.30
4	Code	1.15
5	Documents	0.40
6	Bad fixes	0.20
	Total	2.50
	Total defect potential	2500
Defect Removal Efficiency (DRE)		
1	Pre-test defect removal	85.00%
2	Testing defect removal	90.00%
	Total DRE	98.50%

(Continued)

Table 44.1 (Continued) Example of 1000 Function Point PSP Project with Java Language

	Delivered defects per function point	0.038
	Delivered defects	38
	High-severity defects	6
	Security defects	1

Data Source: Namcook analysis of 600 companies.

Personal Software Process (PSP) Scoring
Best = +10
Worst = –10

Chapter 45

Prince2

Executive summary: *Prince2 ("Projects In Controlled Environments") is a project management standard developed and used by the government of the United Kingdom. Prince2 is in many ways similar to the Project Management Institute's Project Management Body of Knowledge (PMBoK). Both of these methods have published bodies of knowledge, which are the basis for professional certifications. Both require certain levels of education and experience, passing scores on comprehensive exams, and periodic re-certification, including continuing education credits. Prince2 and PMBoK are not software development methods as such, but are used in conjunction with many different methods such as Rational Unified Process (RUP)—they are most commonly used in mid and large size projects, and less frequently with the Agile methods.*

Prince2 is based on seven principles, seven themes, and seven processes.

Principles:

1. Continued business justification
2. Learn from experience
3. Defined roles and responsibilities
4. Manage by stages
5. Manage by exception
6. Focus on products
7. Tailored to suit the project environment

Themes:

1. Business case
2. Organization
3. Quality
4. Plans
5. Risk
6. Change
7. Progress

Processes:

1. Starting up a project (SU): Appoint the project team, create a project brief (charter), establish a project board (oversight group), appoint an executive and project manager, review lessons learned from prior projects, plan the next stage (initiation)
2. Initiating a project (IP): Create the business case, create project and quality plans, identify risks, set up project files, plan the next phase (directing)
3. Directing a project (DP): Define how the project board will control the overall project (review and approvals milestones, quality and status reviews)
4. Controlling a stage (CS): Define and authorize "work packages" (define phases or stages of the project), assessing progress, capturing and resolving project issues, reporting highlights, taking corrective action, escalating issues, accepting completed work packages
5. Managing stage boundaries (SB): Where CS defines what is to be done within a stage, SB defines what is to be done toward the end of a stage—for example, planning the next stage, updating the project plan, business case, and risk register, reporting stage end
6. Managing product delivery (MP): Ensure work on products allocated to teams is authorized and agreed, ensure the team is clear as to what is to be produced and as to estimated effort, cost, and schedule, ensure products are delivered to expectations
7. Closing a project (CP): Decommission the project, identify follow-on actions, conduct project evaluation review (lessons learned)

Work products typically produced:

- Project brief: A short explanation of project need, the management team, project structure and goals
- Business case: Detailed description of need and expected costs and benefits

- Risk register: Risks are named and assigned a risk rating (probability * impact); risks are sorted by rating; risks are assigned precautions and planned response actions if the risk materializes
- Quality register: Updated status of all planned quality activities
- Issues register: A set of notes about problems, complaints, and concerns raised by project participants
- Lessons log: Notes on lessons learned

Prince2 is a generic concept and method that is used for many kinds of projects and not just for software. That being said, software is one of the more common uses.

Table 45.1 shows the scores and results for using Prince2 on a software project of 1000 function points and the Java programming language.

Prince2 is one of the "heavy" methodologies. It is unusual in that it can be used with non-software projects as well as with software, although software is probably the most common use. Prince originated in the United Kingdom and has not traveled to many other locations. It is a marginal "quality-strong" methodology that gives good results for large systems in the 10,000 function point size range. It is probably a bit heavy for small applications below 500 function points where Agile is now the major methodology. It does not seem a good fit with DevOps or container development.

Table 45.1 Example of 1000 Function Point Prince2 Project with Java Language

Project Size and Nature Scores		Prince2	
1	New projects	7	
2	IT projects	7	
3	Small projects <100 FP	6	
4	Medium projects	6	
5	Large projects >10,000 FP	6	
6	Systems/embedded projects	6	
7	Web/cloud projects	6	
8	High-reliability, high-security projects	5	
9	Enhancement projects	4	
10	Maintenance projects	4	
	Subtotal	57	

(Continued)

Table 45.1 (Continued) Example of 1000 Function Point Prince2 Project with Java Language

Project Activity Scores		Prince2	
1	Coding	7	
2	Change control	7	
3	Requirements	6	
4	Requirements changes	6	
5	Architecture	6	
6	Models	6	
7	Design	6	
8	Integration	6	
9	Planning, estimating, measurement	5	
10	Defect prevention	5	
11	Pre-test defect removal	5	
12	Testing defect removal	5	
13	Pattern matching	4	
14	Reusable components	4	
15	Test case design/generation	4	
	Subtotal	82	
	Total	139	
Regional Usage (Projects in 2016)		**Prince2**	
1	North America	100	Canada
2	Pacific	–	
3	Europe	1500	U.K.
4	South America/Central America	–	
5	Africa/Middle East	100	
	Total	1700	

(Continued)

Table 45.1 (Continued) Example of 1000 Function Point Prince2 Project with Java Language

Defect Potentials per Function Point (1000 function points; Java language)		**Prince2**	
1	Requirements	0.35	
2	Architecture	0.30	
3	Design	0.60	
4	Code	1.45	
5	Documents	0.40	
6	Bad fixes	0.30	
	Total	3.40	
	Total defect potential	3400	
Defect Removal Efficiency (DRE)			
1	Pre-test defect removal	55.00%	
2	Testing defect removal	85.00%	
	Total DRE	93.25%	
	Delivered defects per function point	0.23	
	Delivered defects	230	
	High-severity defects	36	
	Security defects	3	
Productivity for 1000 Function Points		**Prince2**	
	Work hours per function point 2016	17.00	
	Function points per month 2016	7.76	
	Schedule in calendar months	16.50	

Data Source: Namcook analysis of 600 companies.

Prince 2 Scoring
Best = +10
Worst = −10

Chapter 46

Product Line Engineering Software Development

Executive summary: *Product line engineering is a development methodology applied to families of related projects rather than to just one project. The idea is to use standard reusable components throughout the entire product line. Product line engineering is somewhat more complex than engineering a single application. However, in the modern world with rapidly increasing numbers of related software packages, this is a methodology that many companies need to understand. Custom designs and manual coding of software applications are intrinsically expensive and error prone. This statement is even more important for families of related applications. Examples of related families of similar applications include private branch exchange (PBX) switching systems, compilers, and many finance and insurance software applications.*

In today's world of 2016, there are many families of related products that need common architecture and common human interface conventions. Examples of these families of related products include the IBM operating system family, Microsoft Office (Word, Excel, PowerPoint, etc.), families of PBX switches by size; flight controls and instruments in a series of military aircraft; and also applications that are planned to run on the same platforms such as Apple i-Phones, android smartphones, and many others.

In today's world, product line engineering is becoming increasingly important and is in daily use by many of the world's top technology companies such as IBM, Microsoft, HP, Sony, Computer Associates (CA).

If you step back and consider what is needed to achieve consistency across a family of perhaps 20 applications that are aimed at the same platform and the same set of users, it is obvious that reuse is an important consideration.

Reuse Considerations for Product Line Families

1. Reusable human interface conventions
2. Reusable software interfaces (API)
3. Reusable color schemes for displays
4. Reusable utilization of physical switches, buttons, and so on
5. Reusable application start and stop commands
6. Reusable data storage formats
7. Reusable architecture
8. Reusable designs
9. Reusable code segments
10. Reusable error message structures
11. Reusable HELP screens
12. Reusable user instructions
13. Reusable test cases and test scripts
14. Reusable customer support training
15. Reusable application distribution methods

Product line engineering is a good attempt at raising the bar on reuse and making it common across multiple applications in the same family.

Most of the development methodologies in this book are aimed at specific individual software applications. Product line engineering on the other hand is aimed at related suites of applications, which might include 20 or more separate software projects. Thus, development has to start at a much higher level than methodologies that only encompass a single application.

Incidentally, the concept of product line engineering does not require that all projects in the line use the same methodology. Assume you are planning a family of 20 related software applications. Once past the high-level planning for all 20, then the individual projects can use whatever methodology is best for their unique technology stack; that is, five might use Agile; five might use the personal software process (PSP)/team software process (TSP), five might use model-based development; and five might use DevOps.

When using product line engineering, the first kinds of requirements, architecture, analysis, and design must, of course, encompass the entire suite of related projects.

Sequence of Product Line Engineering Phases

1. Market analysis for product line
2. Sizing the individual projects and the total scope of the product line
3. Planning the application development sequence for the product line
4. Domain engineering of the product line
5. Analysis of common human interface elements across the product line
6. Analysis of common database structures across the product line
7. Analysis of common HELP screen formats across the product line
8. Analysis of common error messages across the product line
9. Analysis of common tutorial and user information across the product line
10. Analysis of common security features across the product line
11. Analysis of defect prevention methods across the product line
12. Analysis of pre-test defect removal methods across the product line
13. Analysis of standard reusable test cases for use across the product line
14. Definition of common error message formats for use across the product line
15. Analysis of common maintenance approaches across the product line

These 15 high-level activities are in addition to the effort for the specific applications that comprise the product line. A small case study example of product line engineering follows.

Let us assume that a product line will consist of 20 applications each of which is 1000 function points in size. Thus, the total product line will be 20,000 function points.

Let us assume that the high-level product line analysis and planning will take a team of five people and require a total of 2500 hours; 500 hours per team member spread over about 4 calendar months.

Let us assume that each application will be able to use 50% standard reusable materials; that is, 10,000 unique function points and 10,000 generic reusable function points for the entire product line of 20 applications.

Let us assume that the unique function points require 15 work hours per function point and the reusable function points require 3 hours per function point; that is, 150,000 hours of work on custom development and 30,000 hours of work on integrating reusable materials or 180,000 total hours.

If you plan to build all 20 applications simultaneously, the schedule would be about a 12 calendar month time line plus 5 months for product line planning; 17 calendar months in all.

If you plan to build 5 of the 20 applications each calendar year, then a total of just over 4 calendar years will be needed to complete the full set of 20 applications.

The total suite of 20 linked applications in the product line therefore requires 2,500 hours of high-level product line analysis and 180,000 hours of technical development for the applications themselves; 182,500 hours in total.

Thus, the entire work effort for the set of 20 projects in the product line is 182,500 work hours. This is a better than average production rate of 9.125 work hours per function point for the full suite of 20 applications. It can be seen that product lines with significant amounts of reusable materials improve productivity compared with building individual applications with little or no reuse.

Without 50% reuse, the entire set of 20 applications would have needed 300,000 staff hours instead of 182,500 staff hours; a reduction of 117,500 work hours, thanks to software reuse. Thus, product line engineering and 50% reuse saved over 39% of probable work hours for the suite of related applications. Software reuse is the key to overall improvements in software productivity, schedules, and also quality.

Table 46.1 shows the results for building a single application of 1000 function points using the Java programming language. For consistency with the other samples in this book, only a single application out of a product line is illustrated.

Table 46.1 Example of 1000 Function Point Product-Line Project with Java Language

Project Size and Nature Scores		Product Line Engineering
1	Small projects <100 FP	10
2	Medium projects	10
3	Large projects >10,000 FP	10
4	New projects	10
5	IT projects	10
6	High-reliability, high-security projects	10
7	Systems/embedded projects	8
8	Web/cloud projects	8
9	Enhancement projects	5
10	Maintenance projects	5
	Subtotal	86
Project Activity Scores		**Product Line Engineering**
1	Pattern matching	10
2	Reusable components	10

(Continued)

Table 46.1 (Continued) Example of 1000 Function Point Product-Line Project with Java Language

3	Requirements	10
4	Architecture	10
5	Models	10
6	Design	10
7	Coding	10
8	Requirements changes	8
9	Change control	8
10	Integration	8
11	Defect prevention	8
12	Test case design/generation	8
13	Testing defect removal	8
14	Planning, estimating, measurement	7
15	Pre-test defect removal	7
	Subtotal	132
	Total	218
Regional Usage (Projects in 2016)		**Product Line Engineering**
1	North America	500
2	Pacific	400
3	Europe	400
4	South America/Central America	300
5	Africa/Middle East	200
	Total	1800
Defect Potentials per Function Point (1000 function points; Java language)		**Product Line Engineering**
1	Requirements	0.50
2	Architecture	0.20

(*Continued*)

Table 46.1 (Continued) Example of 1000 Function Point Product-Line Project with Java Language

3	Design	0.60
4	Code	1.25
5	Documents	0.35
6	Bad fixes	0.30
	Total	**2.85**
	Total defect potential	**2850**
Defect Removal Efficiency (DRE)		
1	Pre-test defect removal	75.00%
2	Testing defect removal	90.00%
	Total DRE	96.24%
	Delivered defects per function point	0.107
	Delivered defects	**107**
	High-severity defects	**17**
	Security defects	**1**
Productivity for 1000 Function Points		**Product Line Engineering**
	Work hours per function point 2016	14.00
	Function points per month 2016	9.45
	Schedule in calendar months	13.30

Data Source: Namcook analysis of 600 companies.

Product Line Engineering Scores
Best = +10
Worst = −10

Product line engineering is not for everyone and not for every software project. It is a niche methodology and the specific niche is that of suites of similar applications that are not identical. For example, IBM builds dozens of compilers for different programming languages. Telecom companies such as AT&T might build half a dozen different PBX switches with slightly different features. Insurance companies may have similar features but some applications are different enough so that product line engineering could be useful.

Chapter 47

Prototype Development (Disposable)

Executive summary: *Prototypes are not development methodologies themselves, but rather they are partial replicas of larger applications built in order to experiment with key algorithms. Software prototypes come in two flavors—disposable and evolutionary. As the name implies, disposable prototypes are used to examine critical algorithms and are then discarded. Evolutionary prototypes are also used to examine critical algorithms, but are then used to grow and morph into a full application. Because prototypes are usually built in a hurry without formal quality control, evolutionary prototypes can be hazardous. Prototypes are very common in 2016 and have been since the software industry began in the 1950s. One unique feature of disposable prototypes is that they may be built with special tools and languages not planned for the actual product; that is, Mathmatica10 or Excel are useful for mathematical algorithms but may not be suited for the actual working application.*

Software prototypes are not methodologies themselves, but are small replicas of future applications built before a full application in order to try out key algorithms and get a sense of the necessary architecture and design for the full application.

Prototypes are the oldest methodological approach in all of software, and date back to the 1950s. Prototypes are seldom discussed by themselves, but are used in conjunction with all methodologies: Agile, waterfall, Rational Unified Process (RUP), team software process (TSP), iterative, and every one of the 60 in this report with the exception of anti-patterns.

Prototypes come in two distinct flavors: disposable and evolutionary. As the name implies, disposable prototypes are created to test key algorithms and are then discarded. Evolutionary prototypes are also built to test key algorithms, but are not discarded. Indeed, they grow and morph into the full application.

Although evolutionary prototypes are widespread, they have some problems. Because prototypes are usually built in a hurry without rigorous quality control, there may well be hidden flaws in the prototypes and especially hidden security flaws.

Evolutionary prototypes are typically quite a bit larger than disposable prototypes. Table 47.1 shows the approximate differences between the two forms of prototypes.

For smaller projects below 1000 function points in size, disposable prototypes are about 10% as large as the full application while evolutionary prototypes are about 15%. However, both forms of prototype decline in size for really large systems. For example, a major system of 100,000 function points could not possibly build a prototype of 10% because that would require 10,000 function points and a large multi-year project in its own right.

Another interesting difference between disposable and evolutionary prototypes involves the programming languages used. Disposable prototypes are often built in languages designed for fast development speed such as Visual Basic. For mathematical prototypes, sometimes Excel or Mathematica10 or MATLAB might be used.

Evolutionary prototypes, on the other hand, are usually coded in the same language as the full application; Java, C#, or whatever language has been chosen.

Disposable prototypes are the most widely used methodological approach in the history of software, and are probably being used right now for perhaps 275,000 current projects on every continent. By contrast, probably only around 50,000 evolutionary prototypes are in current use.

Disposable prototyping is an early-phase activity that starts during requirements and is finished by the time design is over. Of course, evolutionary prototypes go all the way to delivery.

Table 47.2 shows the results of using a disposable prototype of 90 function points for a full application of 1000 function points. The prototype itself took only

Table 47.1 Comparison of Disposable and Evolutionary Prototypes

Size in Function Points	*Prototype Disposable*	*Prototype Evolutionary*	*Difference*
10	1.00	1.50	0.50
100	10.00	17.00	7.00
1,000	90.00	125.00	35.00
10,000	500.00	1,500.00	1,000.00
100,000	1,500.00	5,000.00	3,500.00

a month and due to the high-level prototyping tools and language, had a productivity rate of about 3.5 work hours per function point or about 38 function points per month. Actual work hours for the prototype were 315.

Disposable prototypes have been useful adjuncts to all software methodologies and to all applications larger than a few hundred function points in size. Prototypes are almost indispensable for large systems in the 10,000 function point size range. Because prototypes are often done in a *fast and dirty* fashion, they are not safe to grow into finished applications. This is why disposable prototypes are somewhat safer to use than evolutionary prototypes.

Historically, prototypes are most often used with methodologies such as waterfall that do a lot of up-front work. Methodologies such as Agile that develop software in small increments may bypass prototypes because each sprint is less than 100 function points.

Table 47.2 Example of 1000 Function Point Project with Disposable Prototype and Java Language

Project Size and Nature Scores		**Prototype Disposable**
1	Large projects >10,000 FP	10
2	High-reliability, high-security projects	10
3	New projects	8
4	IT projects	7
5	Systems/embedded projects	7
6	Web/cloud projects	7
7	Medium projects	6
8	Enhancement projects	3
9	Maintenance projects	2
10	Small projects <100 FP	1
	Subtotal	61
Project Activity Scores		**Prototype Disposable**
1	Defect prevention	10
2	Requirements	9
3	Coding	9

(*Continued*)

Table 47.2 (Continued) Example of 1000 Function Point Project with Disposable Prototype and Java Language

4	Models	7
5	Pre-test defect removal	7
6	Testing defect removal	7
7	Requirements changes	6
8	Planning, estimating, measurement	5
9	Design	5
10	Change control	5
11	Test case design/generation	5
12	Pattern matching	4
13	Reusable components	4
14	Integration	4
15	Architecture	3
	Subtotal	90
	Total	151
Regional Usage (Projects in 2016)		**Prototype Disposable**
1	North America	75,000
2	Pacific	50,000
3	Europe	75,000
4	South America/Central America	50,000
5	Africa/Middle East	25,000
	Total	275,000
Defect Potentials per Function Point (1000 function points; Java language)		**Prototype Disposable**
1	Requirements	0.25
2	Architecture	0.10
3	Design	0.30

(Continued)

Table 47.2 (Continued) Example of 1000 Function Point Project with Disposable Prototype and Java Language

4	Code	1.15
5	Documents	0.40
6	Bad fixes	0.20
	Total	2.40
	Total defect potential	2400
Defect Removal Efficiency (DRE)		
1	Pre-test defect removal	80.00%
2	Testing defect removal	85.00%
	Total DRE	97.00%
	Delivered defects per function point	0.072
	Delivered defects	72
	High-severity defects	11
	Security defects	1
Productivity for 1000 Function Points		
	Work hours per function point 2016	13.68
	Function points per month 2016	9.64
	Schedule in calendar months	12.90

Data Source: Namcook analysis of 600 companies.

Prototype (Disposable) Scoring
Best = +10
Worst = 10

Chapter 48

Prototype Development (Evolutionary)

Executive summary: *Prototypes are not development methodologies themselves, but rather they are partial replicas of larger applications built in order to experiment with key algorithms. Software prototypes come in two flavors—disposable and evolutionary. As the name implies, disposable prototypes are used to examine critical algorithms and are then discarded. Evolutionary prototypes are also used to examine critical algorithms, but are then used to grow and morph into a full application. Because prototypes are usually built in a hurry without formal quality control, evolutionary prototypes can be hazardous. Prototypes are very common and have been since the software industry began.*

Software prototypes are not methodologies themselves, but are small replicas of future applications built before a full application in order to try out key algorithms and get a sense of the necessary architecture and design for the full application.

Prototypes are the oldest methodological approach in all of software, and date back to the 1950s. Prototypes are seldom discussed by themselves, but are used in conjunction with all common methodologies: Agile, waterfall, Rational Unified Process (RUP), team software process (TSP), iterative, and every one of the 60 in this book with the exception of anti-patterns.

Prototypes come in two distinct flavors: disposable and evolutionary. As the name implies, disposable prototypes are created to test key algorithms and are then discarded. Evolutionary prototypes are also built to test key algorithms, but are not discarded. Indeed, they grow and morph into the full application.

Although evolutionary prototypes are widespread, they have some problems. Because prototypes are usually built in a hurry without rigorous quality control, there may well be hidden flaws in the prototypes and especially hidden security flaws.

Evolutionary prototypes are typically quite a bit larger than disposable prototypes. Table 48.1 shows the approximate differences between the two forms of prototypes.

For smaller projects below 1000 function points in size, disposable prototypes are only about 10% as large as the full application while evolutionary prototypes are about 15%. However, both forms of prototype decline in size for really large systems. For example, a major system of 100,000 function points could not possibly build a prototype of 10% because that would require 10,000 function points and a large multi-year project in its own right.

Another interesting difference between disposable and evolutionary prototypes involves the programming languages used. Disposable prototypes are often built-in languages designed for fast development speed such as Visual Basic or Mathematica10. Algorithm and prototypes may be modeled in Excel or MATLAB. Disposable prototypes very often are not coded in the same languages planned for the actual product.

Evolutionary prototypes, on the other hand, are usually coded in the same language as the full application: Java, C#, or whatever language or language combinations have been chosen.

Disposable prototypes are among the most widely used methodological approach in the history of software, and are probably being used right now for

Table 48.1 Comparison of Disposable and Evolutionary Prototypes

Size in Function Points	*Prototype Disposable*	*Prototype Evolutionary*	*Difference*
10	1.00	1.50	0.50
100	10.00	17.00	7.00
1,000	90.00	125.00	35.00
10,000	500.00	1,500.00	1,000.00
100,000	1,500.00	5,000.00	3,500.00

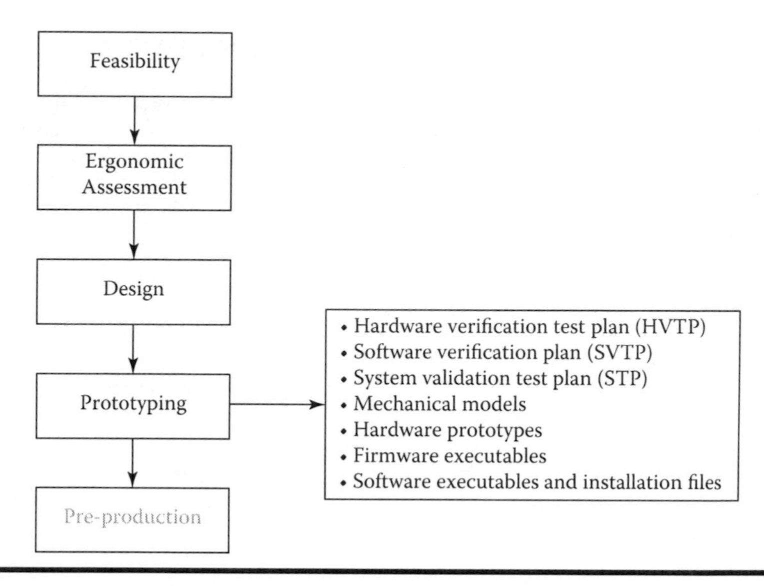

Figure 48.1 Illustration of evolutionary prototype concepts.

perhaps 275,000 current projects on every continent and in every country. By contrast, probably only around 50,000 evolutionary prototypes are in current use.

Disposable prototyping is an early-phase activity that starts during requirements and is finished by the time design is over. Of course, evolutionary prototypes go all the way to delivery.

Another web image shows a hybrid hardware/software prototype, which is very common in the embedded software domain (Figure 48.1).

Table 48.2 shows the results of using an evolutionary prototype of 125 function points for a full application of 1000 function points. The prototype itself took only 6 weeks in Java and had a productivity rate of about 5.5 work hours per function point, which was later subsumed into the full application.

Evolutionary prototypes are fairly common but also fairly troublesome. Since they are often developed in a hurry without formal methodologies and without quality assurance reviews and possibly without formal testing, it is easy to have latent bugs and security flaws in applications that used evolutionary prototypes. The disposable prototype variation is safer, but, of course, some of the work is lost with disposable prototypes.

Table 48.2 Example of 1000 Function Point Project with Evolutionary Prototype and Java Language

Project Size and Nature Scores		Prototype Evolutionary
1	High-reliability, high-security projects	10
2	New projects	8
3	Small projects < 100 FP	7
4	Medium projects	6
5	IT projects	6
6	Systems/embedded projects	6
7	Web/cloud projects	6
8	Enhancement projects	4
9	Maintenance projects	3
10	Large projects > 10,000 FP	–1
	Subtotal	55
Project Activity Scores		**Prototype Evolutionary**
1	Requirements	9
2	Coding	8
3	Defect prevention	8
4	Models	7
5	Pre-test defect removal	7
6	Testing defect removal	7
7	Requirements changes	6
8	Pattern matching	5
9	Reusable components	5
10	Design	5
11	Change control	5
12	Test case design/generation	5
13	Planning, estimating, measurement	4

(*Continued*)

Table 48.2 (Continued) Example of 1000 Function Point Project with Evolutionary Prototype and Java Language

14	Architecture	4
15	Integration	4
	Subtotal	89
	Total	144
Regional Usage (Projects in 2016)		**Prototype Evolutionary**
1	North America	10,000
2	Pacific	10,000
3	Europe	10,000
4	South America/Central America	10,000
5	Africa/Middle East	5,000
	Total	45,000
Defect Potentials per Function Point (1000 function points; Java language)		**Prototype Evolutionary**
1	Requirements	0.25
2	Architecture	0.10
3	Design	0.30
4	Code	1.25
5	Documents	0.40
6	Bad fixes	0.30
	Total	**2.60**
	Total defect potential	2600
Defect Removal Efficiency (DRE)		
1	Pre-test defect removal	70.00%
2	Testing defect removal	85.00%
	Total DRE	95.50%
	Delivered defects per function point	0.117

(*Continued*)

Table 48.2 (Continued) Example of 1000 Function Point Project with Evolutionary Prototype and Java Language

	Delivered defects	117
	High-severity defects	18
	Security defects	2
Productivity for 1000 Function Points		
	Work hours per function point 2016	14.50
	Function points per month 2016	9.10
	Schedule in calendar months	14.90

Data Source: Namcook analysis of 600 companies.

Prototype (Evolutionary) Scores
Best = +10
Worst = 10

Chapter 49

Rapid Application Development (RAD)

Executive summary: *Rapid application development (RAD) is both a general term applied to various alternatives to the traditional waterfall method and also the name of a method attributed to James Martin during the 1980s at IBM. This approach is seen by some as an evolution of the Spiral model defined by Barry Boehm. Martin's method became popular during the reengineering fad and is sometimes associated with it. Rapid prototyping is one of the key elements of this approach. While seldom seen today in its original form, RAD was one of the influences that lead to Agile.*

RAD intended to leverage the iterative concept and to evolve away from the overly rigid sequential notion embodied in the waterfall approach. One of the central techniques used in RAD is the joint application design (JAD) concept originally developed at IBM Canada. RAD starts with a JAD—that is, a facilitated joint session that includes both business users who know the requirements and information technology (IT) specialists familiar with technical implications and legacy system interfaces and dependencies. JAD sessions produce high-level requirements in the first iteration, which are then designed at a high level; prototypes are developed, tested, and reviewed with the user community. This cycle was repeated as required to fully evolve a system that satisfied all requirements.

Early versions of computer-aided software engineering (CASE) tools were sometimes associated with projects using this method, often with very mixed results due to the primitive capabilities of immature tool sets. Both evolutionary and throwaway prototypes were used as circumstances indicated.

RAD is essentially obsolete today, but the JAD element is potentially still relevant and is sometimes used in conjunction with other more modern methods such as the Rational Unified Process (RUP). While not common, it is also possible to use JAD sessions (also sometimes framed as JAR sessions—Joint Application Requirements) as an iterative requirements elicitation technique in conjunction with Agile approaches. JAD, for example, might be an effective technique to use for the quarterly planning cycle associated with extreme programming (XP).

RAD was popular, but like so many methodologies that advertised increased productivity and quality, RAD suffered from a lack of empirical data that proved value. This is not to say that RAD had no value, and indeed some concepts morphed into Agile and other methods. But the software industry has a distressing tendency to bring out one methodology after another, each with vast claims but no proof that the claims are true. Agile is doing the same thing today and pair programming is embarrassingly wasteful.

Table 49.1 shows the results of a RAD project of 1000 function points and the Java language.

Table 49.1 Example of 1000 Function Point RAD Project with Java Language

Project Size and Nature Scores		**RAD**
1	IT projects	9
2	Web/cloud projects	9
3	Small projects <100 FP	7
4	Medium projects	6
5	Large projects >10,000 FP	6
6	New projects	6
7	Systems/embedded projects	5
8	Enhancement projects	4
9	High-reliability, high-security projects	4
10	Maintenance projects	1
	Subtotal	57
Project Activity Scores		**RAD**
1	Requirements	9
2	Requirements changes	8

(Continued)

Table 49.1 (Continued) Example of 1000 Function Point RAD Project with Java Language

3	Coding	8
4	Change control	7
5	Testing defect removal	7
6	Design	6
7	Integration	6
8	Defect prevention	6
9	Models	5
10	Test case design/generation	5
11	Pre-test defect removal	4
12	Pattern matching	3
13	Reusable components	3
14	Planning, estimating, measurement	3
15	Architecture	3
	Subtotal	83
	Total	140
Regional Usage (Projects in 2016)		**RAD**
1	North America	300
2	Pacific	200
3	Europe	200
4	South America/Central America	200
5	Africa/Middle East	100
	Total	1000
Defect Potentials per Function Point (1000 function points; Java language)		**RAD**
1	Requirements	0.35
2	Architecture	0.30
3	Design	0.40

(Continued)

Table 49.1 (Continued) Example of 1000 Function Point RAD Project with Java Language

4	Code	1.45
5	Documents	0.40
6	Bad fixes	0.30
	Total	3.20
	Total defect potential	3200
Defect Removal Efficiency (DRE)		
1	Pre-test defect removal	50.00%
2	Testing defect removal	85.00%
	Total DRE	92.50%
	Delivered defects per function point	0.24
	Delivered defects	240
	High-severity defects	37
	Security defects	3
Productivity for 1000 Function Points		**RAD**
	Work hours per function point 2016	14.00
	Function points per month 2016	9.42
	Schedule in calendar months	13.90

Data Source: Namcook analysis of 600 companies.

Rapid Application Development (RAD) Scoring
Best = +10
Worst = −10

James Martin was an innovator and also an excellent public speaker. His development of the RAD approach contained the seeds of many good ideas, some of which found their way into Agile and other methods. But RAD suffered from the chronic problem of software methodologies and software tools making large claims of success without any solid quantified data to back up those claims. This is not to say that RAD was not effective, but it never produced enough quantified data to support the initial claims of success. (Hopefully, Agile won't do the same thing.)

Chapter 50

Rational Unified Process (RUP)

Executive summary: *The Rational Unified Process (RUP) method was developed about 1996 by well-known software engineering experts Grady Booch, Ivar Jacobsen, Jim Rumbaugh, and others. It was based on a combination of ideas and sub-methods assembled by colleagues at Rational Software. In 2003, IBM acquired Rational and it became an IBM division. RUP is supported by powerful tool suites, among the best for any method. In 2016, RUP is integrating Agile concepts, which it has already been doing for years. Due to IBM's backing, RUP has been one of the most widespread and successful methods for information technology (IT) projects and especially so for large applications in the 10,000 function point size range. But RUP is a heavy methodology that is not as effective for small projects below 500 function points. At the low end, Agile, DevOp, mashups, and other methodologies have taken over from the heavy methodologies such as RUP, whose usage is declining in 2016.*

RUP includes a rich combination of formal training, reference materials, and a tool suite that supports essentially every activity from initial requirements through maintenance.

RUP has its own meta-model that includes:

- Roles of various team members and stakeholders
- Work products such as documents, test materials, code, and so on
- Tasks, which describe how roles and work products interact

The overall RUP concept includes six software engineering disciplines and three support disciplines.

RUP Software Engineering

1. Business modeling
2. Requirements
3. Analysis and design
4. Implementation
5. Test
6. Deployment

The three supporting disciplines are:

RUP Support Disciplines

1. Configuration management
2. Project management
3. Environment

RUP is a mature methodology with a long and generally successful history of major application development. As with many other methodologies, there are image libraries on the web.

Because of IBM's acquisition of this method and IBM's investment in tools that support RUP, it arguably has one of the most complete set of supporting tools of any known software methodology. Not only has IBM funded tools that support RUP, but there are also dozens of third-party tool vendors that also support RUP. RUP has requirements tools such as rational dynamic object-oriented requirements system (DOORS), design tools, static analysis tools, code development tools, test tools, documentation tools, and many more. Probably more than 100 discrete tools support RUP when IBM's own tools and vendor tools are viewed together as a set.

In terms of overall results, RUP has been consistently successful as a methodology for large IT projects. It is not as widely used for systems and embedded applications, or for defense applications. It is probably too heavy for use on smaller projects, where Agile now dominates. However, for large IT applications in the 10,000 function point size range, it is perhaps number one in the world, and more effective than Agile at the top end of the size range for 10,000 function points and above.

The overall results for RUP for a project of 1000 function points and the Java programming language are shown in Table 50.1.

Table 50.1 Example of 1000 Function Point RUP Project with Java Language

Project Size and Nature Meta-Model		**RUP**
1	Large projects >10,000 FP	10
2	New projects	10
3	IT projects	10
4	High-reliability, high-security projects	10
5	Medium projects	9
6	Web/cloud projects	8
7	Small projects <100 FP	7
8	Maintenance projects	7
9	Systems/embedded projects	7
10	Enhancement projects	6
	Subtotal	84
Project Activity Strength Scores		**RUP**
1	Coding	10
2	Requirements changes	9
3	Architecture	9
4	Change control	9
5	Planning, estimating, measurement	8
6	Requirements	8
7	Integration	8
8	Pre-test defect removal	8
9	Testing defect removal	8
10	Design	7
11	Defect prevention	7
12	Test case design/generation	7
13	Pattern matching	6
14	Reusable components	6

(*Continued*)

Table 50.1 (Continued) Example of 1000 Function Point RUP Project with Java Language

15	Models	6
	Subtotal	116
	Total	200
Regional Usage (Projects in 2016)		**RUP**
1	North America	15,000
2	Pacific	10,000
3	Europe	15,000
4	South America/Central America	5,000
5	Africa/Middle East	3,000
	Total	48,000
Defect Potentials per Function Point (1000 function points; Java language)		**RUP**
1	Requirements	0.25
2	Architecture	0.15
3	Design	0.35
4	Code	1.35
5	Documents	0.40
6	Bad fixes	0.40
	Total	2.90
	Total defect potential	3400
Defect Removal Efficiency (DRE)		
1	Pre-test defect removal	80.00%
2	Testing defect removal	90.00%
	Total DRE	98.00%
	Delivered defects per function point	0.068
	Delivered defects	68
	High-severity defects	11
	Security defects	1

(Continued)

Table 50.1 (Continued) Example of 1000 Function Point RUP Project with Java Language

Productivity for 1000 Function Points		
	Work hours per function point 2016	13.31
	Function points per month 2016	9.92
	Schedule in calendar months	14.54

Data Source: Namcook analysis of 600 companies.

Rational Unified Process (RUP) Scores
Best = +10
Worst = 10

RUP is not a new method, having first come out around 1996. It has been successful for large applications, but its usage is now in decline due to the expanding popularity of Agile and the fact that many modern applications are fairly small, while RUP is aimed at large information systems. RUP might be used for systems and embedded software but because of its original target of large information systems, other methods such as the team software process (TSP) have more usage in the systems, embedded, and defense sectors.

Chapter 51

Reengineering

Executive summary: *Software reengineering deals with rebuilding and/or restructuring legacy systems to make them more maintainable and more extensible at a lower cost in time and effort. When documentation of a legacy system does not exist, the reengineering effort may be preceded by a reverse engineering activity. Alternately, the first step in reengineering may be reviewing and updating existing documentation. Reengineering may also involve restructuring of the system architecture and design, and potentially moving to a more modern programming, language, operating system, and so on. Reengineering may also involve changing the data structures associated with the system. Typically, reengineering does not entail changes to the functionality of the system. The term "reengineering" is often applied to the movement known as "business process reengineering" popularized by Hammer, Champy, and others.*

It has been estimated that there are hundreds of billions of lines of legacy code in existence and hundreds of millions of function points, much of it in old languages such as Cobol, Mumps, Chill, Coral, and assembly that are no longer used and are unknown to most current generation programmers.

Recreating all of this code would entail immense cost and substantial risk—hence, reengineering is often an appealing strategy. Reengineering is often cost-effective when the system in question has both high value and high maintenance costs; in addition, it is generally a less risky approach than complete redevelopment.

The principle factors that influence the cost of reengineering include the following:

- The scope of the system: How big it is
- The quality of the existing software: The extent to which it is or is not structured, the degree of cohesion and coupling found in the code

- Tool support for reengineering: Availability tools to ease the process
- The extent of data conversion that may be required: For example, moving the underlying data structure from sequential to relational
- The availability of staff knowledgeable about the application and its technologies

The reengineering process will typically entail the following steps:

- Source code translation: The existing code is converted, ideally by automated methods, from an old to a new version of the same language, or possibly to a different language. In many cases, a degree of manual effort may be unavoidable. There are a number of reasons why code translation may be necessary, including:
  - Hardware platform update: The new platform compliers may not support the older language or version
  - It may be difficult to find staff fluent in the old language (e.g., Cobol)
  - Organizational policy: Licensing cost for older operating systems, compilers, and so on may be prohibitive
- Reverse engineering: This process is described elsewhere and may or may not be necessary.
- Program structure improvement: Many legacy systems exhibit high coupling and low cohesion—the opposite of what is desirable for maintainability. It may be desirable, for example, to apply the concepts of object-oriented development to improve maintainability. Short of that level of change, it may be sufficient to simply improve structure and documentation—much will depend on the skills and experience of the reengineering team.
- Program modularization: This step, similar to structure improvement, may be desirable to eliminate redundancy. This may include:
  - Creating data abstraction hierarchies consistent with object-oriented principles
  - Centralization of hardware control facilities such as device drivers
  - Grouping similar functionality together
  - Grouping business process functionality together
- Data reengineering: For example, moving from sequential to relational data sets. Other data-related consideration that may be addressed include:
  - Adding edits to improve data quality, potentially including scrubbing and conversion of existing data
  - Removing limits included in the original code that are no longer appropriate: For example, field size limitations
  - Architectural considerations: For example, moving from centralized to distributed
- Elimination of hard-coded literals

It is an unfortunate sociological phenomenon that working on old applications is viewed as less glamorous than working on new applications. As a result, most methodologies in this report are aimed almost exclusively at new software and essentially ignore legacy software. This is true for Agile, Crystal, DevOps, extreme programming (XP), and most of the others. This is why *pattern matching* or starting development by examining the results of similar legacy applications is among the most useful in 2016.

Today, in 2016, about 90% of all software projects are not "new" in the sense that they have never been done before. They are either enhancements to legacy applications or attempts to build new and better versions of legacy applications.

An example of this is the patient records system of the U.S. Veterans Administration (VA). Currently, millions of medical records of veteran service people are maintained in a decaying legacy application written in the Mumps programming language more than 25 years ago. Performance is slow and the original specifications and requirements have long vanished. Worse, someone years ago decided to remove comments from the source code. The VA is now spending millions of dollars attempting to build a new, better, and faster version of keeping veteran medical records.

In general, attempting to reengineer an aging legacy system will take longer and cost more than the original version, even if modern languages are used. The problem is that the original requirements, architecture, and design cannot be ignored or left out but these are essentially invisible until some kind of data mining or reengineering takes place.

Table 51.1 shows the results of reengineering an application of 1000 function points in the Java language, using the remains of a legacy application of the same size written in the Mumps language.

Geriatric care for aging legacy software applications is now the most common software activity in the United States. Programmers working on legacy maintenance now outnumber programmers working on new development. Since legacy applications do not age gracefully and need a lot of repairs and many need to be replaced, it is too bad that the majority of software methodologies in this book ignore maintenance and enhancements and focus only on new development. In any case, software maintenance work is expanding faster than development, so research is urgently needed on all forms of legacy work: reengineering, repairs, restructuring, renovation, replacement, and replacement with a commercial off-the-shelf (COTS) package.

Software reengineering is technically challenging due in large part to the fact that legacy applications may be poorly structured, poorly commented, poorly documented, and some are written in antique languages, which are not well understood by maintenance personnel and which may lack tool sets for topics such as data mining.

Table 51.1 Example of 1000 Function Point Reengineering Project with Java Language

Project Size and Nature Scores		**Reengineer**
1	Small projects <100 FP	8
2	Maintenance projects	8
3	IT projects	8
4	Systems/embedded projects	8
5	High-reliability, high-security projects	8
6	Medium projects	7
7	Large projects >10,000 FP	6
8	Enhancement projects	6
9	Web/cloud projects	2
10	New projects	1
	Subtotal	62
Project Activity Scores		**Re Reengineer**
1	Coding	9
2	Integration	9
3	Reusable components	8
4	Change control	8
5	Defect prevention	8
6	Requirements changes	7
7	Design	7
8	Requirements	6
9	Architecture	6
10	Models	6
11	Pattern matching	5
12	Pre-test defect removal	5
13	Test case design/generation	5

(Continued)

Table 51.1 (Continued) Example of 1000 Function Point Reengineering Project with Java Language

14	Testing defect removal	5
15	Planning, estimating, measurement	–5
	Subtotal	89
	Total	151
Regional Usage (Projects in 2016)		**Reengineer**
1	North America	10,000
2	Pacific	10,000
3	Europe	10,000
4	South America/Central America	7,000
5	Africa/Middle East	5,000
	Total	42,000
Defect Potentials per Function Point (1000 function points; Java language)		**Reengineer**
1	Requirements	0.40
2	Architecture	0.55
3	Design	0.45
4	Code	1.30
5	Documents	0.40
6	Bad fixes	0.30
	Total	3.40
	Total defect potential	3400
Defect Removal Efficiency (DRE)		
1	Pre-test defect removal	50.00%
2	Testing defect removal	87.00%
	Total DRE	91.50%
	Delivered defects per function point	0.289
	Delivered defects	289

(Continued)

Table 51.1 (Continued) Example of 1000 Function Point Reengineering Project with Java Language

	High-severity defects	45
	Security defects	4
Productivity for 1000 Function Points		**Reengineer**
	Work hours per function point 2016	17.50
	Function points per month 2016	7.54
	Schedule in calendar months	17.60

Data Source: Namcook analysis of 600 companies.

Legacy Reengineering Scoring
Best = +10
Worst = 10

Chapter 52

Reuse-Based Software Development (85%)

Executive summary: *Custom design and manual coding of software applications are intrinsically expensive and error prone. The future of software engineering lies in moving away from custom development and moving to a model of constructing applications using at least 85% reusable materials. The reusable materials will include reusable requirements, reusable design, reusable code, reusable test cases and test scripts, reusable user documents, and even reusable plans. Pattern-based development and model-based development also include software reuse, but not as the main goal. See also feature-based development and product line engineering, which include reuse too.*

If you have a goal of producing a software application of 1000 function points in size with a productivity rate faster than 100 function points per staff month, how could you do it? Let's assume you also need close to zero defects and you need zero security flaws.

Today's productivity rates for 1000 function points are between about 10 and 20 work hours per function point, even for Agile projects. Today's applications have far too many bugs at delivery and far too many security flaws even for Agile projects.

Custom designs for software applications and manual coding are intrinsically expensive, error prone, and slow regardless of which programming languages and which development methodologies are used. Agile may be a bit faster than waterfall, but it is still slow compared with actual business needs.

The only effective solution for software engineering is to move toward construction of applications using standard reusable materials rather than custom design and development. The idea is to build software more like Ford builds automobiles on an assembly line rather than like the custom design and manual construction of a Formula 1 race car. This analogy is fairly apt because a 10,000 function point software system and a modern passenger car have about the same number of moving parts. If complex devices such as passenger cars can be built from standard parts, then software can eventually be built from standard parts.

An ordinary passenger car and a Formula 1 race car have about the same number of mechanical parts, but the race car costs at least 10 times more to build due to the large volumes of skilled manual labor involved. The schedule would be more than 10 times longer as well. Custom designs and manual construction are intrinsically slow and expensive in every industry.

Because custom designs and manual coding are slow, error prone, and inefficient, it is useful to show the future impacts of varying levels of reusable components. The phrase *reusable components* refers to much more than just reusable source code. The phrase also includes:

Reusable Software Components

1. Reusable requirements
2. Reusable architecture
3. Reusable design
4. Reusable project plans
5. Reusable estimates
6. Reusable source code
7. Reusable test plans
8. Reusable test scripts
9. Reusable test cases
10. Reusable marketing plans
11. Reusable user manuals
12. Reusable training materials
13. Reusable HELP screens and help text
14. Reusable customer support plans
15. Reusable maintenance plans

Figure 52.1 illustrates why software reuse is the ultimate methodology that is needed to achieve high levels of productivity, quality, and schedule adherence at the same time. The figure also illustrates a generic application of 1000 function points coded in the Java language.

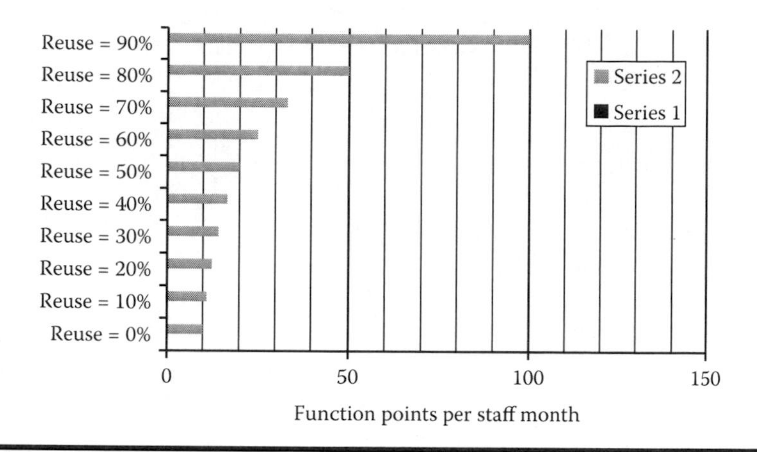

Figure 52.1 Impact of Reuse on Software Productivity.

Table 52.1 Reuse and Software Quality Levels at Delivery

Percentage of Reuse	*Defects per Function Point*	*Defect Removal Percentage*	*Delivered Defects per Function Point*
90	1.00	99.50	0.01
80	1.25	98.00	0.03
70	1.50	95.00	0.08
60	2.00	92.00	0.16
50	2.50	90.00	0.25
40	3.00	88.00	0.36
30	3.75	85.00	0.56
20	4.25	83.00	0.72
10	5.00	81.00	0.95
0	5.75	79.00	1.21

Software reuse is also a top technology for quality improvement. Table 52.1 shows the improvements in defect potentials and defect removal efficiency (DRE) and reuse volumes go up.

Reuse is the best choice for industries where many companies produce and use the same kinds of software. These industries include, but are not limited to, the following 10 examples:

Ten Industries with Substantial Software Reuse in 2016

1. Banks
2. Airlines
3. Insurance companies
4. Telecom operating companies
5. Telecom manufacturing companies
6. State governments
7. Municipal governments
8. Wholesale marketing
9. Retail marketing
10. Publishing

The idea behind reuse is to capture the essential patterns of the common applications that occur many times, and then use those patterns as a key to selecting and utilizing reusable components.

Already today, in 2016, there are dozens of third-party software vendors who sell generic applications to these 10 industries because they do pretty much the same thing.

Unfortunately, there are several gaps in the reuse domain that need to be filled: (1) there is no effective taxonomy of individual reusable features; (2) there are no available catalogs of reusable features that might be acquired from commercial sources; (3) software measurements tend to ignore or omit reusable features, which distorts productivity and quality data; (4) some software estimating tools do not include reuse (although this is a standard feature in the author's Software Risk Master [SRM] estimating tool); (5) much of the literature on reuse only covers code and does not yet fully support reusable requirements, reusable designs, reusable test materials, and reusable user documents.

One major barrier to expanding reuse at the level of specific functions is the fact that there are no effective taxonomies for individual features used in software applications. Current taxonomies work on entire software applications, but are not yet applied to the specific feature sets of these applications. For example, the widely used Excel spreadsheet application has dozens of built-in reusable functions, but there is no good taxonomy for identifying what all of these functions do. The following are some of the elements that would be present in a software application feature set taxonomy:

Taxonomy of Software Application Functions

Software Application Inputs	
1	Sensor based
2	Optical images

(Continued)

(Continued) Taxonomy of Software Application Functions

3	Electrical signals
4	Time signals
5	Pull-down menus
6	Buttons
7	Numerical information
8	Financial information
9	Classified information
10	Legal information
11	Tax information
12	Medical information
13	Text information
14	Graphics information
15	Natural language information
16	Cartographic information
17	Audio information
18	Mixed information
Software Application Processing	
1	Analog processing of electrical signals
2	Logic processing
3	Transformation processing
4	Natural language processing
5	Pattern matching
6	Diagnostic processing
7	Optical signal processing
8	Cartographic processing
9	Mathematical processing
10	Financial processing

(*Continued*)

(Continued) Taxonomy of Software Application Functions

11	Image processing
12	Musical synthesis
13	Text processing
14	Mixed processing
15	Error checking of inputs
16	Security monitoring of inputs
Software Application Data Storage	
1	Ephemeral—no data storage
2	Cloud data storage—formal database
3	Local data storage—disks, thumb drives
4	Remote data storage—disks, thumb drives
5	Remote data storage—formal data base
6	Secure data storage—encryption
7	Inclusion in repository or big data storage
Software Application Outputs	
1	Electrical signals
2	Machine control signals
3	Visual images
4	Audio information
5	Musical outputs
6	Natural language(s)
7	Encrypted outputs
8	Cartographic images
9	Optical information
10	Mathematical data
11	Financial data

(*Continued*)

(Continued) Taxonomy of Software Application Functions

12	Printed information (2-D)
13	Printed information (3-D)
14	Text information
15	Graphics information
16	Mixed information
17	Legal information
18	Tax information
19	Medical information
20	Classified information
21	Big data information

This list shows a top-level feature set taxonomy and each topic would have subcategories. The essential idea is to create an effective working taxonomy of exactly what software applications do, so that reusable components can be created and accessed for the most common functions.

Commercial reuse is a fairly large and growing industry in 2016. For example, hundreds of applications use Crystal Reports. Thousands use commercial and reusable static analysis tools, firewalls, anti-virus packages, and the like. Hundreds of major companies deploy enterprise resource planning (ERP) tools, which attempt reuse at the corporate portfolio level. Reuse is not a new technology, but neither is it yet an industry with proper certification to eliminate bugs and security flaws prior to deployment.

Although reusable materials have major benefits when deployed, the construction of reusable materials and their certification is normally more expensive than ordinary custom development due to the need for rigorous quality control and security flaw prevention (and removal). In general, developing a reusable component takes about 50% longer and is about 75% more expensive than developing the same feature using normal development practices. The sequence for developing standard sets of reusable components includes, but is not limited to, the following:

Development Stages for Certified Reusable Components

1. Sizing of each feature using function points, software non-functional assessment process (SNAP), and logical code size
2. Planning and estimating schedules and costs of feature construction
3. Planning for potential risks and security attacks for each feature
4. Market analysis of reuse potential for each feature (from 10 to 1,000,000 uses)
5. Make or buy analysis for certified reusable materials
6. Tax and government restriction analysis for all reusable materials
7. Mining legacy applications for potential reusable components
8. Patent analysis for unique or valuable intellectual property
9. Formal and reusable requirements creation for each feature
10. Formal and reusable design creation for each feature
11. Formal and reusable code creation for each feature
12. Formal and reusable test plan creation for each feature
13. Formal and reusable test script creation for each feature
14. Formal and reusable test case creation for each feature
15. Formal and reusable user training material for each feature
16. Formal and reusable user documentation for each feature
17. Formal translation into foreign languages, if needed for global sales
18. Formal and reusable HELP text for each feature
19. Formal security inspection for each feature
20. Formal usability inspection for each feature
21. Formal quality inspection for each feature
22. Formal text static analysis
23. Formal running of FOG or FLESCH readability tools on all text
24. Formal code static analysis for all source code
25. Formal mathematical test case construction for each feature's test cases
26. Formal measurement of cyclomatic complexity for all code
27. Formal measurement of test coverage for reusable test cases
28. Formal testing of each reusable code segment for errors and security flaws
29. Formal collection of historical data (quality, costs, etc.) for each feature
30. Formal certification of each feature prior to integration in component library

As can be seen, almost half of the development stages for constructing reusable materials would probably not be performed for ordinary components that are not specifically aimed at being entered into libraries of reusable materials. This is why the development of certified reusable components is slower and more expensive than ordinary development. The high initial costs are, of course, offset by the value and much lower costs of each subsequent reuse of the component.

To demonstrate the economic value of reuse, Table 52.2 shows a side-by-side comparison of the effort involved in building reusable components, and then the

Table 52.2 Development and Deployment of Reusable Software (Work Hours per Function Point for 1000 Function Points; Java)

Activities	*Reuse Develop*	*Reuse Deploy*	*Difference*
Requirements	1.00	0.15	–0.85
Design	2.00	0.10	–1.90
Design inspection	1.25	0.00	–1.25
Coding	5.00	0.15	–4.85
Software quality assurance	1.25	0.10	–1.15
Security inspection	1.50	0.10	–1.40
Testing	6.00	0.50	–5.50
Documentation	3.00	0.36	–2.64
Management	3.00	0.30	–2.70
Total	**24.00**	**1.76**	**–22.24**

effort to deploy them in new applications once they are constructed. Full and formal reuse is the only known methodology that actually does lead to order of magnitude improvements in software work effort.

Because the costs and schedules of creating and certifying reusable software components are much larger than for ordinary development, there is no economic value from creating these components unless they are likely to be reused many times. In general, specific features designed for reuse should have planned deployment in at least 10 applications. The value, of course, increases, and it is possible to envision the same reusable components in over 10,000 applications.

However, once a library of reusable materials is prepared and available, then development schedules, even of major applications, will drop by about 50%–80% compared with 2016 norms. Development schedules in a reusable world for a major system of 10,000 function points would be

Methodologies	**Schedule Months**
Reusable schedule	12
Team software process (TSP) schedule	31
Rational Unified Process (RUP) schedule	33
Agile schedule	36
Waterfall schedule	38

There is no way that any methodology that centers on custom designs and manual coding can even approach the speed and cost-effectiveness of the construction of applications from standard reusable components. The development sequence in a reusable world would resemble the following:

	Start					
	*					
	*					
Place project on formal taxonomy						
	*					
Use taxonomy to match patterns of similar projects						
	*					
Perform initial schedule, risk, cost estimates						
	*					
Extract historical data from similar projects						
1.	Size					
2.	Schedule					
3.	Staffing					
4.	Effort					
5.	Costs					
6.	Risks					
7.	Quality					
8.	Architecture					
9.	Design					
10.	Training materials					
11.	Security					
12.	Tool suites					
13.	Methodologies					
14.	Programming languages					

(*Continued*)

15.	Cyber-attack history					
	*					
Use data from similar projects for current requirements						
	*					
Add unique and custom requirements, if any						
	*					
Inspect custom requirements, if any						
	*					
Use data from similar projects for selecting optimal architecture						
	*					
Use data from similar projects for selecting optimal design						
	*					
Use data from similar projects for selecting optimal quality control						
	*					
Use taxonomy to search for certified reusable components						
	*					
Acquire reusable materials and validate security						
	*					
Use data from similar projects for estimating current project						
	*					
	Divide project into two major segments					
		*				
	*		*			
*				*		
Construction from Reusable Materials 85%				**Custom Design and Manual Coding 15%**		
*				*		

(Continued)

*				Test case design		
*				*		
*				Develop custom code		
*				*		
*				Inspect custom code		
*				*		
Static analysis of reused code				Static analysis of custom code		
*				*		
*				*		
	*		*			
		*				
		*				
		*				
		Test case execution				
		*				
		Defect repairs				
		*				
		Delivery to clients				

The reuse methodology starts about a month earlier than other methods because the first task is placing the proposed project on a standard taxonomy. Then, the taxonomy is used to search for similar projects from benchmark sources such as Namcook Analytics, ISBSG, Q/P Management Group, David's Consulting, Reifer Associates, and several others.

The sum of the requirements and technologies for similar projects is the true starting point for software development. Starting with a blank sheet of paper and assuming every application is unique is like building every automobile in the world by hand using custom designs and hand-crafted parts.

It is likely that some kind of brokerage business might be needed to handle the aggregation and distribution of reusable materials. Specific companies are probably not set up or qualified to market their own reusable materials except very large companies such as IBM and Microsoft.

A reuse clearing house might be created that could purchase reusable software materials at wholesale prices and then remarket the materials at retail prices.

Another form of reuse is also discussed in this book: mashups. This form of reuse is used mainly for web applications and consists of splicing together segments from current web applications to create new ones.

Table 52.3 shows the probable results for a project of 1000 function points and the Java programming language, but utilizing 85% reuse of requirements, design, code, test materials, and project documents.

Table 52.3 Example of 1000 Function Point Project with 85% Reuse and the Java Language

Project Size and Nature Scores		**85% Reuse**
1	Small projects <100 FP	10
2	Medium projects	10
3	Large projects >10,000 FP	10
4	New projects	10
5	IT projects	10
6	Systems/embedded projects	10
7	Web/cloud projects	10
8	High-reliability, high-security projects	10
9	Enhancement projects	5
10	Maintenance projects	5
	Subtotal	90
Project Activity Scores		**85% Reuse**
1	Pattern matching	10
2	Reusable components	10
3	Planning and estimating	10
4	Requirements	10
5	Architecture	10
6	Models	10
7	Design	10

(Continued)

Table 52.3 (Continued) Example of 1000 Function Point Project with 85% Reuse and the Java Language

8	Coding	10
9	Defect prevention	10
10	Test case design/generation	10
11	Testing defect removal	10
12	Requirements changes	8
13	Change control	8
14	Integration	8
15	Pre-test defect removal	7
	Subtotal	141
	Total	231
Regional Usage Reuse		**85% Reuse**
1	North America	300
2	Pacific	195
3	Europe	155
4	South America/Central America	40
5	Africa/Middle East	10
	Total	700
Defect Potentials per Function Point (1000 function points; Java language)		**85% Reuse**
1	Requirements	0.10
2	Architecture	0.10
3	Design	0.20
4	Code	0.40
5	Documents	0.10
6	Bad fixes	0.10
	Total	1.00
	Total defect potential	1000

(Continued)

Table 52.3 (Continued) Example of 1000 Function Point Project with 85% Reuse and the Java Language

Defect Removal Efficiency (DRE)		
1	Pre-test defect removal	95.00%
2	Testing defect removal	95.00%
	Total DRE	99.75%
	Delivered defects per function point	0.002
	Delivered defects	2
	High-severity defects	0
	Security defects	0
Productivity for 1000 Function Points		**85% Reuse**
	Work hours per function point 2016	1.76
	Function point per month	75.00
	Schedule in calendar months	2.82

Data source: Namcook analysis of 600 companies.

Reuse (85%) Scoring
Best = +10
Worst = –10

Moving away from custom designs and manual coding and into the world of construction of software from standard reusable components is definitely the methodology that will dominate the software industry in the future. No other methodology even has a theoretical possibility of coming with 25% of the economic and quality results of using over 90% reusable components.

Although 85% reuse of certified components yields outstanding results, only a few hundred projects have been able to do this up to 2016. The problem, of course, is the lack of sufficient quantities of certified materials to make large-scale reuse possible.

At least 5000 certified reusable components are needed to support all industries and all application types, and the world supply of certified reusable components in 2016 is probably below 1000. At least 600 of the 1000 are proprietary inside companies such as Amazon and Google and are not available anywhere else.

The companies that seem to be leaders in reuse technology in 2016 include Apple, Amazon, AT&T, Google, IBM, Intel, Microsoft, Motorola, Samsung, and Sony. The industries that have more certified reusable components than average include airlines, automobiles, banking, computer games, healthcare and hospitals, hotel chains, insurance, professional sports, real estate, and telecommunications. In all of these industries, every company in each industry carries out similar business functions using similar software packages.

Chapter 53

Reverse Engineering

Executive summary: *Reverse engineering, as it applies to software, may be defined as a process of analyzing an existing system to create representations of that system at a higher level of abstraction. For example, an effort to replicate the functionality of an existing system for which design documentation (and perhaps even source code) does not exist. In a sense, this involves going backwards through a software development life cycle to discover the design and perhaps even the requirements in order to create a faithful replica of the original, perhaps using more modern technology. The intent of a reverse engineering effort is simply to duplicate existing functionality, not necessarily to improve upon it—if the intent is to improve, the project may more properly be categorized as reengineering.*

In both the United States and the EU, most reverse engineering that is intended to create a competitive product is illegal if the software is subject to an End-User License Agreement (EULA). Copyright law in the United States is permitted under the *fair use* doctrine if the software is not covered by an EULA. Efforts to reverse engineer with the intent to enable inter-operability is generally legal—for example, the Open Office suite of programs compatible with Microsoft Office.

Reverse engineering can be achieved using a variety of methods, including:

- Analysis through observation of message sent or received by the software using bus analyzers or packet sniffers, for example, to determine the functionality of device drivers.
- Disassembly of object code using a disassembler tool to create raw machine language. That code would then be subject to further (often time consuming) human effort to deduce the underlying design and perhaps requirements.

- Decompilation using a decompiler to generate source code in a target language from machine language code.

Motivations for reverse engineering may include:

- Military or commercial espionage: May result in a better product or in better countermeasures against an existing product.
- Improved documentation, as in a legacy system that lacks current or complete specifications. Doing so may improve maintainability and extend product life.
- Obsolescence: When, for example, a technology, such as a complier for an old language, is no longer supported.
- Software modernization: Efforts may be necessary to understand the *as is* state as a prerequisite to migration toward a *to be* state.
- Product security analysis: Either to devise countermeasures against hackers or to circumvent access restrictions.
- Bug fixing in software no longer supported.
- Competitive intelligence: Understand what a competitor is actually doing.
- Ethical hacking: Attempts to find vulnerabilities before they are discovered by adversaries.

Often, reverse engineering is attempted as a first step in either replacing or renovating an aging legacy application where the original requirements and specifications are missing or are so out of date as to be useless. Reverse engineering is similar in concept to *data mining* of legacy applications.

Reverse engineering is also used in cyber-attack defenses. For example, viruses and other attack vectors may be reverse engineered in an attempt to develop defense mechanisms.

Table 53.1 shows the results of building an application of 1000 function points in Java by reverse engineering an older legacy application of 1000 function points in basic assembly language.

Reverse engineering is a niche methodology with some valid uses in cyber-attack defenses. It can also be a harmful methodology if it is used for unethical or illegal purposes such as stealing intellectual property or looking inside competitive software projects.

Table 53.1 Example of 1000 Function Point Reverse Engineering Project with the Java Language

Project Size and Nature Scores		**Reverse Engineer**
1	High-reliability, high-security projects	10
2	Small projects <100 FP	8
3	IT projects	8
4	Systems/embedded projects	8
5	Medium projects	6
6	Maintenance projects	6
7	Enhancement projects	4
8	Large projects >10,000 FP	3
9	Web/cloud projects	2
10	New projects	0
	Subtotal	55
Project Activity Scores		**Reverse Engineer**
1	Requirements	10
2	Coding	9
3	Change control	9
4	Integration	9
5	Reusable components	8
6	Defect prevention	8
7	Pre-test defect removal	8
8	Testing defect removal	8
9	Requirements changes	7
10	Design	7
11	Test case design/generation	7
12	Pattern matching	6
13	Architecture	6

(Continued)

Table 53.1 (Continued) Example of 1000 Function Point Reverse Engineering Project with the Java Language

14	Models	6
15	Planning, estimating, measurement	–1
	Subtotal	107
	Total	162
Regional Usage (Projects in 2016)		**Reverse Engineer**
1	North America	1500
2	Pacific	3500
3	Europe	1000
4	South America/Central America	1000
5	Africa/Middle East	1000
	Total	8000
Defect Potentials per Function Point (1000 function points; Java language)		**Reverse Engineer**
1	Requirements	0.4
2	Architecture	0.5
3	Design	0.4
4	Code	1.2
5	Documents	0.4
6	Bad fixes	0.3
	Total	3.2
	Total defect potential	3200
Defect Removal Efficiency (DRE)		
1	Pre-test defect removal	50.00%
2	Testing defect removal	90.00%
	Total DRE	95.00%

(Continued)

Table 53.1 (Continued) Example of 1000 Function Point Reverse Engineering Project with the Java Language

	Delivered defects per function point	0.16
	Delivered defects	160
	High-severity defects	25
	Security defects	2
Productivity for 1000 Function Points		**Reverse Engineer**
	Work hours per function point 2016	16.00
	Function points per month 2016	8.25
	Schedule in calendar months	17.20

Data source: Namcook analysis of 600 companies.

Reverse Engineering Scoring
Best = +10
Worst = –10

Chapter 54

Robotic Software Development Using Standard Parts

Author's note: *All of the other 61 methodologies discussed in this book existed in 2016. Robotic development of software applications does not exist in 2017 either except as an experiment and may not exist for perhaps another 10 years. This chapter is intended to show the precursor steps and preliminary research that will be needed to achieve successful robotic software development. If robotic development did exist in 2016 it would rank number one in software quality, number one in schedules, and number one in productivity.*

Executive summary: *Custom design and manual coding of software applications are intrinsically expensive and error prone no matter which software methodology is used. The future of software engineering lies in moving away from custom development and moving to a model of constructing applications using at least 85% reusable materials and possibly 100%.*

Once a critical threshold of about 95% reuse is reached, then robots might be able to replace human programmers. In many other industries, robots have been able to construct very complex devices so long as the devices are made from standard reusable parts and have a clear construction model. This could be done for software. However, at least 10 more years will be needed before robotic software development is probable. This chapter discusses what must be done to make robotic development possible.

Today, in 2016, robotic development of software seems unlikely. But robotic surgery and robotic manufacturing were unlikely 20 years ago. Today, they are common and effective.

In order to build software via robot, a number of precursor technologies need to be in place:

1. Libraries of standard reusable components are needed for various kinds of software.
2. The libraries need to include test materials and documents as well as code.
3. The libraries of standard parts need to be certified to zero-defect levels.
4. The libraries of standard parts need to be certified free of security flaws.
5. An intelligent agent or a human "composer" needs to select the specific parts needed.
6. The robots need to be programmed to construct various types of applications.
7. Automated static analysis should occur at frequent intervals.
8. Automated testing of the applications is needed at completion.

If you compare the costs and schedules of building an 80-story office building, an 80,000 on cruise ship, and an 80,000 function point software system, the software is much more expensive and also much slower to build than the other two.

When deployed, the software is much less reliable than the other two and has many more defects that interfere with its use than the other two. Worse, the software is much more likely to be attacked by external criminals seeking to steal data or interfere with software operation. In fact, software cyber-attacks in 2016 are about as dangerous as cruising off the coast of Somalia.

These problems are endemic but not impossible to cure. It is technically possible today, in 2016, to build some software applications from standard reusable components. It is also possible to raise the immunity of software to external cyber-attack. (See other chapters on reuse-centered development, container development, and micro services for additional data on software reuse in 2016.)

In the future, more and more standard components will expand the set of applications that can be assembled from certified standard parts. These standard parts should also be free from security vulnerabilities.

Construction from certified components can be more than 10 times faster and cheaper than the best manual methods such as Agile, and also much more secure than today's norms where security vulnerabilities are rampant.

An interesting question is how many standard reusable components are going to be needed for robotic construction to be feasible. The numbers and average sizes of the components vary by industry. Table 54.1 shows an approximation of the reusable components needed for 25 industries with size expressed in terms of International Function Point Users' Group (IFPUG) 4.3 function point metrics.

As already stated, if you have a goal of producing a software application of 1000 function points in size with a productivity rate faster than 100 function points per

Table 54.1 Probable Amounts of Reusable Components (Size Expressed in IFPUG 4.3 Function Points)

	Industry	*Reusable Components*	*Average Size*	*Total Size*
1	Banking/finance	225	50	11,250
2	Computers—operating systems	125	70	8,750
3	Education—primary	90	50	4,500
4	Education—university	150	80	12,000
5	Entertainment—online games	125	80	10,000
6	Government—federal civilian	900	80	72,000
7	Government—federal defense	1250	150	187,500
8	Government—municipal	200	60	12,000
9	Government—state	400	70	28,000
10	Healthcare, hospitals	200	90	18,000
11	Hotels	125	60	7,500
12	Insurance—medical	235	80	18,800
13	Insurance—property	185	60	11,100
14	Intelligence, security	1450	150	217,500
15	Manufacturing—autos	175	50	8,750
16	Manufacturing—electronics	125	50	6,250
17	Online web retailers	250	70	17,500
18	Police, law enforcement	200	80	16,000
19	Real estate	125	50	6,250
20	Social networks	115	60	6,900
21	Stock trading	275	80	22,000
22	Telecommunications—networks	185	80	14,800
23	Transportation—airlines	150	250	37,500
24	Transportation—shipping	135	180	24,300
25	Transportation—trucking	115	240	27,600
	Total	**7510**	**2320**	**806,750**

staff month, how could you do it? Let's assume you also need close to zero defects and you need zero security flaws.

As shown throughout this book, using most methodologies, today's productivity rates for 1000 function points are between about 10 and 20 work hours per function point, even for Agile projects. Today's applications have far too many bugs at delivery and far too many security flaws even for Agile projects.

Custom designs for software applications and manual coding are intrinsically expensive, error prone, and slow regardless of which programming languages and which development methodologies are used.

The only effective solution for software engineering is to move toward the construction of applications using standard reusable materials rather than custom design and development. The idea is to build software more like Ford builds automobiles on an assembly line rather than like the custom design and manual construction of a Formula 1 race car.

An ordinary passenger car and a Formula 1 race car have about the same number of mechanical parts, but the race car costs at least 10 times more to build due to the large volumes of skilled manual labor involved. The schedule would be more than 10 times longer as well. Custom designs and manual construction are intrinsically slow and expensive in every industry.

In today's world, leading automotive producers such as Toyota and Ford use robotic assembly lines. Robots are quicker than humans and if programmed well, make fewer mistakes. What would it take to bring robots to software development?

Reuse has been talked about since the 1960s but has not yet become a major feature of modern software, although, in fact, there are quite a few reusable components in 2016. The barriers to effective reuse include the need to develop an effective taxonomy of reusable features, the high costs of developing secure and safe reusable components, and the need to have a clearing house that can store or access reusable materials developed by a variety of companies and government agencies.

Since reuse is the critical precursor technology, it is probable that the Software Engineering Methods and Theory (SEMAT) approach would be a valuable concept for the construction of reliable and high-quality standard reusable components. It is obvious that components aimed at successful reuse in hundreds of applications need to be well formed, reliable, and essentially free of security flaws. SEMAT is a good starting point for achieving these goals.

Another interesting question about robotic development of software is which company or government organization is most likely to accomplish it first? Of course, there may be hidden research that has yet to be made public. But, from considering the recent track record of innovation, here is a list of possible organizations that might pioneer robotic software development. They are ranked in approximate order of possibly exploring the robotic development of software:

1. Google
2. Amazon

3. Microsoft
4. IBM
5. Huawei
6. Apple
7. Samsung
8. Toshiba
9. Oracle
10. VMWare
11. Sony
12. Docker
13. Box
14. Redhat
15. Intel

There may also be other contenders but these 15 have all demonstrated innovation and all file hundreds or even thousands of patents every year. It is interesting, in 2016, that a Chinese company would be ranked among the top companies in innovation but China is making fast progress as a technology country.

For many years, the first stage of conventional software development, whether Agile, waterfall, the Rational Unified Process (RUP), or something else, has been to gather user requirements. This is the wrong starting place and only leads to later problems.

The first stage of development should be *pattern matching*, or using a formal taxonomy to identify exactly what kind of software application will be needed. Pattern matching is possible because, in 2016, the majority of software applications are not new and novel, but either replacements for existing applications or variations based on existing applications.

The elements of the formal taxonomy used with the author's Software Risk Master (SRM) tool include:

Software Risk Master Application Taxonomy

1. Project nature: New or enhanced or package modification, and so on
2. Hardware platform: Cloud, smartphone, tablet, personal computer, mainframe, and so on
3. Operating system: Android, Linux, Unix, Apple, Windows, and so on
4. Scope: Program, departmental system, enterprise system, and so on
5. Class: Internal, commercial, open source, military, government, and so on
6. Type: Telecom, medical device, avionics, utility, tool, social, and so on
7. Problem complexity: Very low to very high
8. Code complexity: Very low to very high
9. Data complexity: Very low to very high

10. Country of origin: Any country or combination of countries
11. Geographic region: State or province (using telephone codes)
12. Industry: North American Industry Classification (NAIC) codes
13. Project dates: Start date; delivery date (if known)
14. Methodology: Agile, RUP, team software process (TSP), waterfall, Prince2, and so on
15. Languages: C, C#, Java, Objective C, combinations, and so on
16. Reuse volumes: 0%–100%
17. Capability maturity model integrated (CMMI) level: 0 for no use of CMMI; 1–5 for actual levels
18. Software Engineering Method and Theory (SEMAT) use: Using or not using SEMAT
19. Start date: Planned start date for application construction
20. Project age: Calendar months from first deployment to current date
21. Growth rate: Function points from first deployment to current date
22. Custom features: Percentage of total features
23. Reused features: Percentage of total features
24. Custom parts lists: (Added in 2016)
25. Reused parts list: (Added in 2016)

Once the application is placed on the taxonomy, the next step is to dispatch intelligent agents to find out exactly how many applications exist that have the same patterns and what their results have been. The intelligent agents will also report back on the availability of standard reusable materials either from commercial vendors or from an internal library of components.

The taxonomy is also useful in creating benchmarks. Assuming that the intelligent agents find 50 similar existing applications, using data from benchmark sources and web searches, it is then possible to aggregate useful information derived from the set of similar projects.

However, the taxonomy is not the only pattern that needs to be used in order to move toward construction from reusable components and later to move to robotic construction. There are 25 other key patterns:

Important Software Application Patterns

1. Architectural patterns for the overall application structure
2. Design patterns for key application features
3. User interface patterns for optimal human understanding
4. Requirements patterns from all projects with the same taxonomy
5. Data patterns for the information created and used by the application
6. Occupation group patterns (designers, coders, testers, QA, etc.)
7. Development activity patterns (requirements, architecture, design, code, etc.)

8. Growth patterns of new features during development and after release
9. Reuse patterns for standard features from standard components
10. Source patterns for the mix of legacy, commercial off-the-shelf (COTS), reuse, open source, and custom features
11. Code patterns for any custom code, in order to avoid security flaws
12. Risk patterns based on similar applications
13. Security patterns (kinds of attacks noted on similar applications)
14. Regulatory patterns that might impact the software (Food and Drug Administration [FDA], Federal Aviation Authority [FAA], Federal Communications Commission [FCC], etc.)
15. Governance patterns for software dealing with financial data
16. Defect removal patterns for the sequence of inspections, static analysis, and test stages
17. Marketing patterns for the distribution of the software to clients
18. Usage patterns for typical usage scenarios
19. Maintenance patterns for defect repairs after release
20. Support patterns for contacts between customers and support teams
21. Enhancement patterns of future changes after initial deployment
22. Cost and schedule patterns for development and maintenance
23. Value and return on investment (ROI) patterns to compare project costs with long-range value
24. Litigation patterns for patent suits, breach of contract, and so on
25. Life expectancy patterns to show long-range total costs of ownership

These various patterns, combined with growing libraries of standard reusable components, should be able to increase application productivity rates from today's average of below 10 function points per staff month to more than 100 function points per staff month.

In some cases, for fairly small applications productivity could approach or even exceed 200 function points per staff month. The software industry should not be satisfied with custom design and manual coding because these are intrinsically expensive, slow, and error prone.

Because custom designs and manual coding are slow, error prone, and inefficient, it is useful to show the future impacts of varying levels of reusable components. The phrase *reusable components* refers to much more than just reusable source code. The phrase also includes:

Reusable Software Components

1. Reusable requirements
2. Reusable architecture
3. Reusable design

4. Reusable project plans
5. Reusable estimates
6. Reusable source code
7. Reusable test plans
8. Reusable test scripts
9. Reusable test cases
10. Reusable marketing plans
11. Reusable user manuals
12. Reusable training materials
13. Reusable HELP screens and help text
14. Reusable customer support plans
15. Reusable maintenance plans

Software reuse is also a top technology for quality improvement. Table 54.2 shows the improvements in defect potentials and defect removal efficiency (DRE) and reuse volumes go up.

Once software achieves the technical capability of constructing software from standard components aided by robotic development, today's high software risks will be greatly reduced. Table 54.3 shows the probable risk patterns for 95% reusable software.

Table 54.2 Reuse and Software Quality Levels at Delivery

Percentage of Reuse	*Defects per Function Point*	*Defect Removal (%)*	*Delivered Defects per Function Point*
90	1.00	99.50	0.01
80	1.25	98.00	0.03
70	1.50	95.00	0.08
60	2.00	92.00	0.16
50	2.50	90.00	0.25
40	3.00	88.00	0.36
30	3.75	85.00	0.56
20	4.25	83.00	0.72
10	5.00	81.00	0.95
0	5.75	79.00	1.21

Table 54.3 Risk Reductions from Robotic Software Construction (Assumes 10,000 Function Points)

	Risk Analysis from Similar Projects	*Normal Odds (%)*	*Robotic Odds (%)*
1	Optimistic cost estimates	35.00	1.00
2	Inadequate quality control using only testing	30.00	0.50
3	Excessive schedule pressure from clients, executives	29.50	2.00
4	Technical problems hidden from clients, executives	28.60	2.00
5	Executive dissatisfaction with progress	28.50	1.00
6	Client dissatisfaction with progress	28.50	1.00
7	Poor quality and defect measures (omits > 10% of bugs)	28.00	1.00
8	Poor status tracking	27.80	0.10
9	Significant requirements creep (> 10%)	26.30	4.00
10	Poor cost accounting (omits > 10% of actual costs)	24.91	2.00
11	Schedule slip (> 10% later than plan)	22.44	0.25
12	Feature bloat and useless features (> 10% not used)	22.00	3.00
13	Unhappy customers (> 10% dissatisfied)	20.00	2.00
14	Cost overrun (> 10% of planned budget)	18.52	1.00
15	High warranty and maintenance costs	15.80	0.50
16	Cancellation of project due to poor performance	14.50	0.50
17	Low reliability after deployment	12.50	0.10
18	Successful cyber-attacks after deployment	12.00	2.50%

(Continued)

Table 54.3 (Continued) Risk Reductions from Robotic Software Construction (Assumes 10,000 Function Points)

	Risk Analysis from Similar Projects	*Normal Odds (%)*	*Robotic Odds (%)*
19	Negative ROI due to poor performance	11.00	1.00
20	Litigation (patent violation)	9.63	1.00
21	Security vulnerabilities in software	9.60	1.50
22	Theft of intellectual property	8.45	2.00
23	Litigation (breach of contract)	7.41	1.00
24	Toxic requirements that should be avoided	5.60	1.00
25	Low team morale	4.65	2.00
	Average risks for this size and type of project	**19.25**	**1.36**
	Financial risk: (cancel; cost overrun; negative ROI)	**44.27**	**3.12**

Reuse and robotic construction are the best choice for industries where many companies produce and use the same kinds of software. These industries include, but are not limited to, the following 20 examples:

Industries with Substantial Software Reuse in 2017

1. Airlines
2. Aircraft manufacturing
3. Automobile manufacturing
4. Banks
5. Commercial software vendors
6. Computer game companies
7. Federal government (civilian)
8. Federal government (defense)
9. Hotels

10. Insurance companies
11. Municipal governments
12. Open-source software vendors
13. Publishing (books, periodicals)
14. Retail marketing
15. Software outsource vendors
16. State governments
17. Stock brokerages
18. Telecom manufacturing companies
19. Telecom operating companies
20. Wholesale marketing

The idea behind reuse and robotic development is to capture the essential patterns of the common applications that occur many times, and then use those patterns as a key to selecting and utilizing reusable components.

Already today, in 2016, there are dozens of third-party software vendors who sell generic applications to these 20 industries because they do pretty much the same thing.

Unfortunately, there are several gaps in the reuse domain that need to be filled:

- There is no effective taxonomy of individual reusable features.
- There are no available catalogs of reusable features that might be acquired from commercial sources.
- Software methodologies ignore reuse and do not include a *reuse acquisition* phase.
- Software measurements tend to ignore or omit reusable features, which distorts productivity and quality data.
- Some software estimating tools do not include reuse (although this is a standard feature in the author's SRM estimating tool).
- Much of the literature on reuse only covers code and does not yet fully support reusable requirements, reusable designs, reusable test materials, and reusable user documents.

One major barrier to expanding reuse at the level of specific functions is the fact that there are no effective taxonomies for individual features used in software applications. Current taxonomies work on entire software applications, but are not yet applied to the specific feature sets of these applications.

For example, the widely used Excel spreadsheet application has dozens of built-in reusable functions, but there is no good taxonomy for identifying what all of these functions do. The following are some of the elements that would be present in a software application feature set taxonomy:

Taxonomy of Software Application Functions

Software Application Inputs	
1	Sensor based
2	Optical images
3	Electrical signals
4	Time signals
5	Pull-down menus
6	Buttons
7	Numerical information
8	Financial information
9	Classified information
10	Legal information
11	Tax information
12	Medical information
13	Text information
14	Graphics information
15	Natural language information
16	Cartographic information
17	Audio information
18	Mixed information
Software Application Processing	
1	Analog processing of electrical signals
2	Logic processing
3	Transformation processing
4	Natural language processing
5	Pattern matching
6	Diagnostic processing
7	Optical signal processing

(*Continued*)

(Continued) Taxonomy of Software Application Functions

8	Cartographic processing
9	Mathematical processing
10	Financial processing
11	Image processing
12	Musical synthesis
13	Text processing
14	Mixed processing
15	Error checking of inputs
16	Security monitoring of inputs
Software Application Data Storage	
1	Ephemeral—no data storage
2	Cloud data storage—formal database
3	Local data storage—disks, thumb drives
4	Remote data storage—disks, thumb drives
5	Remote data storage—formal database
6	Secure data storage—encryption
7	Inclusion in repository or big data storage
Software Application Outputs	
1	Electrical signals
2	Machine control signals
3	Visual images
4	Audio information
5	Musical outputs
6	Natural language(s)
7	Encrypted outputs
8	Cartographic images
9	Optical information

(Continued)

(Continued) Taxonomy of Software Application Functions

10	Mathematical data
11	Financial data
12	Printed information (2-D)
13	Printed information (3-D)
14	Text information
15	Graphics information
16	Mixed information
17	Legal information
18	Tax information
19	Medical information
20	Classified information
21	Big data information

A top-level feature set taxonomy and each topic would have sub-categories. The essential idea is to create an effective working taxonomy of exactly what software applications do so that reusable components can be created and accessed for the most common functions.

Commercial reuse is a fairly large and growing industry in 2016. For example, hundreds of applications use Crystal Reports. Thousands use commercial and reusable static analysis tools, firewalls, anti-virus packages, and the like. Hundreds of major companies deploy enterprise resource planning (ERP) tools, which attempt to reuse at the corporate portfolio level. Reuse is not a new technology, but neither is it yet an industry with proper certification to eliminate bugs and security flaws prior to deployment.

Although reusable materials have major benefits when deployed, the construction of reusable materials and their certification is normally more expensive than ordinary custom development due to the need for rigorous quality control and security flaw prevention (and removal). In general, developing a reusable component takes about 50% longer and is about 75% more expensive than developing the same feature using normal development practices. The sequence for developing standard sets of reusable components includes, but is not limited to, the following:

Development Stages for Certified Reusable Components

1. Sizing of each feature using function points, software non-functional assessment process (SNAP), and logical code size
2. Planning and estimating schedules and costs of feature construction
3. Planning for potential risks and security attacks for each feature
4. Market analysis of reuse potential for each feature (from 10 to 1,000,000 uses)
5. Make or buy analysis for certified reusable materials
6. Tax and government restriction analysis for all reusable materials
7. Mining legacy applications for potential reusable components
8. Patent analysis for unique or valuable intellectual property
9. Formal and reusable requirements creation for each feature
10. Formal and reusable design creation for each feature
11. Formal and reusable code creation for each feature
12. Formal and reusable test plan creation for each feature
13. Formal and reusable test script creation for each feature
14. Formal and reusable test case creation for each feature
15. Formal and reusable user training material for each feature
16. Formal and reusable user documentation for each feature
17. Formal translation into foreign languages, if needed for global sales
18. Formal and reusable HELP text for each feature
19. Formal security inspection for each feature
20. Formal usability inspection for each feature
21. Formal quality inspection for each feature
22. Formal text static analysis
23. Formal running of FOG or FLESCH readability tools on all text
24. Formal code static analysis for all source code
25. Formal mathematical test case construction for each feature's test cases
26. Formal measurement of cyclomatic complexity for all code
27. Formal measurement of test coverage for reusable test cases
28. Formal testing of each reusable code segment for errors and security flaws
29. Formal collection of historical data (quality, costs, etc.) for each feature
30. Formal certification of each feature prior to integration in component library

As can be seen, almost half of the development stages for constructing reusable materials would probably not be performed for ordinary components that are not specifically aimed at being entered into libraries of reusable materials. This is why the development of certified reusable components is slower and more expensive than ordinary development. The high initial costs are, of course, offset by the value and much lower costs of each subsequent reuse of the component.

One key topic is not addressed in the software methodology literature. When designing software applications, the correct place to start after examining legacy applications is with the outputs for the new application. Outputs are the key feature of all software and need to be pinned down first. Then, it is easy to develop suitable input patterns to match the outputs.

If you try and do it the other way, and define the inputs first, you will invariably either have gaps that don't support needed outputs or you will have more inputs than the application actually needs. Designing the outputs first cuts down on later rework and redesign when the inputs and outputs are synchronized.

The actual processing stages that transform the inputs to the outputs should be done after both the inputs and outputs are fully defined and carefully designed. It is surprising that this basic fact is almost never mentioned in the software methodology literature.

Incidentally, a software "robot" will not look like an industrial robot or a surgical robot with moving arms and complex physical motion. Because a software robot only deals with intangible objects, it will probably look like just another server or a large desktop computer.

The operating sequence for a software robot will start with an analysis of the total set of reusable components needed, probably in the range of 100–350 components. Then, the components will be acquired from the component library and the construction of the application will take place sequentially. The sequence will probably be

1. Collect and test the application outputs
2. Collect and test the applications inputs
3. Collect and test the application security features
4. Collect and test each processing step
5. Insert any manual or custom features into the application
6. Final assembly of all reusable components
7. Final test and certification of all components
8. Application delivered to clients

To demonstrate the economic value of reuse, Table 54.4 shows a side-by-side comparison of the effort involved in building reusable components, and then the effort to deploy them into new applications once they are constructed. Full and formal reuse is the only known methodology that actually does lead to order of magnitude improvements in software work effort.

Because the costs of and schedules for creating and certifying reusable software components are much larger than for ordinary development, there is no economic value from creating these components unless they are likely to be reused many times. In general, specific features designed for reuse should have planned deployment in at least 10 applications. The value, of course, increases, and it is possible to envision the same reusable components in over 10,000 applications.

Table 54.4 Development and Deployment of Reusable Software

(Work Hours per Function Point for 1000 Function Points; Java)			
Activities	*Reuse Develop*	*Reuse Deploy*	*Difference*
Requirements	1.00	0.15	–0.85
Design	2.00	0.10	–1.90
Design inspection	1.25	0.00	–1.25
Coding	5.00	0.15	–4.85
Software quality assurance	1.25	0.10	–1.15
Security inspection	1.50	0.10	–1.40
Testing	6.00	0.50	–5.50
Documentation	3.00	0.36	–2.64
Management	3.00	0.30	–2.70
Total	**24.00**	**1.76**	**–22.24**

However, once a library of reusable materials is prepared and available, then the development schedules of major applications will drop by about 50%–80% compared with 2016 norms.

Examining the requirements and technologies used for similar existing projects is the true starting point for software development. Starting with a blank sheet of paper and assuming every application is unique is like building every automobile in the world by hand using custom designs and hand-crafted parts.

It is likely that some kind of brokerage business might be needed to handle the aggregation and distribution of reusable materials. Specific companies are probably not set up or qualified to market their own reusable materials except very large companies such as IBM and Microsoft.

A reuse clearing house might be created that could purchase reusable software materials at wholesale prices and then remarket the materials at retail prices.

Another form of reuse is also discussed in this report: mashups. This form of reuse is used mainly for web applications and consists of splicing together segments from current web applications to create new ones.

Table 54.5 shows the probable results for a project of 1000 function points and the Java programming language, but utilizing 95% reuse of requirements, design, code, test materials, and project documents and robotic construction.

Table 54.5 Example of 1000 Function Point Robotic Development Project with the Java Language

Project Size and Nature Scores		Robotic Reuse
1	Small projects <100 FP	10
2	Medium projects	10
3	Large projects >10,000 FP	10
4	New projects	10
5	IT projects	10
6	Systems/embedded projects	10
7	Web/cloud projects	10
8	High-reliability, high-security projects	10
9	Enhancement projects	5
10	Maintenance projects	5
	Subtotal	90
Project Activity Scores		**Robotic Reuse**
1	Pattern matching	10
2	Reusable components	10
3	Planning and estimating	10
4	Requirements	10
5	Architecture	10
6	Models	10
7	Design	10
8	Coding	10
9	Defect prevention	10
10	Test case design/generation	10
11	Testing defect removal	10
12	Requirements changes	10
13	Change control	10

(*Continued*)

Table 54.5 (Continued) Example of 1000 Function Point Robotic Development Project with the Java Language

14	Integration	10
15	Pre-test defect removal	7
	Subtotal	147
	Total	237
Regional Usage (Robotic Development 2016)		**Robotic Reuse**
1	North America	0
2	Pacific	0
3	Europe	0
4	South America/Central America	0
5	Africa/Middle East	0
	Total	0
Defect Potentials (1000 function points, Java programming language)		
1	Requirements	0.10
2	Architecture	0.10
3	Design	0.10
4	Code	0.30
5	Documents	0.10
6	Bad fixes	0.00
	Total	0.70
	Total defect potential	700
Defect Removal Efficiency (DRE)		
1	Pre-test defect removal	95.00%
2	Testing defect removal	95.00%
	Total DRE	99.75%
	Delivered defects per function point	0.002

(*Continued*)

Table 54.5 (Continued) Example of 1000 Function Point Robotic Development Project with the Java Language

	Delivered defects	2
	High-severity defects	0
	Security defects	0
Productivity for 1000 Function Points		**Robotic Reuse**
	Work hours per function point 2016	1.50
	Function points per month 2016	88.00
	Schedule in calendar months	2.00

Data source: Namcook analysis of 600 companies.

Robotic Scoring
Best = +10
Worst = +10

Moving away from custom designs and manual coding and into the world of construction of software from standard reusable components followed by robotic construction is definitely the methodology that will dominate the software industry in the future. No other methodology even has a theoretical possibility of coming within 15% of the economic and quality results of using over 95% reusable components and robotic construction.

Unfortunately, this is not an easy technology to master. However, robots have benefited dozens of industries and also surgical procedures. No doubt, software will eventually receive similar benefits but probably not for another 10 years or more.

Chapter 55

Service-Oriented Architecture

Executive summary: *Service-Oriented Architecture (SOA) is a vendor and technology-independent approach to software design in which discrete software elements provide "services" to other elements via a defined protocol. Each service is a self-contained unit of functionality independent of other units—a "black box" whose internals are hidden. This approach achieves both high cohesion and loose coupling, both highly desirable characteristics. SOA embodies many of the same ideas found in object-oriented development, carried to the level of whole systems. There are two distinct aspects of SOA: (1) creating the services initially, which is conventional development; (2) assembling new applications from services, which takes reuse to the same level as mashups in terms of ease and speed.*

SOA developers commonly use the Web Service Description Language (WSDL) to describe the services themselves and the Simple Object Access Protocol (SOAP) to describe the communications protocols.

SOA Framework

SOA-based solutions achieve business objectives by building enterprise-level systems whose architecture consists of five horizontal layers:

- Consumer interface layer: End-user interfaces or apps
- Business process layer: Sets of services that implement specific use cases

- Services: An inventory that totals the complete set of services available to the enterprise
- Service components: The individual units of code, data, and protocols used to compose services
- Operational systems: Contains data models, data repository, and technical platforms

In addition, there are four vertical layers that are applied to and supported by the horizontal layers:

- Integration layer: Protocol support, data integration, service integration, and application integration to produce enterprise-level applications supporting business-to-business (B2B) and business-to-consumer (B2C)
- Quality of service: Security, availability, and so on configured to meet required service-level agreement (SLAs) and operation-level agreements (OLAs)
- Informational: Provide business information
- Governance: Policy enforcement

SOA Principles

While there are no generally accepted industry standards governing SOA, various industry sources have published principles, including the following:

- Standardized service contract: All services adhere to published communications agreements
- Loose coupling: Dependencies between services are carefully minimized
- Service abstraction: All services adhere to the principle of information hiding—that is, internal logic is not exposed
- Reusability: Services are defined to promote reuse
- Autonomy: Services retain control over the logic they encapsulate, both design-time and run-time
- Discoverability: Services are supplemented by meta-data to facilitate discovery and understanding
- Composability: Services are designed to be used together in highly flexible arrangements
- Granularity: Design to achieve an appropriate level of cohesion consistent with loose coupling
- Normalization: Decomposition to minimize redundancy
- Optimization: High quality is a primary goal
- Location transparency: The ability to invoke a service independent of its actual location in a network

SOA has two kinds of tasks: (1) building the services initially; (2) constructing applications from completed services.

Initial service construction is conventional software development and would probably use a methodology such as the Rational Unified Process (RUP) or the team software process (TSP) that is quality strong. Using services once they are completed is another excellent example of reuse, and in some cases even beyond the 85% reuse level discussed in the chapter on reuse. The mashup chapter has a similar theme and similar results.

Because SOA is a new and rather complex topic, it is suggested that a Google search be used to achieve in-depth understanding of the approach. A full tutorial is outside the scope of this chapter. There are numerous articles on the web and some books about SOA from Amazon and other e-book stores.

Also, SOA depends to a large degree upon meta-data, which is not fully standardized as of 2014. The methods leading to full deployment of SOA are in rapid evolution.

As with Agile, an "SOA manifesto" was published in 2009 by, coincidentally, a set of 17 SOA practitioners (the Agile Manifesto also had 17 participants). Among the key points in the SOA manifesto are

- Business value over technical strategy
- Strategic goals over project-specific benefits
- Intrinsic interoperability over custom integration
- Shared services over specific-purpose implementation
- Flexibility over optimization
- Evolutionary refinement over initial perfection

SOA is, of course, built on the idea of common and standard application program interfaces (API) of which there are quite a few in 2016. SOA is an emerging technology in 2016 but not a mature one.

Examples of the kinds of industries where SOA seems beneficial would include:

- Banking services
- Insurance services
- State governments
- Manufacturing
- Wholesale distribution

Examples of the kinds of industries and applications where SOA is probably not a good choice in 2016 include:

- Medical devices such as pacemakers and cochlear implants
- Aircraft navigation packages
- Automotive brakes and fuel injection

- Weapons systems such as shipboard target acquisition
- Telephone switching systems

It is clear that as SOA expands, all of the services need to be placed on a standard taxonomy, and the services also need to be available in a catalog to make it easy to select the appropriate set.

Table 55.1 shows the scoring and the results of constructing an SOA application of 1000 function points. Because the SOA services are independently developed, there is no easy way of knowing what programming languages they were coded in, nor does it matter unless SOA maintenance is your responsibility. If you are using SOA services from third parties or your own company, you may not need to know

Table 55.1 Example of 1000 Function Point Service-Oriented Architecture Project with the Java Language

Project Size and Nature Scores		**Service Oriented**
1	Small projects <100 FP	10
2	Medium projects	10
3	New projects	10
4	IT projects	10
5	Web/cloud projects	10
6	Large projects >10,000 FP	8
7	Enhancement projects	2
8	Maintenance projects	2
9	Systems/embedded projects	–1
10	High-reliability, high-security projects	–1
	Subtotal	60
Project Activity Scores		**Service Oriented**
1	Pattern matching	10
2	Reusable components	10
3	Requirements changes	10
4	Models	10

(Continued)

Table 55.1 (Continued) Example of 1000 Function Point Service-Oriented Architecture Project with the Java Language

5	Coding	10
6	Integration	10
7	Change control	9
8	Pre-test defect removal	9
9	Requirements	8
10	Testing defect removal	8
11	Defect prevention	7
12	Design	5
13	Planning, estimating, measurement	4
14	Architecture	4
15	Test case design/generation	4
	Subtotal	118
	Total	178
Regional Usage (Projects in 2016)		**Service Oriented**
1	North America	1500
2	Pacific	1200
3	Europe	1000
4	South America/Central America	700
5	Africa/Middle East	300
	Total	4700
Defect Potentials per Function Point (1000 function points; Java language)		**Service Oriented**
1	Requirements	0.10
2	Architecture	0.25
3	Design	0.50
4	Code	0.90

(Continued)

Table 55.1 (Continued) Example of 1000 Function Point Service-Oriented Architecture Project with the Java Language

5	Documents	0.40
6	Bad fixes	0.25
	Total	2.40
	Total defect potential	2400
Defect Removal Efficiency (DRE)		
1	Pre-test defect removal	70.00%
2	Testing defect removal	85.00%
	Total DRE	95.50%
	Delivered defects per function point	0.108
	Delivered defects	108
	High-severity defects	17
	Security defects	2
Productivity for 1000 Function Points		**Service Oriented**
	Work hours per function point 2016	7.81
	Function points per month 2016	16.90
	Schedule in calendar months	7.00

Data Source: Namcook analysis of 600 companies.

Service-Oriented Architecture Scoring
Best = +10
Worst = −10

the languages so long as the APIs and features are well defined. This example is using the SOA services for a new application; not building the SOA services in the first place.

SOA is a niche development methodology aimed primarily at cloud applications. However, that niche is rapidly expanding and may soon overtake other platforms for a variety of applications. This means that SOA should have increasing usage at perhaps 15% per year for at least the next 10 years.

Chapter 56

Specifications by Example Development

Executive summary: *The term "specifications by example" refers to a useful form of gathering requirements by using teams to create specific requirements of actual features. The requirements are then turned into "executable requirements" that feed into test cases and into the application itself. This is already a useful method for requirements collection. But it could evolve into something even better if two conditions are met: (1) start with pattern matching and examining the requirements from similar existing applications by means of a formal taxonomy; (2) use the formal taxonomy so that general requirements from the current application can be added to the growing library of reusable requirements for future applications. Most requirements are not truly unique but have occurred in many similar applications and will occur in many other applications in the future. The software industry needs to stop considering every application as novel and unique and recognize the essential similarities among applications in the same industries such as banking, insurance, energy, and many others.*

As we all know, software requirements have been troubling and difficult since the industry began. They are ambiguous, incomplete, have errors, and are often hard to understand. They are also difficult to migrate into design, code, and test cases. Further, requirements "creep" can run from less than 1% per calendar month to more than 5% per calendar month.

The following is a list of the most common problems for software requirements based on more than 26,000 projects and 600 companies:

Overview of Common Software Requirements Problems in Rank Order

1. Toxic requirements that should not be implemented
2. Poor client understanding of their own requirements
3. Disputes between stakeholders about specific requirements
4. Conflicting and mutually contradictory requirements
5. Failure to include security requirements
6. Failure to include quality requirements
7. Arbitrary schedule requirements by clients or executives
8. Requirements defects >1 per function point or 0.5 per page
9. Requirements gaps and omissions (goes up with size)
10. Requirements creep (>1% per calendar month)
11. Requirements volume (may exceed lifetime reading speeds)
12. Inadequate change control by stakeholders and developers
13. Inadequate requirements templates that omit key topics
14. Superfluous requirements not needed by users
15. Inadequate requirements inspection methods
16. Inadequate requirements test methods
17. Inadequate requirements change control methods
18. Inadequate requirements gathering methods
19. Inadequate requirements analysis methods
20. Lack of an effective taxonomy for specific software features
21. Lack of automatic function point generation from requirements
22. Lack of effective, certified reusable requirements
23. Inclusion of potentially harmful uncertified reusable requirements
24. Requirement to include commercial off-the-shelf (COTS) packages without due diligence
25. Requirement to include open-source packages without due diligence
26. Poor communication between stakeholders and team members
27. Poor requirements communication among team members
28. Deferred requirements omitted from planned release
29. Poor traceability of requirements to other materials
30. Static representations of dynamic phenomena
31. Poor readability of text portions of requirements
32. Failure to update requirements after initial release
33. Failure to remove canceled requirements
34. Obscure graphics and visual elements of requirements
35. Evolutionary prototypes may have quality and security flaws
36. Prototypes >10% of key features may be cumbersome

As can be seen, requirements problems are numerous and endemic. From studying thousands of projects in hundreds of companies, it is possible to show the kinds of software projects that have the most complete and least complete requirements:

Ranked List of Requirements Completeness by Origin

1. Applications invented by one person
2. Applications with certified reusable requirements
3. Applications <100 function points in size
4. Replacement of a legacy application with no new features
5. Applications using requirements modeling
6. Small projects with embedded users in team (Agile)
7. Projects with stable requirements (any size)
8. Applications with quality function deployment (QFD)
9. Applications requiring Food and Drug Administration (FDA), Federal Aviation Authority (FAA), or government certification
10. Replicating competitive features in commercial packages
11. Medium project with joint application design (JAD)
12. Large system with JAD
13. Applications requiring strict governance
14. Replacing a legacy application plus new features
15. Large commercial package with focus groups
16. Embedded application with mixed software/hardware requirements
17. Commercial applications with requirements by marketers
18. Applications associated with frequent hardware changes (defense)
19. Informal interviews with users
20. Applications with complex data (taxation, health insurance)
21. Projects with unstable requirements due to business changes
22. Applications with unpredictable government mandates
23. Applications with clients who disagree about features
24. Applications with >100,000 users
25. Massive applications >100,000 function points

Because requirements have been troublesome since the software industry began, quite a few methods have been developed to bring requirements under control. Among the first of these was JAD, developed by IBM in the 1970s and first used for financial and banking applications in Canada. JAD then spread worldwide based on its success.

Other methods were also developed including QFD from Japan, user stories, use cases, the Unified Modeling Language (UML), model-based requirements, and more recently *specification by example*.

Table 56.1 shows the overall comparative results for eight requirements methods for a typical application of 1000 function points in size. Note that all of the rows in Table 56.1 are sensitive to size and change significantly for larger or smaller applications. Table 56.1 is sorted on the column of "requirements completeness."

Table 56.1 Requirements Comparisons for a 1000 Function Point Software Application

Requirement Method	*Pages*	*Words*	*Diagrams*	*Bugs*	*Percentage Complete*	*Monthly Change (%)*
Pattern based	300.00	120,000	60.00	60	99.00	0.50
Specs by example	322.00	128,800	62.00	60	99.00	0.80
Requirement model	380.00	152,000	76.00	105	97.00	0.90
QFD	450.80	180,320	90.00	220	96.00	1.30
JAD	289.00	115,600	46.00	330	94.00	1.00
Use cases	338.00	135,200	74.00	360	93.20	1.70
Conventional text	275.00	110,000	55.00	400	91.40	2.20
User stories	234.00	93,600	46.00	300	89.00	4.50
Average	**323.60**	**129,440**	**63.63**	**234**	**94.70**	**1.61**

The specification by example approach was first described by Ward Cunningham (inventor of technical debt) in a paper in 1996. However, the name *specification by example* was first used by Martin Fowler in 2002.

The concept is allied with the Agile family of methods and extreme programming (XP) but that is not intrinsic; the same approach would work quite well with team software process (TSP), model-based development, DevOps, Microsoft, and most other development methodologies.

The specification by example approach seems to be successful based on anecdotes and subjective opinions. But, like the whole Agile family, the method lacks solid quantitative data based on 50 or more actual projects.

The specific techniques used by specification by example are highly collaborative. The requirements team includes one or more users, one or more developers, and one or more testers. Possibly other participants such as business analysts, architects, or quality assurance might also participate.

The idea is to create an *executable specification* that can feed directly into development and also into automated test case generation. Further, unlike conventional text and diagrammatic requirements, the executable specifications are intended as living documents that are constantly updated perhaps for more than 10 years as the software ages in use.

The full details of specification by example are somewhat complex so a Google search is recommended. There is also an extensive literature so acquiring one or more of the published books on the approach is easily done. As with many contemporary methods, the ads are impressive, but actual data that demonstrates value is sparse.

Figure 56.1 illustrates the technical flow of information when using specifications by example.

Despite the genuine merits of the specifications by example approach, it shares several weaknesses with the other Agile methods and indeed with software engineering in general:

1. The method makes the false assumption that every project is new and unique.
2. The process should start with placing the project on a formal taxonomy.
3. The taxonomy should be used to find similar requirements from similar projects.
4. User requirements and non-functional requirements are needed.
5. Requirements should be divided into truly new and generic from historical projects.
6. The method ignores the fact that certifiable reuse is the best overall methodology.
7. The method should place new requirements onto a formal taxonomy.
8. Many new requirements can become reusable requirements for future software.
9. Historical data should be collected for every project >100 function points in size.

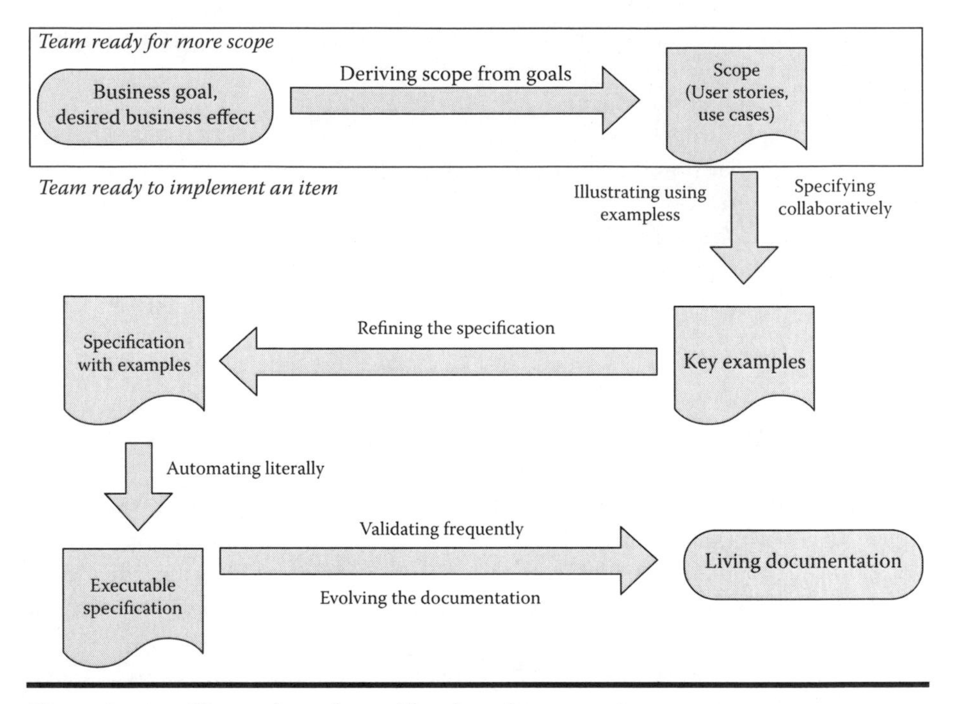

Figure 56.1 Illustration of specifications-by-example.

10. Quality should be measured using defect potentials and defect removal efficiency (DRE).
11. Planning and estimating should predict the whole project as well as the next sprint.

In general, specifications by example are a useful and fast-growing approach. The method will probably be used on more than 25,000 projects within a few years.

Specifications by example would also benefit from improved representation techniques. It is painfully obvious that text and black and white static diagrams are inadequate for software, which is the world's fastest-moving product when it is operating and software also changes fairly often over time as new features are added. Clearly animated, 3-D, full color holograms are the ultimate method for specifying software features.

Table 56.2 shows the results of using specifications by example on an application of 1000 function points and the Java programming language as predicted by Software Risk Master (SRM).

Specifications by example are not yet in the mainstream but have demonstrated success for a number of complex technical applications. The method is

Table 56.2 Example of 1000 Function Point Project Using Specifications by Example with the Java Language

Project Size and Nature Scores		**Specs by Example**
1	Small projects <100 FP	10
2	Medium projects	10
3	New projects	10
4	IT projects	10
5	High-reliability, high-security projects	10
6	Large projects >10,000 FP	9
7	Systems/embedded projects	9
8	Web/cloud projects	9
9	Enhancement projects	6
10	Maintenance projects	3
	Subtotal	86
Project Activity Scores		**Specs by Example**
1	Requirements	10
2	Requirements changes	10
3	Design	10
4	Coding	10
5	Integration	10
6	Defect prevention	10
7	Pre-test defect removal	10
8	Test case design/generation	10
9	Testing defect removal	10
10	Pattern matching	8
11	Reusable components	8
12	Planning, estimating, measurement	2
13	Architecture	8

(*Continued*)

Table 56.2 (Continued) Example of 1000 Function Point Project Using Specifications by Example with the Java Language

14	Models	8
15	Change control	8
	Subtotal	132
	Total	218
Regional Usage (Projects in 2016)		**Specs by Example**
1	North America	3,000
2	Pacific	2,500
3	Europe	2,500
4	South America/Central America	1,500
5	Africa/Middle East	500
	Total	10,000
Defect Potentials per Function Point (1000 function points; Java language)		**Specs by Example**
1	Requirements	0.10
2	Architecture	0.25
3	Design	0.35
4	Code	1.15
5	Documents	0.30
6	Bad fixes	0.20
	Total	2.35
	Total defect potential	2350
Defect Removal Efficiency (DRE)		
1	Pre-test defect removal	84.00%
2	Testing defect removal	90.00%
	Total DRE	98.38%
	Delivered defects per function point	0.376

(Continued)

Table 56.2 (Continued) Example of 1000 Function Point Project Using Specifications by Example with the Java Language

	Delivered defects	38
	High-severity defects	6
	Security defects	1
Productivity for 1000 Function Points		**Specs by Example**
	Work hours per function point 2016	14.45
	Function points per month 2016	9.13
	Schedule in calendar months	14.80

Data source: Namcook analysis of 600 companies.

Specifications by Example Scoring
Best = +10
Worst = −10

a bit complicated to learn and use so it needs more training than some of the other methodologies. In the future, if software develops animated, 3-D, full color diagrams then specifications by example would be very powerful, and probably a precursor to full software reuse and later to robotic development using standard parts.

Chapter 57

Spiral Development

Executive summary: *The "Spiral" methodology is one of the classic software engineering methodologies. It was first published in 1986 by the famous software author Dr. Barry Boehm of the University of Southern California. Because of Dr. Boehm's work at the Defense Advanced Research Projects Agency (DARPA) and with the defense community, Spiral was used early on by many defense projects. Spiral includes early risk analysis, which is a valuable feature.*

The idea of Spiral development was among the first to move away from waterfall development and try a new direction. According to Dr. Boehm's recent publications, the Spiral model centers on risk analysis and is methodology neutral.

Each major feature is examined and then an appropriate method is selected based on size, complexity, and feature risk profiles. The selected methodology can be iterative, waterfall, Agile, team software process (TSP), or whatever seems best suited for the specific component.

Of course, this assumes that the development teams will know which method is best, which is unlikely. It is more common for teams to look around and merely select the most popular methodology of the day, which would be Agile in 2016. Methodology selection is a weak link in software engineering.

Although Spiral development is simple in concept, in real life it is fairly complicated due to the fact that development teams need to evaluate risks, select appropriate methodologies, and also satisfy stakeholder demands.

There are a number of books and journal articles on Spiral development, and those considering Spiral should do extensive literature analysis.

Because Dr. Boehm is the developer of the well-known Constructive Cost Model (COCOMO) software cost estimating tool, Spiral projects often use COCOMO for feature and system estimation. However, COCOMO is only 1 of 10 available parametric cost estimating tools.

The available parametric tools in alphabetical order include:

1. COCOMO III
2. CostXpert
3. ExcelerPlan
4. KnowledgePlan
5. R2Estimator
6. Reliability, Availability, Serviceability, Security (RASS) Estimate (United Kingdom)
7. SEER
8. Software Lifecycle Management (SLIM)
9. Software Risk Master (SRM) by the author of this book
10. True Price

Note that because COCOMO algorithms have been published and placed in the public domain, there are at least half a dozen COCOMO clones available mainly as open-source tools. Some of these include COSTAR, COSMOS, and CoolSoft, but there are many others.

Spiral development has many images on the web. In fact, the main image of Spiral development is arguably the most widely published methodology image in the history of software, as shown in Figure 57.1.

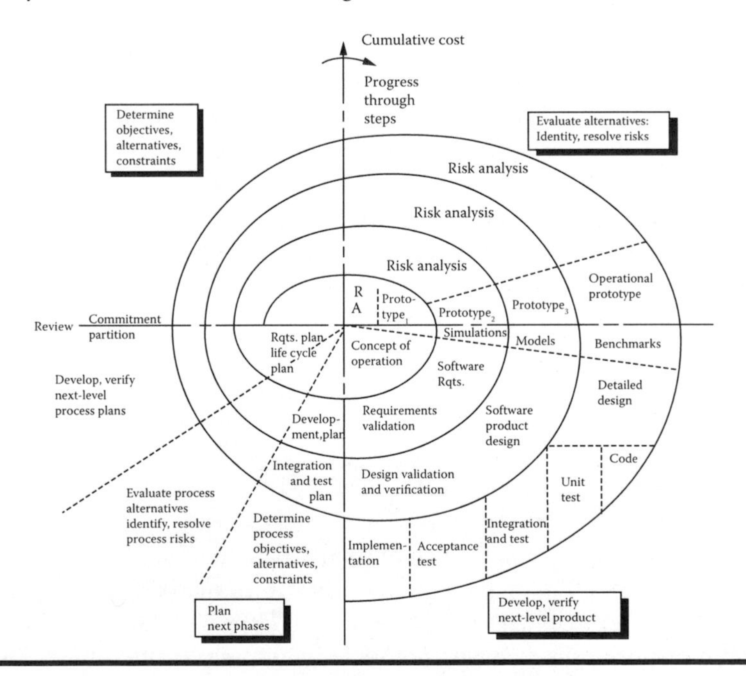

Figure 57.1 Illustration of Spiral development methodology.

The choice of what appears to be a classic Archimedes spiral gives Spiral development the appearance of being elegant and even beautiful. This is probably due more to the intrinsic appeal of the Spiral illustration than to the virtues of the actual methodology.

The Archimedes spiral is known to embody the *golden ratio*, which is widely used in both ancient and modern art.

Although Spiral has been in continuous use since the 1980s, it does not include formal measurement of quality and productivity unlike TSP, which does include both measures. As a result, there is a shortage of empirical Spiral results.

Table 57.1, shows the Namcook results for Spiral based on a project of 1000 function points and the Java programming language.

Table 57.1 Example of 1000 Function Point Spiral Project with the Java Language

Project Size and Nature Scores		Spiral
1	Enhancement projects	**10**
2	Systems/embedded projects	**10**
3	High-reliability, high-security projects	**10**
4	Small projects < 100 FP	9
5	Medium projects	9
6	Large projects > 10,000 FP	9
7	New projects	9
8	IT projects	9
9	Web/cloud projects	8
10	Maintenance projects	5
	Subtotal	88
Project Activity Strength Scores		**Spiral**
1	Defect prevention	10
2	Coding	9
3	Change control	9
4	Integration	9
5	Pattern matching	8
6	Planning, estimating, measurement	8

(*Continued*)

Table 57.1 (Continued) Example of 1000 Function Point Spiral Project with the Java Language

7	Pre-test defect removal	8
8	Testing defect removal	8
9	Requirements	7
10	Requirements changes	7
11	Design	7
12	Test case design/generation	7
13	Architecture	6
14	Models	6
15	Reusable components	5
	Subtotal	114
	Total	202
Regional Usage (Projects in 2016)		**Spiral**
1	North America	7,500
2	Pacific	5,000
3	Europe	5,000
4	South America/Central America	4,000
5	Africa/Middle East	1,000
	Total	22,500
Defect Potentials per Function Point (1000 function points; Java language)		**Spiral**
1	Requirements	0.30
2	Architecture	0.30
3	Design	0.35
4	Code	1.45
5	Documents	0.40
6	Bad fixes	0.40

(*Continued*)

Table 57.1 (Continued) Example of 1000 Function Point Spiral Project with the Java Language

	Total	3.20
	Total defect potential	3200
Defect Removal Efficiency (DRE)		
1	Pre-test defect removal	70.00%
2	Testing defect removal	85.00%
	Total DRE	95.50%
	Delivered defects per function point	0.144
	Delivered defects	144
	High-severity defects	22
	Security defects	2
Productivity for 1000 Function Points		**Spiral**
	Work hours per function point 2016	16.4
	Function points per month 2016	8.05
	Schedule in calendar months	14.6

Data source: Namcook analysis of 600 companies.

Spiral Scoring
Best = +10
Worst = –10

Spiral development is one of the older software methodologies but it is still in use, mainly for government and systems software. Spiral provided concepts that were later adopted by other methodologies including Agile.

Today, in 2016, spiral development is in competition with a number of newer methodologies including Agile, container development, DevOps, iterative development, model-based development, and pattern-based development.

Chapter 58

Structured Development (Design, Programming)

Executive summary: *Other than cowboy development, the "structured development" methodology is the oldest in this book. It originated around 1968 in a famous letter from Edsger Dijkstra to the "Communications of the ACM" on "Go to statements considered harmful." The first and oldest form of structured development was structured programming. Then, the concept expanded to structured analysis and structured design and eventually to the blanket term "structured development." The method was adopted by large corporations such as IBM in the 1970s when they were struggling to control the costs and quality of applications that were growing to 10,000 function points or 1,000,000 logical code statements. Structured development replaced and was better than "cowboy programming" and somewhat better than the later waterfall method. However, in time, other methodologies such as iterative development, object-oriented development, and spiral development would begin to displace structured development. Today, many newer methodologies are in competition: Agile, container development, continuous development, DevOps, model-driven development, and many more. However, the fundamental principles of structured development are still valid in 2016 and implicitly assumed by other methodologies, even if not acknowledged.*

In the 1960s, software projects began to increase rapidly in number, size, and complexity. Development teams necessarily increased with application size. In the 1950s, a majority of software applications had only one person, a programmer. By the end of the 1960s, team sizes had grown to about 8 on average, and large systems sometimes had teams greater than 50.

Further, occupation groups also expanded from basic programming to include dozens of specialties such as architects, business analysts, configuration control, software quality assurance, software technical writers, and many more. (Today, in 2016, there are a total of 126 software occupation groups.)

Traditional cowboy programming by one person rapidly proved to be inadequate for large systems, which routinely ran late, exceeded their budgets, and were filled with bugs when, and if, they were released.

Software researchers were dissatisfied with the way software was being built and tried to find better alternatives. In 1968, Edsger Dijkstra wrote a famous open letter about the harm caused by "go to statements." His idea was that branches or go to statements raised complexity to such high levels that it was difficult to debug and test programs. His concepts proved successful and led to a variety of structured programming languages.

However, not everyone agreed that branches were harmful, and other researchers such as Donald Knuth opposed structured programming. Even today, there are some debates about the pros and cons of go to statements. The current prevailing opinion is that judicious use of go to statements can simplify overall logic and are not necessarily harmful. However, everybody agrees that too many go to statements or using them carelessly raises cyclomatic complexity to hazardous levels well above the safe range of about 10.

(Cyclomatic complexity was invented by Tom McCabe and is basically a quantification of branches. Code with no branches at all has a cyclomatic complexity of 1. As cyclomatic complexity goes up, branches and paths increase, which means that test cases needed for 100% coverage go up geometrically. In real life, software test coverage declines as cyclomatic complexity goes up because it is not possible to create enough test cases for high coverage with cyclomatic complexity above about 20. Structured programming has the effect of reducing cyclomatic complexity.)

Within another 10 years, the structure concept had shifted from code to include structured analysis and design. Eventually, several formal methods were described with structured analysis and design (SADT) being one of the most popular.

The structured methods were soon supported by a variety of graphical design methods such as structure charts, Warnier/Orr diagrams, and Hierarchical plus Input, Process, Output (HIPO) diagrams from IBM, and others. James Martin wrote an interesting book that cited 40 different kinds of design languages, many of which are still in use.

Eventually, the concepts of structured development found their way into a growing family of software development methods including but not limited to the Jackson method, the Yourdon method, Information Engineering, Structured Systems Analysis and Design Method (SSADM) created by the U.S. Department of Commerce, and a number of others.

In general, from observations of about 500 unstructured and 500 structured applications when at IBM in the 1970s, the author did find better quality and productivity as a result of the family of structured methodologies. These results were published in several books by the author including *Applied Software Measurement*, *Estimating*

Software Costs, and *Assessment and Control of Software Risks* (which coincidentally is patterned after a medical text called *Control of Communicable Diseases in Man*.)

The results of using structured development (SADT) for an application of 1000 function points and the Java language is shown in Table 58.1.

Although structured development originated in the 1960s, the essential principles are still valid in 2016. Most software development methodologies today include structured development concepts, whether they acknowledge this or not.

Table 58.1 Example of 1000 Function Point Structured Development Project with the Java Language

Project Size and Nature Scores		Structured Development
1	Small projects < 100 FP	10
2	High-reliability, high-security projects	10
3	Medium projects	9
4	New projects	9
5	IT projects	9
6	Systems/embedded projects	9
7	Large projects > 10,000 FP	8
8	Web/cloud projects	8
9	Enhancement projects	5
10	Maintenance projects	5
	Subtotal	82
Project Activity Scores		**Structured Development**
1	Design	10
2	Coding	9
3	Integration	9
4	Defect prevention	9
5	Pre-test defect removal	8
6	Testing defect removal	8
7	Requirements	7

(*Continued*)

Table 58.1 (Continued) Example of 1000 Function Point Structured Development Project with the Java Language

8	Requirements changes	7
9	Test case design/generation	7
10	Architecture	6
11	Models	6
12	Pattern matching	5
13	Reusable components	5
14	Planning, estimating, measurement	5
15	Change control	5
	Subtotal	106
	Total	188
Regional Usage (Projects in 2016)		**Structured Development**
1	North America	10,000
2	Pacific	7,500
3	Europe	9,000
4	South America/Central America	6,500
5	Africa/Middle East	3,000
	Total	36,000
Defect Potentials per Function Point (1000 function points; Java language)		**Structured Development**
1	Requirements	0.40
2	Architecture	0.30
3	Design	0.50
4	Code	1.30
5	Documents	0.40
6	Bad fixes	0.30
	Total	3.20
	Total defect potential	3200

(Continued)

Table 58.1 (Continued) Example of 1000 Function Point Structured Development Project with the Java Language

Defect Removal Efficiency (DRE)		
1	Pre-test defect removal	65.00%
2	Testing defect removal	85.00%
	Total DRE	94.75%
	Delivered defects per function point	0.168
	Delivered defects	168
	High-severity defects	26
	Security defects	2
Productivity for 1000 Function Points		
	Work hours per function point 2016	15.00
	Function points per month 2016	8.80
	Schedule in calendar months	16.30

Data source: Namcook analysis of 600 companies.

Structured Development Scoring
Best = +10
Worst = –10

Structured development, in 2016, is more or less like aspirin: a common medicine that reduces the pain of software development. Also like aspirin, which reduces the odds of a heart attack, structured development can reduce the odds of poor quality and having latent security flaws. Structured development is not a panacea, but it is an effective palliative for a number of software development problems.

Chapter 59

Test-Driven Development (TDD)

Executive summary: *Test-driven development (TDD) is a methodology "in the small"—it deals with the work of programmers but does not address a full development life cycle, which includes architecture, design, requirements, and other non-code topics. TDD is typically used as an element of Agile methods, but may in principle be used in conjunction with any method wherein code is developed incrementally. The central concept of TDD is that a test is written before related functional code is developed. Both tests and code are developed in very small increments. Some advocates of this approach view it as, in essence, a specification technique that simply has the side effect of ensuring thorough testing at the function and unit levels. As each test is developed, it is added to the set of all prior tests and the total set is executed against each new increment of functional code. This approach tends to ensure that a new test or a new element of functionality is not only correct in itself but also does not "break" a previous element. TDD requires the availability of suitable automated test tools and assumes each developer writes their own tests. TDD is often combined with a related approach known as Acceptance TDD (ATDD). See also the chapter on extreme programming (XP), which also uses TDD.*

Despite TDD being a nominal quality approach, other than this book there is almost no quantified data published on defect removal efficiency (DRE). As readers will see at the end of the chapter, DRE for TDD is about 95.00%, which is better than waterfall (87.00%) and Agile (92.50%) but below team software process (TSP) (98.50%), which also uses inspections.

> *The lack of TDD quantitative data is typical of the way the software industry works. The medical concept of validation and checking for side effects prior to the release of a therapy is totally absent from software where vast claims are made with zero supporting data.*

TDD Process

The process consists of five steps:

1. Write a test to check the outcome of a small unit of functionality. Developing the test forces the developer to fully understand the requirements of the functional element to be developed—necessarily, if we don't know how to test it, we don't really understand what is expected. Once developed, the new test is added to the test suite using a tool appropriate to the technology being used.
2. Run *all* tests to determine if the new one, as expected, fails. As the functional code associated with the new test has not been written, it will always be expected to fail. This step ensures the following:
   a. The new test does in fact fail as expected (if it passes, something is clearly wrong)
   b. It demonstrates that the test harness is functioning as expected
   c. It confirms that the new test fails for the reason expected
3. Write some code to implement the intended functionality in a way that is expected to pass the previously written test.
4. Run the full set of tests developed to date. If all test cases pass, the developer gains confidence that the new code meets requirements and does not break earlier code. If the full suite of tests does not pass, the developer makes any further necessary refinements to the functional code. Note that due to the small increments involved in each iteration of this process, the debugging process is much more simplified as the "search space" is necessarily small.
5. Clean up (*refactor*) code—for example, object, class, module, variable, and method names—may need to change to more accurately reflect the current level of functionality to facilitate readability and maintainability. Inheritance hierarchies may need to be revised to enhance modularity. Whenever refactoring is done, the test suite is re-executed to ensure new defects have not been introduced.

This five-step process is iterated until all desired functionality has been built and tested.

Critical Success Factors

- Developers must understand testing techniques: Using this approach relies on generalists rather than specialists.

- Automated test tools are essential: Hence, developers must also have sufficient understanding of the tools to be used.
- Tests must be fast: Slow execution of test suites will lead to reduced frequency of use. This means that the test will generally not access external networks or databases and will rely on mock responses from external interfaces. This means that additional layers of integration and system-level testing may be required to supplement function and unit-level tests built by developers.
- The test suite requires a level of attention and care equal to that devoted to the functional code. It must be maintainable and extensible, and may itself require periodic refactoring.

There are several flavors of TDD, and the chapter on XP is another. Table 59.1 shows the results of a TDD project of 1000 function points in the Java language.

Table 59.1 Example of 1000 Function Point Test-Driven Development Project with the Java Language

Project Size and Nature Scores		**Test Driven**
1	Small projects < 100 FP	10
2	New projects	10
3	Web/cloud projects	9
4	IT projects	7
5	Medium projects	5
6	Systems/embedded projects	5
7	High-reliability, high-security projects	5
8	Enhancement projects	4
9	Maintenance projects	1
10	Large projects > 10,000 FP	**–1**
	Subtotal	55
Project Activity Scores		**Test Driven**
1	Coding	9
2	Defect prevention	9
3	Testing defect removal	8

(Continued)

Table 59.1 (Continued) Example of 1000 Function Point Test-Driven Development Project with the Java Language

4	Change control	7
5	Test case design/generation	7
6	Integration	6
7	Pre-test defect removal	6
8	Requirements	5
9	Requirements changes	5
10	Design	5
11	Models	4
12	Pattern matching	3
13	Reusable components	3
14	Architecture	3
15	Planning, estimating, measurement	**−1**
	Subtotal	79
	Total	134
Regional Usage (Projects in 2016)		**Test Driven**
1	North America	8,000
2	Pacific	6,000
3	Europe	8,000
4	South America/Central America	5,000
5	Africa/Middle East	3,000
	Total	30,000
Defect Potentials per Function Point (1000 function points; Java language)		**Test Driven**
1	Requirements	0.30
2	Architecture	0.40
3	Design	0.50
4	Code	1.25

(Continued)

Table 59.1 (Continued) Example of 1000 Function Point Test-Driven Development Project with the Java Language

5	Documents	0.40
6	Bad fixes	0.30
	Total	3.15
	Total defect potential	3150
Defect Removal Efficiency (DRE)		
1	Pre-test defect removal	50.00%
2	Testing defect removal	90.00%
	Total DRE	95.00%
	Delivered defects per function point	0.158
	Delivered defects per function point	158
	High-severity defects	24
	Security defects	2
Productivity for 1000 Function Points		**Test Driven**
	Work hours per function point 2016	13.60
	Function points per month 2016	9.70
	Schedule in calendar months	14.30

Data source: Namcook analysis of 600 companies.

Test-Driven Development Scoring
Best = +10
Worst = −10

The basic concept of TDD is a good one. Writing test cases prior to writing code makes sense and is relatively effective. However, TDD has a very narrow focus and completely ignores defects that originate in requirements and design. TDD also ignores static analysis, formal inspections, requirements models, and other kinds of pre-test defect removal activities.

Worse, TDD has not yet measured DRE although it is shown to be about 95% in this book based on client results.

This failure to provide quantitative data prior to releasing a methodology would never happen in medical practice, but it is the normal way of releasing software development methodologies, almost none of which were validated prior to release, except Watts Humphrey's TSP and personal software process (PSP).

Chapter 60

Team Software Process (TSP) and Personal Software Process (PSP)

Executive summary: *The team software process (TSP) and personal software process (PSP) methodologies are among the most effective for large software projects above 10,000 function points in size. TSP/PSP methods are also effective for systems software, embedded software, and for software that needs certification by government groups such as the Food and Drug Administration (FDA) and the Federal Aviation Authority (FAA). TSP/PSP are also effective for applications covered by Sarbanes–Oxley requirements. Both TSP/PSP are endorsed by the Software Engineering Institute (SEI) and are therefore used on many defense and military software projects.*

The TSP and PSP methodologies were developed by the late Watts Humphrey. The two methodologies are designed to be used together, with students learning PSP first before learning TSP. These two methods are based in part on IBM's internal development methods for building systems software such as operating systems.

Watts Humphrey was IBM's corporate director of programming technology before he left to join SEI. At SEI, Watts developed the essential concepts for the famous Capability Maturity Model® or CMM®.

TSP and PSP are supported by rigorous training. PSP and TSP users of the methodologies are expected to complete formal courses prior to utilization. TSP and PSP are also supported by a growing library of books including several excellent books authored by Watts Humphrey himself. The first publication on TSP was in 2000 when a monograph by Watts Humphrey was published by the U.S. Department of Defense. A Google or Amazon search on "TSP" will find dozens of books and articles.

The web has an image collection of interesting PSP and TSP diagrams, including the one shown in Figure 60.1.

Due to Watts' heading up the SEI appraisal program and writing several books on TSP and PSP while employed at SEI, these methods are strongly endorsed and supported by the SEI. Among Namcook clients, the majority of TSP users are in organizations that are ranked at capability maturity model integrated (CMMI) Levels 3 or 5. In other words, TSP is strongly associated with fairly sophisticated companies. This may be one of the reasons why TSP quality is among the best of any software methodology.

Due to the background of these methods for large and complex systems software such as operating systems, TSP and PSP are among the best for systems software, embedded software, telecommunications software, medical devices, and defense software for weapons systems and other technical software projects.

TSP and PSP are "quality-strong" development methods that include rigorous inspections prior to testing and, more recently, static analysis prior to testing. TSP and PSP are also very strong on quality and productivity measurements, and keep much more data than many methodologies. In particular, TSP and PSP keep much more data than Agile, which is very sparse in keeping either quality or productivity data.

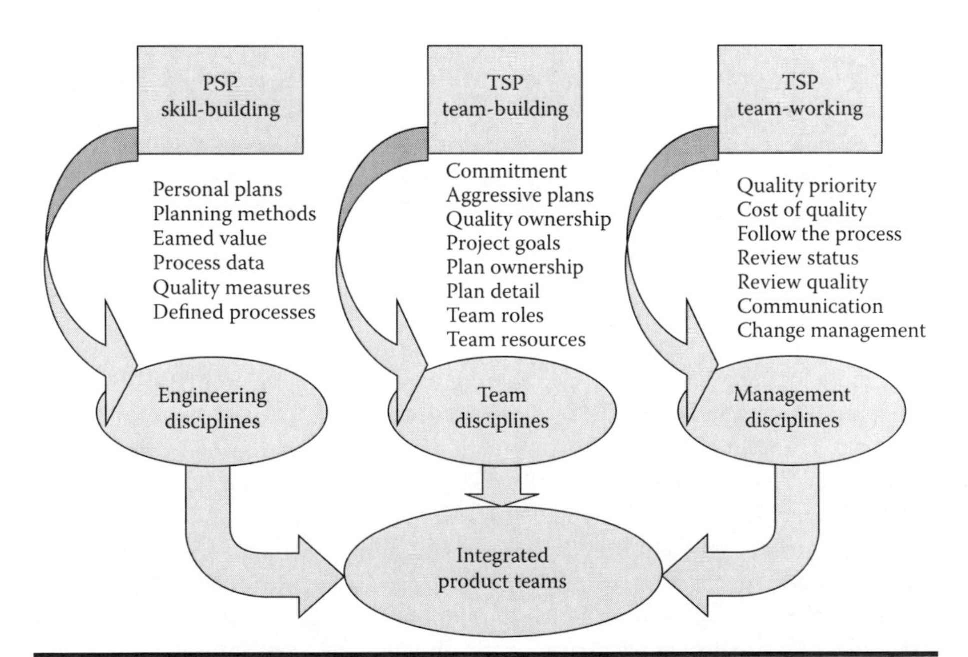

Figure 60.1 Illustration of the synergy between the TSP and PSP methodologies.

Measures of TSP and PSP quality show an interesting and unusual trend validated by Watts Humphrey personally. TSP and PSP projects show almost steady high-quality results even for large projects that approach 100,000 function points. Most software methodologies have elevated defect potentials and reduced defect removal efficiency (DRE) for large systems, but TSP seems to have overcome that endemic problem.

Because of the careful records kept by TSP projects, they seem to be somewhat "heavy" for small projects of only a few hundred function points in size. TSP and PSP will work, of course, but the overhead is quite a bit more than Agile or other lighter methods.

TSP and PSP show their strengths for larger and more complex software projects >5000 function points. They are also a good choice for projects that need FDA or FAA certification, or Sarbanes–Oxley governance.

Because TSP and PSP originated inside IBM before function point metrics were the major software metric, some of the measures were originally based on lines of code (LOC). Function points are a much better choice than LOC for both quality and productivity measurements with TSP in 2016.

TSP and PSP were selected by the government of Mexico for internal software projects, and the method is also used elsewhere in Central and South America. Because of the strong endorsement of the SEI, TSP and PSP are also used on many military and defense software projects.

TSP and PSP have annual conferences with hundreds of attendees. The method is not as popular as Agile, but then embedded and systems software projects are not as numerous as web and information technology (IT) projects.

Table 60.1 shows the Namcook meta-model and quantitative results for TSP in 2016. The quantitative results assume 1000 function points and the Java programming languages.

The combination of TSP with PSP comprises one of the most "quality strong" of all available software methodologies. However, this combination is also one of the "heavy" software methodologies and is best for large and critical applications in the 10,000 function point size range where quality, reliability, and security are key goals. TSP/PSP might be too heavy for small applications of about 100 function points, although PSP itself is useful for small applications. Overall, TSP/PSP has proved useful for many large systems and the results have better than average quality and reliability.

TSP has been endorsed by SEI. This is not surprising since Watts Humphrey was head of the SEI assessment program. However, the endorsement was based on technical merits.

Table 60.1 Example of 1000 Function Point TSP/PSP Project with the Java Language

Project Size and Class Scores		**TSP/PSP**
1	Large projects > 10,000 FP	10
2	New projects	10
3	Systems/embedded projects	10
4	High-reliability, high-security projects	10
5	Medium projects	8
6	IT projects	8
7	Web/cloud projects	8
8	Small projects < 100 FP	7
9	Enhancement projects	7
10	Maintenance projects	6
	Subtotal	84
Project Activity Scores		**TSP/PSP**
1	Coding	10
2	Pre-test defect removal	10
3	Planning and estimating	9
4	Change control	9
5	Defect prevention	9
6	Requirements changes	8
7	Architecture	8
8	Integration	8
9	Testing defect removal	8
10	Requirements	7
11	Models	7
12	Design	7
13	Test case design/generation	7
14	Pattern matching	6

(Continued)

Table 60.1 (Continued) Example of 1000 Function Point TSP/PSP Project with the Java Language

15	Reusable components	6
	Subtotal	119
	Total	203
TSP Usage in 2016		**TSP/PSP**
1	North America	7,500
2	South and Central America	3,500
3	Pacific	2,500
4	Europe	1,500
5	Africa/Middle East	500
	Total	15,000
Defect Potentials per Function Point (1000 function points; Java language)		**TSP/PSP**
1	Requirements	0.20
2	Architecture	0.10
3	Design	0.30
4	Code	1.15
5	Documents	0.40
6	Bad fixes	0.20
	Total	2.35
	Total defect potential	2350
Defect Removal Efficiency (DRE)		**TSP/PSP**
1	Pre-test defect removal	85.00%
2	Testing defect removal	90.00%
	Total DRE	98.50%
	Delivered defects per function point	0.035
	Delivered defects	35

(*Continued*)

Table 60.1 (Continued) Example of 1000 Function Point TSP/PSP Project with the Java Language

	High-severity defects	5
	Security defects	0
Productivity for 1000 Function Points		**TSP/PSP**
	(Assumes Java programming language)	
	Work hours per function point 2016	11.34
	Function points per month 2016	11.65
	Schedule in calendar months	14.29

Data source: Namcook analysis of 600 companies.

Team Software Process (TSP) Scoring
Best = +10
Worst = −10

Chapter 61

T-VEC Development

Executive summary: *T-VEC is a U.S. company that applies meta-models to requirements, design, and verification of software. The T-VEC tool suite is often used in avionics and defense applications, including the Mars Rover. T-VEC is not a full development methodology, but rather a model-driven set of tools to handle requirements, architecture, and design. T-VEC tools also support test case generation from models. Actual code development is not part of the T-VEC suite. However, T-VEC requirements can be input into any of several application generators. See also the chapters on model-driven development and IntegraNova.*

Model-based development is often used for formal requirements and design that feed into some form of application generator. The T-VEC concept is to ensure very high quality and reliability of front-end requirements, architecture, and design. T-VEC can also help in test planning and test case design. However, T-VEC is not a full development methodology but only a front-end methodology. Actual code development is not part of the current T-VEC tool suite.

The idea of software development from abstract models is a fairly old concept. Model-driven development and computer-aided software engineering (CASE) had a burst of popularity in the 1980s, almost like Agile is having today. Some of the companies involved include Bachman Systems, Higher-Order Software, Cadre Technologies, and Logic Works. More recently, companies such as T-VEC and IntegraNova in Spain are bringing out operational application generators based on modeling.

T-VEC has been used on a fairly wide variety of real-time, embedded, and defense applications and apparently these uses have been perceived as valuable by T-VEC clients.

In model-based development, requirements changes are handled by going back to the model and regenerating the application: actual manual code changes are discouraged. This is because manual code changes deviate from the underlying model and quickly lead to errors that are hard to find and hard to correct.

Once the code is generated and goes into production, it will fairly soon need to be updated both to repair bugs and to add various enhancements or modifications based on external changes to business situations. If you go back and recreate the model then everything stays synchronized. But if you change the code itself, then the code and the models are no longer in synch and all future changes will have to be done to the code with probably incomplete documentation.

One of the gaps in the model-driven literature is that of actual productivity and quality results. This was a shortcoming of CASE in the 1980s and is a chronic shortcoming of Agile today, in 2016.

T-VEC does not measure productivity but its measures of quality and defect removal efficiency (DRE) are fairly good and indicate high DRE levels.

Sample results for a T-VEC model-driven development for an application of 1000 function points and Java code are shown in Table 61.1.

Table 61.1 Example of 1000 Function Point T-VEC Project with the Java Language

Project Size and Nature Scores		**T-VEC**
1	New projects	10
2	Systems/embedded projects	10
3	High-reliability, high-security projects	10
4	Small projects < 100 FP	9
5	Medium projects	9
6	Large projects > 10,000 FP	9
7	IT projects	8
8	Web/cloud projects	8
9	Enhancement projects	2
10	Maintenance projects	2
	Subtotal	77
Project Activity Scores		**T-VEC**
1	Pattern matching	10
2	Reusable components	10

(Continued)

Table 61.1 (Continued) Example of 1000 Function Point T-VEC Project with the Java Language

3	Requirements changes	10
4	Design	10
5	Defect prevention	10
6	Pre-test defect removal	10
7	Test case design/generation	10
8	Testing defect removal	10
9	Architecture	9
10	Models	9
11	Requirements	8
12	Coding	8
13	Change control	7
14	Integration	7
15	Planning, estimating, measurement	6
	Subtotal	134
	Total	211
Regional Usage (Projects in 2016)		**T-VEC**
1	North America	600
2	Pacific	300
3	Europe	200
4	South America/Central America	100
5	Africa/Middle East	50
	Total	1250
Defect Potentials per Function Point (1000 function points; Java language)		**T-VEC**
1	Requirements	0.15
2	Architecture	0.15
3	Design	0.30

(Continued)

Table 61.1 (Continued) Example of 1000 Function Point T-VEC Project with the Java Language

4	Code	0.40
5	Documents	0.40
6	Bad fixes	0.10
	Total	1.50
	Total defect potential	1500
Defect Removal Efficiency (DRE)		
1	Pre-test defect removal	85.00%
2	Testing defect removal	90.00%
	Total DRE	98.50%
	Delivered defects per function point	0.023
	Delivered defects	23
	High-severity defects	3
	Security defects	0
Productivity for 1000 Function Points		**T-VEC**
	Work hours per function point 2016	13.20
	Function points per month 2016	10.00
	Schedule in calendar months	15.00

Data source: Namcook analysis of 600 companies.

T-VEC Scoring
Best = +10
Worst = −10

The T-VEC approach is not a full methodology but rather a front-end that offers sophisticated requirement and design tools and concepts. These are then fed into an application generator or are developed using traditional programming languages.

Overall, T-VEC has been shown to benefit quality, although T-VEC the company does not publish quantitative data. Because T-VEC is mainly used for complex systems and defense applications, schedules and productivity are not usually issues.

Chapter 62

Virtual Reality Global Development

Executive summary: *About 30% of large software applications in 2016 are built in more than one country. Popular outsource destinations with good software personnel and lower costs than mainland Europe and the United States include India, Ireland, Malaysia, Peru, and the Philippines. Long distances over many time zones make physical travel expensive. Conference calls or video conferences are inconvenient due to spanning many time zones. It is technically possible to build a virtual reality development environment that can improve coordination for global software development.*

Software in 2016 is a global business. Many large software applications are fully or partially outsourced from high-cost countries such as the United States and Germany to lower-cost countries such as India or Peru. About 30% of large software applications in 2016 are built in more than one country.

Long distances spanning many time zones make physical travel expensive and time consuming. Coordination via conference calls, webinars, and video conferences are cheaper substitutes, but not particularly convenient. Further, they are usually narrow in focus and related only to one or a few topics and have no continuing value after the immediate conference is finished.

It would be technically feasible to construct a virtual reality software environment that would make coordination and cooperation across long distances easier than it is in 2016. If done well, the virtual reality software development environment might also benefit domestic software if more than one city is involved. It would also be useful for large defense software projects, some of which may include over a dozen sub-contractors.

The software industry has built very sophisticated design tools and development environments for other industries such as aircraft and automotive construction, architectural construction, ship construction, medical device construction, and a number of others. But software itself lags in sophisticated software development and project management tools and especially so for large distributed projects that include multiple development teams located in different cities or different countries.

Prototype Virtual Development Methodology

The author has seen a prototype of a working tool similar to the one discussed here so the technologies are available in 2016, even if they are not yet commercially available. The prototype was being developed by a large company that sold both hardware products and software applications globally. It was an impressive tool that had features far beyond the current crop of software project management tools. The prototype supported a rich set of project management functions.

The purpose of the prototype was to provide up-to-date information on key technology topics to the corporation's chairman, the board of directors, the president, C-level executives, and selected customer executives. The company was a multinational high-tech corporation and wanted to impress clients with its overall technical excellence. The prototype certainly benefited the corporation since few, if any, other companies even know how many software and hardware projects and personnel are on board or how much software they own and are developing.

The prototype could be used with any or all software development methodologies but was most beneficial for quality-strong methodologies such as the team software process (TSP). Agile projects seldom collect enough data to be useful for benchmarks or future estimates.

The first screen of the prototype tool displayed a rotating map of the world with all of the company's development laboratories highlighted as glowing colored dots by the names of each city that had a development laboratory (26 cities were shown in 10 countries on the prototype).

The image looked somewhat like the one shown in Figure 62.1 from the Shutterstock image library, only this image does not highlight software development laboratories.

If you zoomed in on the United States in the prototype, the image would show all of the U.S. software and hardware development laboratories, more or less like the image in Figure 62.2 from Shutterstock.

This image shows network connections but if you imagine that each dot represented a software development laboratory you can see what the prototype looked like in operation.

A sub-screen showed total global hardware and software personnel, total volumes of deployed software, total world software budgets, and total numbers of new applications under development in all locations combined.

Figure 62.1 Starting point of a virtual software project map.

Figure 62.2 Illustration of a national software project map.

(The deployed software data showed internal projects, commercial off-the-shelf [COTS] packages, and open-source packages. The corporate portfolio was about 7,500,000 function points with roughly 50% being COTS packages. The corporate software staff comprised about 10,000 employees and roughly 3500 contractors and outsource personnel.)

If you clicked on a city with one of the company's development laboratories, a new screen opened up that showed all of the current software packages owned

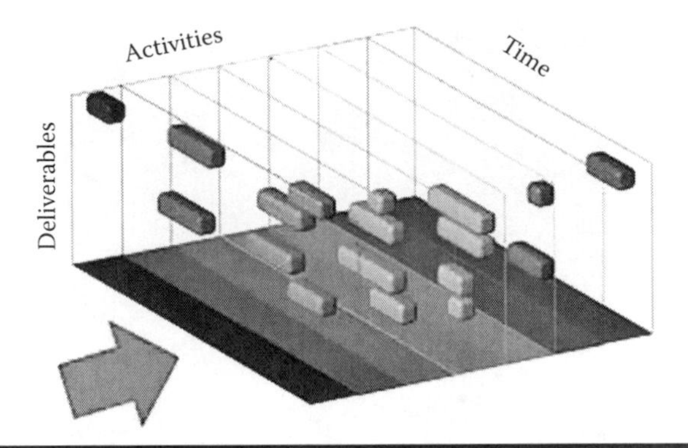

Figure 62.3 Illustration of 3-D projects time line.

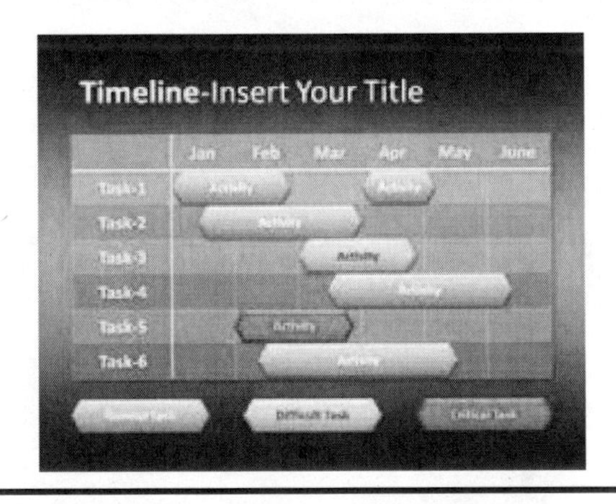

Figure 62.4 Illustration of a timeline or Gantt chart.

by the specific location, and all of the new software applications that were under development. A related sub-screen showed total software staffing and current software budgets for the location. (Hardware products were also shown, but this book will discuss only the software application side and not the hardware side.) Software application development progress for the location was shown for all projects using a chart somewhat like the chart in Figure 62.3.

Clicking on one of the current software projects in development brought up a new screen that showed key data about each specific software project including start date, projected delivery date, size of the application in function points and lines of code (LOC), current budget for the project, funds expended to date, cost to complete, time to complete, and numbers of software personnel assigned to the project and current status; that is, in design, coded, being tested, and so on. (Progress was shown via a moving Gantt chart that highlighted progress to date and work remaining, which somewhat resembled the sample in Figure 62.4.)

A sub-screen showed software occupation groups working on the project, such as business analysts, software engineers, test personnel, quality assurance, integration, and so on. A chronological progress sub-screen showed any changes in dates such as schedule slips. A quality sub-screen showed bugs found to date and projected bugs remaining.

The prototype was a good starting point for a sophisticated software management planning and control tool, but it could be expanded to also support all development work and additional quality control methods.

Building a Virtual Reality Methodology Package

The starting point for a virtual reality software development and planning tool would be to license a virtual reality engine from one of the computer game companies. That would be used to build a dynamic moving software development support environment.

The virtual reality engine would be used to build an artificial software development laboratory. One of the screens would be what appears to be a circular conference room with a series of large high-definition wall panels running around the room. Each of the panels would display current data, and users of the virtual reality tool could move from screen to screen to see important information shown in real time. Since we are dealing with virtual reality and not a real room, the room itself could circle to the screens of current interest. The total number of screens would probably be about 30 to show both project management and technical work.

Using Google Glasses or one of the virtual reality headsets would probably be better than a regular computer screen. The virtual screens would look like they were about 4 feet by 6 feet in size, and some screens would include animated, full color, 3-D images.

The virtual software development environment would show sets of screens devoted to project management, quality control, development, and post-release maintenance and enhancement. This is why at least 30 primary screens and many sub-screens will be part of the virtual development environment.

The virtual reality screens would show the following kinds of information:

Software Project Management Screens

Benchmark Screen

The starting point for every software project would be a benchmark screen that showed numbers of completed projects similar to the project under development. The historical benchmark data could come under one or more of the software benchmark organizations such as the International Software Benchmarking Standards Group (ISBSG), Davids' Consulting, Namcook Analytics, Q/P Management Group, and so on.

Each completed project benchmark example would show start and end dates, size in function points and LOC, costs of the projects, and quality data using defect potentials and defect removal efficiency (DRE).

The benchmark screen would also identify special topics such as litigation filed against one or more of the precursor projects for breach of contract, patent violations, or any other serious issue.

(An example of the litigation data would be lawsuits between state governments and software vendors for projects like child-support software applications or motor-vehicle support applications, both of which have had frequent litigation. All 50 states have both kinds of software and many have been late, over budget, and had litigation between states and their software vendors.)

A major gap with the current generation of software project management tools is that none of them includes historical benchmark data, although they all should do this. You can hardly be successful in controlling a new software project if you don't have any knowledge of the results of similar or identical projects already completed.

Software project management tools should include historical benchmarks. Further, every project using a project management tool should have its own data added to the benchmark collection so that the software industry has a steadily growing suite of accurate benchmarks instead of a spotty collection.

To do this means that all software project management tools should support function point metrics. (Some 35 out of 37 benchmark organizations only support function point metrics.) All forms of software should be benchmarked: web applications, embedded software, medical devices, defense software, big data applications, and even computer games.

Corporate Portfolio Screen

Most companies have no idea how much software they own, how many software personnel are employed, or other quantitative facts. An effective corporate software management tool would provide data such as that shown in Table 62.1 for an actual Fortune 500 company studied by the author.

Very few companies and almost no government organizations have this kind of data available as of 2016. But every large company should have this kind of software demographic data available since C-level executives should know these basic facts.

Competitive Analysis Screen

For commercial software that is sold or leased, one of the screens would show the major competitive products that are similar to the one being developed. For building

Table 62.1 Data for an Actual Fortune 500 Company

Corporate Software Development				
Software development laboratories	26			
Software development countries	10			
Overall Portfolio Size				
Initial internalapplications in January	4,310			
Average language level	3.25			
Average size in FP	1,100			
Average size in software non-functional assessment process (SNAP) points	189			
Average size in LOC	108,308			
Initial size of portfolio in FP	4,015,000			
Average size in SNAP points	648,423			
Initial size of portfolio in LOC	395,323,077			
Latent defects in portfolio	28,105			
Latent security flaws in portfolio	1,827			
Outsourced Activities				
Development (%)	10.00			
Enhancement (%)	10.00			
Maintenance (%)	75.00			
Customer support (%)	85.00			

(*Continued*)

Table 62.1 (Continued) Data for an Actual Fortune 500 Company

Portfolio Size Distribution		**Average Age (Years)**		
>10,000 function points; 1,000 SNAP points	35	15.00		
5,000–10,000 FP	200	8.75		
1,000–5,000 FP	475	6.50		
500–1,000 FP	1,790	2.25		
<500 function points	1,810	2.00		
Total	4,310	6.90		
Portfolio Application Types	**Internal**	**COTS**	**Total**	**Percentage**
Information systems	1,460	96	1,556	28.35
Systems software	913	125	1,038	18.90
Embedded applications	548	125	673	12.25
Tools and support software	365	433	798	14.54
Manufacturing applications	365	20	385	7.02
Web applications	335	345	680	12.39
Cloud applications	125	34	159	2.90
End-user applications	200	0	200	3.64
Total	4,310	1,178	5,488	100.00
Portfolio Languages				
Information systems	COBOL, PL/I, SQL, QBE, ASP			
Systems software	Chill, ESPL/1, Coral, Bliss, C, Objective C			
Embedded applications	Forth, Java, Objective C, J2ME			
Tools and support software	Objective C, C, C++, C#, PERL, Ruby			
Manufacturing applications	Assembly, Objective C, C, C++, C#			

(*Continued*)

Table 62.1 (Continued) Data for an Actual Fortune 500 Company

COTS packages	Multiple			
Open-source packages	Java, C, C++, Ruby			
End-user applications	Visual Basic, Java, Ruby, Perl			
Portfolio Platforms	**Internal**	**COTS**	**Total**	**Percentage**
Mainframe (IBM primarily)	2190	343	2533	46.16
Client-Server	730	224	954	17.38
Windows-based PC	335	236	571	10.40
Custom/embedded	548	120	668	12.16
SaaS	365	224	589	10.72
Cloud	125	13	138	2.51
Handheld/tablets	17	19	36	0.66
Total	4310	1178	5488	100.00

a new kind of quality or static analysis tools, results would be shown from companies such as Coverity, KlocWork, OptiMyth, CAST Software, and so on. For building a new telecom product, results would be shown from other telecom companies such as Alcatel, AT&T, Huawei, SPRINT, Nippon Telephone, Siemens, and so on.

It is probable that the virtual reality tool would bring in additional data about companies from sources such as Hoover's Guides and additional country data from sources such as the Central Intelligence Agency (CIA) fact book. In fact, one of the benefits of a virtual reality tool would be access to much larger collections of data than are currently available. Of course, copyrights would be honored and the data acquired under licenses from the owners. No doubt the software benchmark companies would be willing to provide their data for normal fees.

Planning and Estimating Screen

The next large screen would be the planning and estimating screen that would show dynamic moving program evaluation review technique (PERT) and generalized activity normalization time table (Gantt) charts for the current project. This screen would also show application size in terms of function points and LOC. The plans and estimates themselves would probably come from one of the commercial parametric estimation tools such as Software Evaluation and Estimation of Resources (SEER),

Software Lifecycle Management (SLIM), Software Risk Master (SRM), or possibly all three used concurrently. (Manual estimates tend to be distressingly optimistic.)

Work completed to date and planned effort from today to completion would be shown in real time. For defense software projects, earned value analysis (EVA) would be displayed.

This screen would also show any potential "red flag" topics that might cause a schedule delay or a cost overrun. In fact, red flags would trigger immediate alarms to all project managers and C-level sponsors of the current application. There would be no more hiding or concealing problems until too late to solve them, which is far too common in 2016 with today's project management tools!

Requirements creep would also be shown on this screen. (Software requirements have been measured to grow at rates that may top 1% per calendar month.) Effort predictions would be based on the activity-based cost concept as shown in Table 62.2, which shows a large system of 10,000 function points coded in Java.

Phase-level estimates or single-point estimates that only show complete projects are useless for serious project management and effective project controls.

Project Risk Screen

The risk screen would show the full set of possible risks identified for the project under development. Software is a high-risk industry and this screen is important and should be displayed and analyzed for every major software project in the world.

Software projects have over 210 known risks, which is more than any other industry except medicine. The author's SRM estimating tool predicts the major risks and assigns probabilities for each risk based on historical data from similar applications (Table 62.3).

As can be seen, large system risk factors are both numerous and severe. It is important to know these risks early before starting the project. It is also important to take effective steps to remediate them during development.

Quality Control Screen

Quality control is a weak link in software and has been for over 50 years. Since poor quality is a key reason for schedule delays and cost overruns, this quality control screen would include defect potentials, defect prevention, pre-test defect removal, testing, and probable delivered defect volumes. The quality data would use a combination of metrics developed by IBM in the 1970s that show defect potentials and DRE. The kinds of quality data needed for successful larger projects are shown in Table 62.4 for an application of 2500 function points.

The data shown here for software quality control are the normal quality estimates from SRM.

Table 62.2 Activity-Based Cost Example

	Development Activities	*Work Hours per Function Point*	*Burdened Cost per Function Point ($)*
1	Business analysis	0.01	0.42
2	Risk analysis/sizing	0.00	0.14
3	Risk solution planning	0.00	0.21
4	Requirements	0.29	23.33
5	Requirement—inspection	0.24	19.09
6	Prototyping	0.38	30.00
7	Architecture	0.05	4.20
8	Architecture—inspection	0.04	3.00
9	Project plans/estimates	0.04	3.00
10	Initial design	0.66	52.50
11	Detail design	0.88	70.00
12	Design inspections	0.53	42.00
13	Coding	6.60	525.00
14	Code inspections	3.30	262.50
15	Reuse acquisition	0.00	0.14
16	Static analysis	0.01	0.70
17	COTS package purchase	0.01	0.42
18	Open-source acquisition	0.00	0.21
19	Code security audit	0.07	5.25
20	Ind. verif. & valid	0.01	1.05
21	Configuration control	0.03	2.10
22	Integration	0.02	1.75
23	User documentation	0.26	21.00
24	Unit testing	1.06	84.00
25	Function testing	0.94	75.00

(*Continued*)

Table 62.2 (Continued) Activity-Based Cost Example

	Development Activities	*Work Hours per Function Point*	*Burdened Cost per Function Point ($)*
26	Regression testing	1.47	116.67
27	Integration testing	1.06	84.00
28	Performance testing	0.26	21.00
29	Security testing	0.38	30.00
30	Usability testing	0.22	17.50
31	System testing	0.75	60.00
32	Cloud testing	0.06	4.38
33	Field (Beta) testing	0.03	2.63
34	Acceptance testing	0.03	2.10
35	Independent testing	0.02	1.75
36	Quality assurance	0.18	14.00
37	Installation/training	0.03	2.63
38	Project measurement	0.01	1.11
39	Project office	0.24	19.09
40	Project management	1.76	140.00
	Cumulative results	21.91	1743.08

The quality control screen would not show hazardous metrics that distort reality such as "cost per defect" other than as a warning to clients to avoid bad metrics and start using good metrics such as defect removal cost per function point.

The quality control screen would also include a suite of quality tools for things like using cause-effect graphs for test case design, automatic proofs of correctness, requirements models, quality function deployment (QFD), and other quality control techniques.

It would also include a complete bug tracking facility that would record 100% of bugs found via every form of defect removal such as pre-test inspections, static analysis, and all forms of testing. Post-release bugs would also be tracked. The virtual tool would calculate DRE 90 days after the software project is deployed. If developers found 900 bugs and users report 50 bugs in the first 3 months, DRE is 95%.

Table 62.3 Example of 1000 Function Point Virtual-Reality Project with the Java Language

	Risk Analysis from Similar Projects	*Project Odds (%)*
1	Optimistic cost estimates	35.00
2	Inadequate quality control using only testing	30.00
3	Excessive schedule pressure from clients, executives	29.50
4	Technical problems hidden from clients, executives	28.60
5	Executive dissatisfaction with progress	28.50
6	Client dissatisfaction with progress	28.50
7	Poor quality and defect measures (omits >10% of bugs)	28.00
8	Poor status tracking	27.80
9	Significant requirements creep (>10%)	26.30
10	Poor cost accounting (omits >10% of actual costs)	24.91
11	Schedule slip (>10% later than plan)	22.11
12	Feature bloat and useless features (>10% not used)	22.00
13	Unhappy customers (>10% dissatisfied)	20.00
14	Cost overrun (>10% of planned budget)	18.52
15	High warranty and maintenance costs	15.80
16	Cancellation of project due to poor performance	14.50
17	Low reliability after deployment	12.50
18	Negative return on investment (ROI) due to poor performance	11.00
19	Litigation (patent violation)	9.63
20	Security vulnerabilities in software	9.60
21	Theft of intellectual property	8.45
22	Litigation (breach of contract)	7.41
23	Toxic requirements that should be avoided	5.60
24	Low team morale	4.65
25	High personnel turnover	3.50
	Average risks for this size and type of project	18.44
	Financial risk (cancel, cost overrun, negative ROI)	44.02

Table 62.4 Quality Data Needed for Successful Larger Projects

Software Quality				
Defect Potentials		**Potential**		
Requirements defect potential		1366		
Design defect potential		1743		
Code defect potential		2692		
Document defect potential		262		
Total defect potential		6063		
Per function point		2.43		
Per thousands of lines of code (KLOC)		45.47		
Defect Prevention	**Efficiency (%)**	**Remainder**	**Bad Fixes**	
Joint application design (JAD)	25	4578	30	
QFD—not used	0	4607	0	
Prototype	20	3685	18	
Models—not used	0	3704	0	
Subtotal	**39**	3704	47	
Pre-test Removal	**Efficiency (%)**	**Remainder**	**Bad Fixes**	
Desk check	26	2741	55	
Pair programming—not used	0	2795	56	
Static analysis	55	1283	26	
Inspections	90	131	3	
Subtotal	**96**	133	139	
				Test

(Continued)

Table 62.4 (Continued) Quality Data Needed for Successful Larger Projects

Software Quality				
Test Removal	**Efficiency (%)**	**Remainder**	**Bad Fixes**	**Cases**
Unit	31	92	2	1987
Function	34	62	2	2198
Regression	13	56	1	989
Component	31	39	2	1319
Performance	12	36	1	659
System	35	24	1	2308
Acceptance	16	21	0	484
Subtotal	84	21	9	9945
Defects delivered		21		
High-severity defects		3		
Security flaws		1		
High severity (%)		13.05		
Delivered per FP		0.009		
High severity per FP		0.001		
Security flaws per FP		0.001		
Delivered per KLOC		0.160		
High severity per KLOC		0.021		
Security flaws per KLOC		0.009		
Cumulative (%)		99.65	Excellent	
Removal efficiency				

Table 62.4 shows a small subset of the quality information predicted by the author's SRM tool. Every software project should measure defect potentials and DRE and also measure defect removal costs per function point.

Note that the *technical debt* metric published by Ward Cunningham has become popular. It is a good metaphor but unfortunately, not yet a good metric. There are no International Organization for Standardization (ISO) or Object Management Group (OMG) standards and every company measures technical debt differently. Worse, it does not even address some of the major software quality issues such as litigation for poor quality or consequential damages to customers.

User Cost Screen

For internal information systems and especially for those using the Agile methodology, users are active participants in software development projects. Their costs

Table 62.5 User Costs for an Internal Project Coded in Java

User Activities During Development					
	User Activities	*Effort Hours*	*Hours per Function Point*	*Costs ($)*	*$ per Function Point*
1	User requirements team	1,447.62	1.45	109,669	109.67
2	User architecture team	723.81	0.72	54,834	54.83
3	User planning/ estimating team	868.57	0.87	$65,801	65.80
4	User prototype team	1,520.01	1.52	115,152	115.15
5	User design review team	1,650.29	1.65	125,022	125.02
6	User change control team	1,737.15	1.74	131,602	131.60
7	User governance team	434.29	0.43	32,901	32.90
8	User document review team	1,230.48	1.23	9,322	9.32
9	User acceptance test team	1,302.86	1.30	98,702	98.70
10	User installation team	506.67	0.51	38,384	38.38
	Total user costs	11,421.76	11.42	781,388	781.39

should be captured and used for planning future projects since for some projects user costs are almost as large as development costs. Table 62.5 shows user costs for a 1000 function point internal project coded in Java.

User costs are seldom estimated and seldom tracked, but they are a large component of internal software projects and they should be captured. The virtual software development environment would include both user cost estimates and collecting user historical data. The author's SRM tool predicts user costs as an optional feature for internal projects. User costs have no relevance for commercial software or embedded applications or medical device software.

Progress to Date Screen

An important part of software project management is accumulating data that shows the progress of software projects. This data includes schedules to date, costs to date, and defect removal to date. The data would be displayed in a series of Gantt charts plus accumulated historical data. Of course, time to complete and cost to complete would be shown too and updated dynamically as the project grows and changes.

Table 62.6 Total Cost of Ownership Estimates (TCO time period is from calendar year 2011 through year end 2015)

	Staffing	*Effort*	*Costs ($)*	*$ per Function Point at Release*	*Percentage of TCO*
Development	7.48	260.95	3,914,201	1565.68	46.17
Enhancement	2.22	79.75	897,169	358.87	10.58
Maintenance	2.36	85.13	877,561	351.02	10.35
Customer support	0.34	12.29	58,062	23.22	0.68
User costs	4.20	196.69	2,722,773	1089.11	32.12
Additional costs			7,500	3.00	0.09
Cyber-attacks (if any)	0.00	0.00	0	0.00	0.00
Litigation (if any)	0.00	0.00	0	0.00	0.00
Total TCO	16.60	634.81	8,477,266	3390.91	100.00

Total Cost of Ownership (TCO) Screen

Software economic analysis should be based on TCO. These costs include development, customer support, maintenance, enhancements, and cyber-attack recovery costs. As of 2016, the software industry lacks reliable and accurate TCO data, and the virtual software development environment would change that. Table 62.6 shows a sample TCO estimate from the author's SRM estimation tool for an application of 2500 function points.

Note that today, in 2016, TCO estimates and benchmarks should include cyber-attack costs and litigation costs, if either or both occurred.

Application Backlog Screen

All corporations and government groups have backlogs of projects waiting to start. An effective methodology and management tool will show the backlog as well as in-progress software. Table 62.7 shows the actual backlog for a Fortune 500 company but sanitized to avoid giving away the name of the company.

Application Usage and Work Value Screen

Companies and government agencies build software to improve their performance. But very few actually measure or quantify the benefits. An effective methodology and management tool suite will show overall usage of software by the complete company or agency and project the work value of the software (Table 62.8).

Table 62.7 Actual Backlog for a Fortune 500 Company

Application Backlog	
Applications in backlog	250
Average size in function points	1,500
Average size in SNAP points	263
Backlog size in function points	375,000
Backlog size in SNAP points	65,625
Language level	4.00
Backlog size in LOC	30,000,000
Average age of backlog (months)	12
Oldest backlog application (months)	24

Table 62.8 Example of 1000 Function Point Virtual-Reality Global Project with the Java Language

Application Usage	
Total corporate employment	250,000
Number of application users	75,000
Annual usage growth (%)	10.00
Users per application	21
Applications per user	15
Usage per work day (hours)	2.5
Total hours per work day	187,500
Total hours per year	41,250,000
Annual software usage ($)	5,478,515,625
Annual usage cost per user ($)	73,047
Portfolio Work Value	
Effort reduction for key tasks (hours)	4.50
Effort reduction per work day (hours)	337,500
Total hours saved per year	74,250,000
Annual effort reduction ($)	9,861,328,125
Annual value per user ($)	131,484
Net annual value per user ($)	58,438
Net annual ROI ($)	1.80

Portfolio Value Analysis Screen

Software has both tangible financial value and intangible value. Intangible value includes topics such as security benefits or medical benefits for medical devices. Value should be shown in a full software corporate methodology and management tool set. The value table in Table 62.9 is based on a Fortune 500 company with a portfolio of about 4300 applications.

Value is seldom considered at the corporate level, but to understand software economics in full both complete costs and complete value need to be measured.

Table 62.9 Example of Portfolio Value Analysis Screen

Portfolio Tangible Value	
Software and embedded sales ($)	3,500,000,000
Warranty cost reduction ($)	400,000,000
Manufacturing efficiency ($)	500,000,000
Improved customer service ($)	75,000,000
Reduced scrap and rework ($)	50,000,000
Tool, service acquisition ($)	25,000,000
Annual tangible value ($)	4,550,000,000
Net annual tangible value ($)	1,333,373,321
Net annual tangible ROI ($)	1.41
Portfolio Intangible Value	
National defense ($)	0
Human life or safety ($)	0
Market share increase ($)	1,000,000,000
Customer satisfaction ($)	350,000,000
Competitive advantages ($)	300,000,000
Security and encryption ($)	125,000,000
Enterprise prestige ($)	10,000,000
Employee morale ($)	50,000,000
Reduced employee turnover ($)	10,000,000
Overall intangible value ($)	1,845,000,000
Net annual intangible value ($)	−1,371,626,679
Net annual ROI ($)	0.57
Tangible+intangible value ($)	6,395,000,000
Tangible+intangible net value ($)	3,178,373,321
Tangible+intangible ROI ($)	1.99
Work value ROI ($)	1.80

(Continued)

Table 62.9 (Continued) Example of Portfolio Value Analysis Screen

Tangible value ROI ($)	1.41
Intangible value ROI ($)	0.57
Enterprise value ROI ($)	3.79
Portfolio Replacement Value	
Value of applications owned ($)	6,825,500,000
Replacement dollar per function point	1,700
Replacement dollar per SNAP point	1,768

Cyber-Attack and Security Analysis Screen

Because cyber-attacks and cyber-crime are increasing rapidly, the virtual development tool would have links to all of the major government cyber-attack databases. This screen would highlight attacks against similar projects and would also show the kinds of security defense steps planned for the current project, such as security inspections, security testing, security static analysis, using ethical hackers, and many more. Cyber-attack costs, if any, should be included in TCOs as they are in the author's SRM tool.

Litigation Cost Analysis Screen

If there is litigation between a software developer and clients for breach of contract, or litigation between two software companies for issues such as theft of intellectual property, these costs should be included in TCO.

Unbudgeted Cost Screen

It is an unfortunate fact of life that not all expenses can be planned and budgeted. For example, a lawsuit against a company can be filed at any time. A hurricane can shut down work and destroy buildings. It is useful to show past unbudgeted expenses by fiscal year (Table 62.10).

The unbudgeted expense screen shows a number of real-life problems that should be monitored and included in project management tools and also dealt with in a full life-cycle methodology package.

Table 62.10 Past Unbudgeted Expenses by Fiscal Year

Unbudgeted Costs 2015	
Denial of service attacks ($)	35,000,000
Virus and spyware recovery ($)	30,000,000
Canceled software projects ($)	80,277,778
Software budget overruns—accrued costs ($)	88,305,556
Software schedule delays—lost revenues ($)	20,000,000
Unanticipated requirements changes—new ($)	20,000,000
Unanticipated updates —legacy ($)	15,000,000
Unanticipated regulatory changes ($)	5,000,000
IRS tax litigation: software assets ($)	20,000,000
Patent violation litigation ($)	15,000,000
Breach of contract litigation ($)	15,000,000
Intellectual property litigation ($)	12,000,000
Sarbanes–Oxley compliance litigation ($)	5,000,000
Total unbudgeted costs ($)	360,583,333
Unbudgeted dollar per application	87,831
Unbudgeted dollar per user	4,274
Unbudgeted dollar per staff worker	25,414
Unbudgeted dollar per function point	79.85
Unbudgeted dollar per SNAP point ($)	81.84
Unbudgeted dollar per LOC	0.81
Canceled Projects	
Projects canceled	5
Average size in function points	8,500
Average size in SNAP points	1,240
Total size in function points	42,500
Total size in SNAP points	6,201

(*Continued*)

Table 62.10 (Continued) Past Unbudgeted Expenses by Fiscal Year

Total size in LOC	4,184,615
Staff on canceled projects	315
Effort on canceled projects (months)	2,916
Costs for canceled projects	$80,277,778
Schedule at cancellation point	48
Schedule slip at cancellation	12
Total effort on canceled projects	15,111
Total cost for canceled projects	$321,111,111
Total cost per function point	$1,888.89
Total cost per SNAP point	$1,912.00
Total cost per LOC	$19.18
Budget overruns	$88,305,556
Overrun dollars per function point	2,077.78
Overrun dollars per SNAP point	2129.72
Overrun dollars per LOC	21.10

Virtual Reality Software Development Methodology

It is painfully obvious that text and static black and white diagrams are totally inadequate to display the reality of software, which is the fastest known product in the world when running and also changes rapidly over time at more than 1% per month during development and more than 8% per year after release. Software is dynamic and needs dynamic representation methods: not just text and static black and white diagrams.

Software design and software project controls need animated, full color, 3-D illustrations to simulate software growth and performance, and also to model cyber threats. This is why a virtual reality environment will eventually replace today's antique software development methods, some of which are over 40 years old. The following are the essential features of a virtual reality software development methodology that would encompass development, quality, and post-release maintenance and enhancements, which are omitted from many methodologies.

Requirements Screen

The requirements screen would use sophisticated data mining to show requirements from similar legacy applications as a starting point for developing new requirements. Both functional and non-functional requirements would be included. The function point metric would size user requirements and the SNAP metric would size non-functional requirements. This screen would also link to the test and quality screens to be sure that all requirements were inspected and tested. The requirements feature would also include running the FOG and Flesch readability indexes against all requirements to ensure that they were complete and comprehensible. This screen would show projected requirements growth during development and after deployment.

Architecture Screen

The architecture screen is one that would need animated, 3-D, full color graphics to show the architecture of the application and also to show the application fit into the cloud and links to related applications.

Table 62.11 Example of Document Types and Sizes

	Document Sizing				
	Document Types	*Pages*	*Words*	*Percentage of Pages*	*Percentage Complete*
1	Requirements	613	245,037	7.46	86.27
2	Architecture	141	56,234	1.71	87.28
3	Initial design	737	294,915	8.98	82.14
4	Detail design	1361	544,306	16.57	85.35
5	Test plans	324	129,575	3.94	82.41
6	Development plans	138	55,000	1.67	85.28
7	Cost estimates	141	56,234	1.71	88.28
8	User manuals	600	239,979	7.31	90.87
9	HELP text	482	192,980	5.88	91.25
10	Courses	363	145,000	4.41	90.68
11	Status reports	209	83,557	2.54	87.28
12	Change requests	477	190,791	5.81	91.27
13	Bug reports	2628	1,051,061	32.00	88.08
	Total	8212	3,284,669	100.00	87.42

Design Screen

The design screen is another screen based on animated, 3-D, full color graphics. It would show the overall structure of the application as a hologram with input and output screens moving in real time, or slowed down to investigate performance. A small sample of a possible 3-D image is shown in Table 62.11.

Today, in 2016, there are over a dozen graphics packages that can produce 3-D graphics such as Wolfram Alpha, MATLAB, and many others. The widespread use of 3-D printers and the emerging use of holographic displays make it possible to envision both animation and three dimensions as future software design tools.

Document Status Screen

Not many people know that for large software projects, and especially for defense software projects, the costs of producing paper documents are larger than the costs of producing source code. The virtual reality status room would show the current status of all major documents being produced for the software application under development as shown in Table 62.11 in a sample for 2500 function points.

Software document costs are not well understood and are sometimes left out of software cost estimates. But for the whole software industry, document costs are a major cost driver and should be included in cost estimates and benchmarks. For civilian projects, documents can total more than 30% of all software costs; for defense projects, more than 40%. Department of Defense standards call for way too many documents, most of which have no technical value.

Software Occupation Group Screen

Software has a total of 126 occupations based on a study performed by the author and funded by AT&T. Most benchmarks and few estimates actually show the occupations that are working on software projects, but this data should be a standard feature of estimates and benchmarks, as it is with SRM (Table 62.12).

Standards and Certification Screen

The standards and certification screen would show all of the ISO and OMG standards that apply to the application. It would also show any government mandates such as the Federal Aviation Authority (FAA) or the Federal Communications Commission (FCC) certification or Sarbanes–Oxley. It would also show the number of certified test personnel, quality personnel, and certified project managers involved with the project. It would also show the capability maturity model integrated (CMMI) level for the organization building the software. This screen would also discuss the specific methodologies used for the project; that is, Agile, DevOps, container development, RUP, TSP, waterfall, or whatever (Table 62.13).

Table 62.12 Example of Project Occupation Groups

	Occupation Groups and Specialists on Project	*Normal Staff*	*Peak Staff*
	Occupation Groups (alphabetical order)		
1	First-line managers	1.0	2.0
2	Second-line managers	0.4	0.6
3	Third-line managers	0.1	0.1
4	Administrative support	0.6	0.7
5	Agile coaches	0.1	0.7
6	Architects	0.1	0.2
7	Business analysts	3.1	4.8
8	C-level (CIO/CTO/CRO, etc.)	0.1	1.0
9	Configuration control	0.3	0.5
10	Customer support	4.0	6.0
11	Database administration	0.6	0.8
12	Designers	3.1	5.1
13	Embedded user personnel	0.6	3.0
14	Estimating specialists	0.2	0.3
15	Function point counters	0.1	0.1
16	Graphic designers	0.1	0.1
17	Human factors specialists	0.1	0.1
18	Maintenance team	2.0	4.0
19	Management consultants	0.1	0.1
20	Patent attorneys	0.1	0.1
21	Performance specialists	0.1	0.1
22	Programmers	6.6	9.9
23	Project librarians	0.3	0.4
24	Project office staff	0.3	0.7
25	Quality assurance	1.3	2.1

(*Continued*)

Table 62.12 (Continued) Example of Project Occupation Groups

	Occupation Groups and Specialists on Project	*Normal Staff*	*Peak Staff*
26	Scrum masters	0.1	0.7
27	Security specialists	0.1	1.0
28	Technical writers	1.4	2.0
29	Testers	6.0	8.0
30	Translators/international projects	0.1	0.1
	Totals	33.0	55.2

Table 62.13 Examples of International Standards Used for Project

ISO and Other Standards Used for Project	
IEEE 610.12-1990	Software engineering terminology
IEEE 730-1999	Software assurance
IEEE 12207	Software process tree
ISO/IEC 9001	Software quality
ISO/IEC 9003	Software quality
ISO/IEC 12207	Software engineering
ISO/IEC 25010	Software quality
ISO/IEC 29119	Software testing
ISO/IEC 27034	Software security
ISO/IEC 20926	Function point counting
OMG Corba	Common Object Request Broker Architecture
OMG Models	Meta-models for software
OMG function points	Automated function points (legacy applications)
UNICODE	Globalization and internationalization

Table 62.14 Examples of Tools Used for Project

	Tool Information for Project	
	Tasks	*Tools Utilized*
1	Architecture	QEMU
2	Automated test	HP QuickTest Professional
3	Benchmarks	ISBSG, Namcook SRM
4	Coding	Eclipse, Slickedit
5	Configuration	Perforce
6	Cost estimate	SRM
7	Cost tracking	APO, Microsoft Project
8	Cyclomatic	BattleMap
9	Debugging	GHS probe
10	Defect tracking	Bugzilla
11	Design	Projects Unlimited
12	Earned value	DelTek Cobra
13	ERP	Microsoft Dynamics
14	Function points 1	SRM
15	Function points 2	Function point workbench
16	Function points 3	CAST automated function points
17	Graphics design	Visio
18	Inspections	SlickEdit
19	Integration	Apache Camel
20	ISO tools	ISOXpress
21	Maintenance	Mpulse
22	Manual test	DevTest
23	Milestone track	KIDASA Software Milestone Professional
24	Progress track	Jira
25	Project management	APO

(*Continued*)

Table 62.14 (Continued) Examples of Tools Used for Project

26	Quality estimate	SRM
27	Requirements	Rational Doors
28	Risk analysis	SRM
29	Source code size 1	SRM
30	Source code size 2	Unified code counter (UCC)
31	SQA	NASA Goddard ARM tool
32	Static analysis	OptiMythKiuwin, CAST, Coverity
33	Support	Zendesk
34	Test coverage	Software Verify suite
35	Test library	DevTest
36	Value analysis	Excel and Value Stream Tracking

Tool Usage Screen

Large software projects use many kinds of tools for purposes such as cost estimating (i.e., SLIM or SRM); requirements (i.e., Rational Doors), architecture, design, code development, test case design, project management (i.e., automated project office [APO], Microsoft projects), quality tracking, and so on. This screen would show the full suite of tools that is being used on the current project. This information is of use when planning new projects, but is seldom included in software benchmarks (Table 62.14).

Code Development Screen

The code development screen would support all current programming languages. It would also show sources for reused code as well as providing support such as debugging and static analysis for custom manual code. Cyclomatic complexity would be calculated automatically for all modules and code segments. Links would be available to test libraries and static analysis tools.

Software Quality Assurance Screen

The software quality assurance (SQA) screen would support SQA personnel in reviewing all critical deliverables. It would support inspection results and have links to all bugs or defects identified via any channel such as desk checks, inspections, static analysis modeling, or testing.

Reusable Component Acquisition Screen

This screen would show the sources and volumes of all reusable materials utilized by the application. A probable mix might be 75% reuse from similar internal software applications and 25% reuse from external vendors of reusable materials such as currency exchange rate or interest calculation rates.

Defect Prevention Screen

This screen would identify and show the results of all defect prevention methods used, such as JAD, QFD, requirements models, and cause-effect graphing. It might also show the root causes of defects found in similar legacy applications.

Pre-test Defect Removal Screen

This screen would show the quantified results for desk checks, formal inspections, automated proofs of correctness, and static analysis used for quality control before testing begins. It would also have links to things like static analysis tools that were licensed for use by the project.

Test Case Design and Execution Screen

The most effective ways of developing software test cases today use cause–effect graphs or are based on design of experiments. This screen would show numbers of test cases and test scripts, and also effort and costs for test preparation. As testing is performed, defect counts would be accumulated. Also, accumulated would be testing costs.

Defect Reporting and Tracking Screen

Software defect tracking is woefully incomplete and sometimes omits over 75% of actual bugs. Many companies don't start tracking bugs until after unit test so all early bugs are invisible. Finding and fixing bugs is the number one cost driver for the entire software industry and it won't get much better until all bugs are tracked and given root cause analysis. Bugs occur not only in source code but also in requirements, in architecture, in designs, in the code itself, and in user documents. About 7% of bug repairs contain new bugs called "bad fixes." Bugs are not randomly distributed but clump in a small number of error-prone modules (EPM). Failure to track and analyze software defect origins and removal efficiency is an endemic weakness of the software industry.

Table 62.15 Example of 1000 Function Point Global Virtual Reality Project with the Java Language

Project Size and Nature Scores		**Global Virtual**
1	Small projects <100 FP	**5**
2	Medium projects	**7**
3	Large projects >10,000 FP	**10**
4	New projects	**10**
5	IT projects	**10**
6	Systems/embedded projects	**10**
7	Web/cloud projects	**10**
8	High-reliability, high-security projects	**10**
9	Enhancement projects	7
10	Maintenance projects	6
	Subtotal	85
Project Activity Scores		**Global Virtual**
1	Pattern matching	10
2	Reusable components	10
3	Planning and estimating	10
4	Requirements	10
5	Architecture	10
6	Models	10
7	Design	9
8	Coding	9
9	Defect prevention	10
10	Test case design/generation	9
11	Testing defect removal	9
12	Requirements changes	10
13	Change control	10

(*Continued*)

Table 62.15 (Continued) Example of 1000 Function Point Global Virtual Reality Project with the Java Language

14	Integration	9
15	Pre-test defect removal	10
	Subtotal	145
	Total	230
Regional Usage (Virtual Global 2016) Virtual		**Global Virtual**
1	North America	5
2	Pacific	1
3	Europe	1
4	South America	0
5	Africa/Middle East	0
	Total	7
Defect Potentials per Function Point (1000 function points; Java language)		**Global Virtual**
1	Requirements	0.25
2	Architecture	0.10
3	Design	0.50
4	Code	0.90
5	Documents	0.10
6	Bad fixes	0.15
	Total	2.00
	Total defect potential	2000
Defect Removal Efficiency (DRE)		
1	Pre-test defect removal	90.00%
2	Testing defect removal	92.00%
	Total DRE	99.20%
	Delivered defects per function point	0.016

(Continued)

Table 62.15 (Continued) Example of 1000 Function Point Global Virtual Reality Project with the Java Language

	Delivered defects	16
	High-severity defects	2
	Security defects	0
Productivity for 1000 Function Points		**Global Virtual**
	Work hours per function point 2016	9.04
	Function points per month 2016	14.60
	Schedule in calendar months	10.90

Data source: Namcook analysis of 600 companies.

Virtual Global Scores
Best = +10
Worst = −10

Maintenance and Customer Support Screen

Large software applications live for over 10 years and some have lived for over 30 years. It is important to show post-release maintenance. Since the virtual reality tool discussed here is intended to be kept up after release, it will show predicted maintenance costs for at least 3 years and measured maintenance costs as well once the application is deployed.

Enhancement Screen

Once software is deployed, it continues to grow at about 8% per calendar year for as long as there are active users. It is important to estimate future enhancements for at least 3 years, and to track enhancement costs on a year-by-year basis. The virtual reality tool would include both estimates and measures of software enhancements.

As can be seen, a full methodology with project management support and corporate reporting integrated will be a large superset of any such tools and methods available in 2016. But software is one of the most critical industries in world history and it should be a world leader in economic and quality analysis, instead of a laggard as it is today, in 2016.

Table 62.15 shows the projected scores for the virtual development environment combined with results from working prototypes. There are no commercial versions of a virtual reality tool set available as of 2016, but among the author's technology clients about half a dozen U.S. companies are working on tools that resemble the feature set shown in Table 62.15, as are a few in Europe and the Pacific regions.

The software industry has many useful tools but these are not integrated into a single development environment. Worse, there is little quantitative data available in 2016 on tool results in terms of defect potential reduction, DRE improvement, or work hours per function point. The virtual reality tool set and methodology would provide a seamless integrated work space for software projects and full collection of quality and cost data for benchmark purposes.

The methodology and tool suite described here could be used from before requirements are fully defined all the way through development and continue to be used for 10 or more years after deployment to show continuous changes in software applications over time.

The overall purpose is to provide software with the most accurate estimates and the most complete benchmark data of any industry, rather than the current set of inaccurate manual estimates and skimpy partial benchmarks that are insufficient for serious economic analysis.

A further goal is to make international multi-location software projects more cost-effective than today, and achieve higher quality and productivity as well in spite of the difficulty of international cooperation across multiple countries and multiple time zones.

Chapter 63

V-Model

Executive summary: *The V-Model methodology was developed in Germany and is (was?) mandatory for governmental and military projects there. It is a full life-cycle methodology that emphasizes the importance of testing—"develop tests simultaneously with other steps." This method is an older approach that may be considered an extension of the waterfall model. Although it is unlikely that anyone would adopt this approach for a current era project, it nonetheless includes ideas that are valuable and certainly applicable to other more modern approaches such as the Rational Unified Process (RUP). The central notion, as illustrated by Figure 63.1, is that tests are designed in parallel with each phase of the life cycle. The value of early detection of defects is well understood and is one of the key elements necessary to achieve minimum cost and schedule and maximum delivered quality, regardless of the methodology chosen. The early design of tests in parallel with each development phase or iteration can be an effective basis for early defect detection and removal.*

The V-Model is potentially applicable in circumstances in which requirements are well defined and stable, and the technology to be used is well understood. It is not well adapted to high rates of requirements change—hence, it does not work well for very large projects in which significant requirements evolution is almost inevitable. The V-Model is subject to all of the criticism of waterfall.

The model is described as consisting of distinct verification and validation phases. (Note, this terminology is inconsistent with the way these terms are commonly used in U.S. government procurements.)

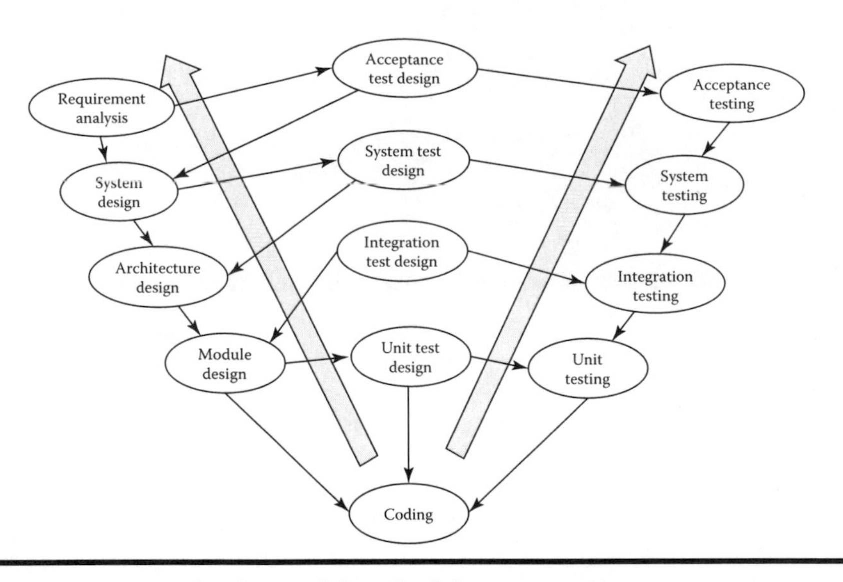

Figure 63.1 Example of V-model methodology concepts.

Verification phases: Requirements analysis, systems design, architecture design, module design

Validation phases: Unit testing, integration testing, system testing, user acceptance testing

A sample of V-Model development for an application of 1000 function points in the Java language is shown in Table 63.1.

The V-Model approach had some good quality concepts but was one of the "heavy" methodologies that tended to be displaced by newer "light" and "lean" methodologies such as Agile, DevOps, mashups, and so on. Its European origin is why V-Model was not widely used in the United States, except for some defense projects.

Table 63.1 Example of 1000 Function Point V-Model Project with the Java Language

Project Size and Nature Scoring		**V-Model**
1	Small projects <100 FP	7
2	Medium projects	7
3	Large projects >10,000 FP	7
4	New projects	7
5	Enhancement projects	7
6	Web/cloud projects	6
7	High-reliability, high-security projects	6
8	Systems/embedded projects	5
9	Maintenance projects	4
10	IT projects	3
	Subtotal	59
Project Activity Scoring		**V-Model**
1	Coding	7
2	Change control	7
3	Integration	7
4	Requirements changes	6
5	Models	6
6	Design	6
7	Planning, estimating, measurement	5
8	Requirements	5
9	Architecture	5
10	Defect prevention	5
11	Pre-test defect removal	5
12	Test case design/generation	5
13	Testing defect removal	5

(Continued)

Table 63.1 (Continued) Example of 1000 Function Point V-Model Project with the Java Language

14	Pattern matching	4
15	Reusable components	3
	Subtotal	81
	Total	140
Regional Usage (Projects in 2016)		**V-Model**
1	North America	200
2	Pacific	100
3	Europe	3500
4	South America/Central America	100
5	Africa/Middle East	100
	Total	4000
Defect Potentials per Function Point (1000 function points; Java language)		**V-Model**
1	Requirements	0.40
2	Architecture	0.35
3	Design	0.60
4	Code	1.45
5	Documents	0.40
6	Bad fixes	0.30
	Total	3.50
	Total defect potential	3500
Defect Removal Efficiency (DRE)		
1	Pre-test defect removal	50.00%
2	Testing defect removal	85.00%
	Total DRE	92.50%
	Delivered defects per function point	0.27
	Delivered defects	270

(Continued)

Table 63.1 (Continued) Example of 1000 Function Point V-Model Project with the Java Language

	High-severity defects	42
	Security defects	4
Productivity for 1000 Function Points		**V-Model**
	Work hours per function point 2016	18.5
	Function points per month 2016	7.14
	Schedule in calendar months	19.5

Data Source: Namcook Analysis of 600 companies.

V-Model Scores
Best = +10
Worst = −10

Chapter 64

Waterfall Development

Executive summary: *This method is the second oldest method after cowboy. It originated in the late 1960s when software projects grew large enough to require teams and therefore became cumbersome. A similar earlier method was defined in 1956 for use on the Semi-Automatic Ground Environment (SAGE) defense system. Historically, waterfall development was used on more than 2,000,000 projects between about 1970 and 2014, including some major applications that are still in use. Waterfall is still widely used, although many other methods such as Agile, team software process (TSP), and Rational Unified Process (RUP) have supplanted waterfall. The quality and productivity results of waterfall are not as good as the better replacement methods.*

The name *waterfall development* is due to the fact that a visual representation of this method resembles a stream flowing over a series of waterfalls, as shown in Figure 64.1.

Despite the name *waterfall*, in real life the waterfall phases are not actually completed before the next phase begins. For example, requirements are usually only about 50% complete when design starts; design is only about 60% complete when coding starts; coding is only about 35% complete when testing starts. These overlapping phases make project planning and estimating tricky because the sum total of the phases is not equal to the sum total of the project schedule. Phase overlaps are shown in Figure 64.2.

Another historical attribute of waterfall development is attempting to design the full system at the beginning. This is contrasted to the Agile approach of having only a rough approximation of the full system and doing formal designs at the beginning of each sprint.

Analysis
Requirement signification
Design
Development
Testing and Integration
Implementation/ deployment

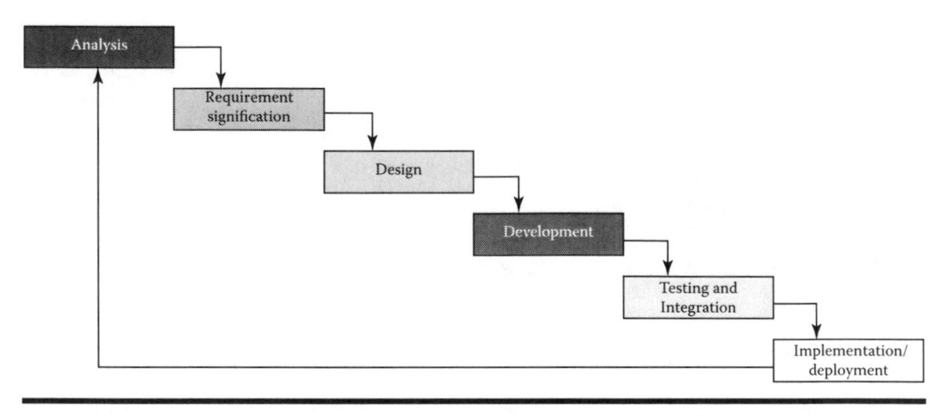

Figure 64.1 Illustration of software waterfall development methodology.

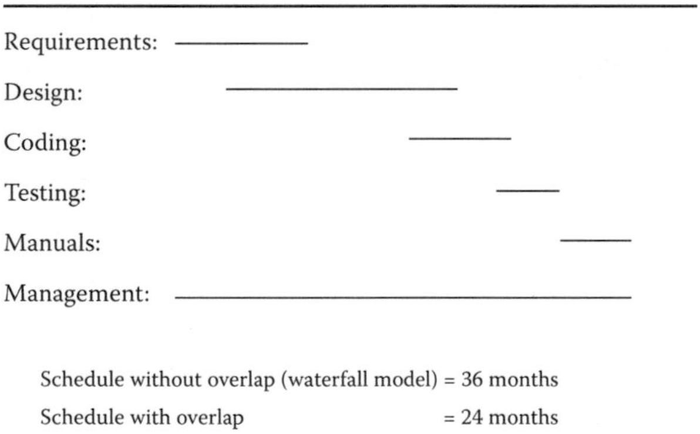

Requirements:
Design:
Coding:
Testing:
Manuals:
Management:
Schedule without overlap (waterfall model) = 36 months
Schedule with overlap = 24 months

Figure 64.2 Illustration of software timeline or Gantt chart.

Historically, a number of major systems were developed using the waterfall approach so it does have some proof of success. Among these major systems built using waterfall can be found:

Examples of Waterfall Development

1. SAGE air defense system
2. SABRE airline reservation system
3. IBM OS/360 and other operating systems

4. AT&T ESS/1 switching system
5. ITT System/12 switching system
6. Microsoft Windows
7. SAP
8. Oracle
9. State driver's license systems
10. IRS taxation packages

In today's world of 2016, waterfall is regarded as something of an antique that is slowly fading from use, despite many current usages in many countries. For large systems, the more popular replacements to waterfall include RUP and TSP. For smaller projects, the more popular replacements to waterfall include Agile, DevOps, and extreme programming (XP).

For those companies and projects that still use waterfall, it can be augmented by modern tools and approaches such as static analysis and automated testing. Formal inspections of key deliverable topics such as requirement, design, and code inspections are also useful in a waterfall context.

The results of waterfall for a project of 1000 function points using the Java programming language are shown in Table 64.1.

The venerable waterfall approach is often criticized today, in 2016. But it provided over 30 years of continuous service and was used to successfully develop hundreds of major applications. In fact, waterfall is still used in 2016 for applications in the 10,000 function point size range where Agile development is not the best choice.

Waterfall has also been used in hybrid approaches, with waterfall combined with Agile being surprisingly successful. Waterfall has also been used with the Software Engineering Institute (SEI) capability maturity model integrated (CMMI) approach. The combination of waterfall and the higher CMMI scores such as Level 3 or Level 5 has also been successful for defense software.

The main criticism of waterfall, which seems to be justified, is that waterfall tries to accomplish too much too soon, such as the development of full requirements before starting design. In real life, requirements are seldom more than about 50% firm when design starts.

In any case, waterfall is supported by all parametric estimating tools and has more valid benchmark data than any other methodology. This means that estimating and measuring waterfall projects is fairly easy to do.

In fact, since Agile is quirky and hard to measure, there are probably 100 times more valid waterfall benchmarks than there are Agile benchmarks. It is typical of the software industry that new methodologies such as Agile go into service with zero empirical data or any proof that they even work.

Waterfall is slow in the beginning but once code and testing start, the last half of software development is no worse than Agile or other newer methodologies.

Table 64.1 Example of 1000 Function Point Waterfall Project with the Java Language

Project Size and Scoring		**Waterfall**
1	New projects	7
2	Small projects <100 FP	6
3	IT projects	6
4	Systems/embedded projects	6
5	Medium projects	5
6	Enhancement projects	5
7	Maintenance projects	5
8	Large projects >10,000 FP	4
9	High-reliability, high-security projects	2
10	Web/cloud projects	–2
	Subtotal	44
Project Activity Scoring Waterfall		
1	Requirements	6
2	Requirements changes	6
3	Architecture	5
4	Design	5
5	Integration	5
6	Testing defect removal	5
7	Pattern matching	4
8	Reusable components	4
9	Planning, estimating, measurement	4
10	Change control	4
11	Test case design/generation	4
12	Coding	3
13	Pre-test defect removal	3

(*Continued*)

Table 64.1 (Continued)— Example of 1000 Function Point Waterfall Project with the Java Language

14	Models	–1
15	Defect prevention	–1
	Subtotal	56
	Total	100
Regional Usage (Projects in 2016)		**Waterfall**
1	North America	100,000
2	Pacific	100,000
3	Europe	110,000
4	South America/Central America	60,000
5	Africa/Middle East	15,000
	Total	385,000
Defect Potentials per Function Point (1000 function points; Java language)		**Waterfall**
1	Requirements	0.35
2	Architecture	0.30
3	Design	0.60
4	Code	1.60
5	Documents	0.40
6	Bad fixes	0.50
	Total	3.75
	Total defect potential	3750
Defect Removal Efficiency (DRE)		
1	Pre-test defect removal	40.00%
2	Testing defect removal	80.00%
	Total DRE	88.00%
	Delivered defects per function point	0.45

(Continued)

Table 64.1 (Continued)— Example of 1000 Function Point Waterfall Project with the Java Language

	Delivered defects	450
	High-severity defects	70
	Security defects	6
	Productivity for 1000 function points	
	Work hours per function point 2016	16.85
	Function points per month 2016	7.83
	Schedule in calendar months	17.20

Data source: Namcook Analysis of 600 companies.

Waterfall Scoring
Best = +10
Worst = −10

When waterfall is used for defense software, it suffers under the huge burden of defense paperwork, which totals about three times the volumes of civilian projects of the same size and type.

Surprisingly, waterfall is not a bad choice for enhancing legacy applications where the bulk of the work involves understanding existing software. Although this book does not break out enhancements as a separate topic, waterfall and Agile are about equal for legacy updates.

Chapter 65

Summary and Conclusions about Software Methodologies

An unanswered sociological question is why the software industry has developed so many variations of so many things. Is there really a need for over 3000 programming languages, for more than 60 software development methodologies, for more than 50 static analysis tools, and for some 25 software size metrics including about a dozen flavors of function point metrics?

The fundamental reason for all these variations seems to be that the software industry does not know how to measure either productivity or quality. In the absence of any empirical data about effectiveness, researchers simply develop something new if they feel it might be helpful. Once developed, the new methodology or new language is put into service without any kind of validation or proof of effectiveness. Software methodology selection and software programming language selection in 2016 are closer to joining religious cults than to making informed technical decisions.

This is not to say that many of the 60 methodologies are not useful. Some of them such as structured development, Agile, and information engineering introduced new and beneficial concepts that should be widely understood and widely deployed.

But until bad metrics such as lines of code, cost per defect, story points, and use-case points are replaced by effective metrics such as function points and defect removal efficiency (DRE), the software industry will continue to move like a drunkard's walk, with some progress but also some retrograde movements.

No matter what software development methodology is selected, it is obvious that custom designs and manual coding are intrinsically expensive and error prone. The methodologies in this book that deal with constructing software from standard reusable components are the ones to consider and explore for the next 20 years or so.

When considering methodologies, readers are urged to evaluate the kinds of software applications that are to be built. No *one size fits all* methodology is a good fit for all sizes and types of software applications.

This book provides some guidance about the effectiveness of the various methodologies for

- Small projects <1000 function points
- Large systems >10,000 function points
- Systems, embedded, web, cloud, and other types of software
- Methodology emphasis on quality control
- Methodology focus on all activities or narrow focus on only code development

Many of the methodologies in this book are useful and have added value to large numbers of software applications. But no methodology in this book is a panacea that is useful for all sizes, types, and flavors of software applications.

References and Readings on Software Methodologies and How to Measure Them

This bibliography includes three important books that are not about software. Paul Starr's book The Social Transformation of American Medicine *won a Pulitzer Prize in 1982 and provides an excellent model for transforming a craft into a true profession. Software would do well to follow the same path as medicine. The psychologist Dr. Leon Festinger's book* Cognitive Dissonance *provides important insights about resistance to change and about rejecting a new and often better concept such as continuing with lines of code metrics instead of using function point metrics. It also gives insights into why methodologies are selected, like joining religious cults rather than informed technical decisions. Phil Crosby's widely cited book* Quality is Free *proves that high quality saves time and money, which is true for software as well as for manufacturing.*

Most of the data in this book did not come from the software literature, but rather from onsite studies with clients while performing benchmark data collection studies or estimating their major new software projects. Because software methodologies are changing and evolving, it is always useful to augment published books with search engines to find the latest available data.

Abran, Alain and Robillard, Pierre N.; Function point analysis, an empirical study of its measurement processes; *IEEE Transactions on Software Engineering*, Vol. 22, No. 12; Dec. 1996; pp. 895–909.

Austin, Robert D.; *Measuring and Managing Performance in Organizations*; Dorset House Press, New York; 1996; ISBN 0-932633-36-6; 216 pages.

Beck, Kent; *Test-Driven Development*; Addison Wesley, Boston, MA; 2002; ISBN 10: 0321146530; 240 pages.

Black, Rex; *Managing the Testing Process: Practical Tools and Techniques for Managing Hardware and Software Testing*; Wiley, Indianapolis, IN; 2009; ISBN-10 0470404159; 672 pages.

Brown, William K; Malveau Ralphael C.; McMormick III, Hays W.; Mowbray, Thomas J.; *AntiPatterns: Refactoring Software Architectures and Projects in Crisis*; John Wiley & Sons, Chichester; 1998.

Cohen, Lou; *Quality Function Deployment: How to Make QFD Work for You*; Prentice Hall, Upper Saddle River, NJ; 1995; ISBN 10: 0201633302; 368 pages.

Crosby, Philip B.; *Quality is Free*; New American Library, Mentor Books, New York; 1979; 270 pages. *Note*: This book is not about software but about the economic value of high-quality levels. Crosby was the ITT vice president of quality. This book is true for software quality as well as for manufacturing quality.

Curtis, Bill, Hefley, William E., and Miller, Sally; *People Capability Maturity Model*; Software Engineering Institute, Carnegie Mellon University, Pittsburgh, PA; 1995.

Festinger, Leon; *A Theory of Cognitive Dissonance*; Stanford University Press, Stanford, CA; 1962. *Note*: This book is not about software but about opinion formation and resistance to change. Festinger's theory of cognitive dissonance plays a part in software methodology selections and also about resistance to new and better metrics such as function points.

Gilb, Tom and Graham, Dorothy; *Software Inspections*; Addison Wesley, Reading, MA; 1993; ISBN 10: 0201631814.

Grady, Robert B.; *Successful Process Improvement*; Prentice Hall PTR, Upper Saddle River, NJ; 1997; ISBN 0-13-626623-1; 314 pages.

Humphrey, Watts S.; *Managing the Software Process*; Addison Wesley Longman, Reading, MA; 1989.

Humphrey, Watts S.; *Introduction to Team Software Process (TSP)*; Addison Wesley, Reading, MA; 1999 (Kindle e-book).

Humphrey, Watts S.; *PSP: A Self Improvement Process for Software Engineers*; Addison Wesley, Reading, MA; 2005.

Jacobsen, Ivar, Griss, Martin, and Jonsson, Patrick; *Software Reuse Architecture, Process, and Organization for Business Success*; Addison Wesley Longman, Reading, MA; 1997; ISBN 0-201-92476-5; 500 pages.

Jones, Capers; *Assessment and Control of Software Risks*; Prentice Hall, Englewood Cliffs, NJ; 1994; ISBN 0-13-741406-4; 711 pages.

Jones, Capers; *Patterns of Software System Failure and Success*; International Thomson Computer Press, Boston, MA; December 1995; ISBN 1-850-32804-8; 292 pages.

Jones, Capers; *Software Quality: Analysis and Guidelines for Success*; International Thomson Computer Press, Boston, MA; 1997; ISBN 1-85032-876-6; 492 pages.

Jones, Capers; *The Economics of Object-Oriented Software*; Software Productivity Research, Burlington, MA; April 1997; 22 pages.

Jones, Capers; *Software Assessments, Benchmarks, and Best Practices*; Addison Wesley Longman, Boston, MA; 2000; 600 pages.

Jones, Capers; *Estimating Software Costs*; 2nd edition; McGraw-Hill, New York; 2007.

Jones, Capers; *Applied Software Measurement*; 3rd edition; McGraw-Hill, New York; 2008.

Jones, Capers; *Software Engineering Best Practices*; 1st edition; McGraw-Hill, New York; 2010.

Jones, Capers; *The Technical and Social History of Software Engineering*; Addison Wesley, Englewood Cliffs, NJ; 2014.

Jones, Capers and Bonsignour, Olivier; *The Economics of Software Quality*; Addison Wesley, Boston, MA; 2011; ISBN 978-0-13-258220-9; 587 pages.

Kan, Stephen H.; *Metrics and Models in Software Quality Engineering*; 2nd edition; Addison Wesley Longman, Boston, MA; 2003; ISBN 0-201-72915-6; 528 pages.

Keys, Jessica; *Software Engineering Productivity Handbook*; McGraw-Hill, New York; 1993; ISBN 0-07-911366-4; 651 pages.

Land, Susan K.; Smith, Douglas B.; Walz, John Z.; *Practical Support for Lean Six Sigma Software Process Definition: Using IEEE Software Engineering Standards*; Wiley Blackwell, Hoboken, NJ; 2008; ISBN 10: 0470170808; 312 pages.

Love, Tom; *Object Lessons*; SIGS Books, New York; 1993; ISBN 0-9627477 3-4; 266 pages.

McMahon, Paul; *15 Fundamentals for Higher Performance in Software Development*; PEM Systems; 2014.

Multiple authors; *Rethinking the Software Process*; (CD-ROM); Miller Freeman, Lawrence, KS; 1996. (This is a new CD-ROM book collection jointly produced by the book publisher, Prentice Hall, and the journal publisher, Miller Freeman. This CD-ROM disk contains the full text and illustrations of five Prentice Hall books: *Assessment and Control of Software Risks* by Capers Jones; *Controlling Software Projects* by Tom DeMarco; *Function Point Analysis* by Brian Dreger; *Measures for Excellence* by Larry Putnam and Ware Myers; and *Object-Oriented Software Metrics* by Mark Lorenz and Jeff Kidd.)

Multiple authors; DevOps; Wikipedia article; 2007 but continuously updated.

Multiple authors; Git; Wikipedia article; 2013 but continuously updated.

Nandyal, Raghav; *Making Sense of Software Quality Assurance*; Tata McGraw-Hill Publishing, New Delhi, India; 2007; ISBN 0-07-063378-9; 350 pages.

Paulk Mark et al.; *The Capability Maturity Model; Guidelines for Improving the Software Process*; Addison Wesley, Reading, MA; 1995; ISBN 0-201-54664-7; 439 pages.

Putnam, Lawrence H. and Myers, Ware; *Industrial Strength Software: Effective Management Using Measurement*; IEEE Press, Los Alamitos, CA; 1997; ISBN 0-8186-7532-2; 320 pages.

Radice, Ronald A.; *High Quality Low Cost Software Inspections*; Paradoxicon Publishing Andover, MA; 2002; ISBN 0-9645913-1-6; 479 pages.

Reifer, Don; Annual Report on Agile Software Development; Reifer Associates; Phoenix AZ; 2016.

Royce, Walker E.; *Software Project Management: A Unified Framework*; Addison Wesley Longman, Reading, MA; 1998; ISBN 0-201-30958-0.

Rubin, Howard; *Software Benchmark Studies for 1997*; Howard Rubin Associates, Pound Ridge, NY; 1997.

Rubin, Howard (Editor); *The Software Personnel Shortage*; Rubin Systems, Inc.; Pound Ridge, NY; 1998.

Shore, James and Worden Shane; *The Art of Agile Development*; O'Reilly Media, Sebastapol, CA; 2007.

Starr, Paul; *The Social Transformation of American Medicine*; Basic Books; Perseus Group, New York; 1982; ISBN 0-465-07834-2. *Note*: This book won a Pulitzer Prize in 1982 and is highly recommended as a guide for improving both professional education and professional status. There is much of value for the software community in Starr's book. It shows how medicine evolved from a low-status craft to the most respected of all learned professions.

Strassmann, Paul; *Information Payoff*; Information Economics Press, Stamford, CT; 1985.

Strassmann, Paul; *Information Productivity*; Information Economics Press, Stamford, CT; 1999.

Strassmann, Paul; *Governance of Information Management: The Concept of an Information Constitution*; 2nd edition; (eBook); Information Economics Press, Stamford, CT; 2004.

Thayer, Richard H. (editor); *Software Engineering and Project Management*; IEEE Press, Los Alamitos, CA; 1988; ISBN 0-8186-075107; 512 pages.

Weinberg, Gerald M.; *The Psychology of Computer Programming*; Van Nostrand Reinhold, New York; 1971; ISBN 0-442-29264-3; 288 pages.

Weinberg, Gerald M.; *Becoming a Technical Leader*; Dorset House; New York; 1986; ISBN 0-932633-02-1; 284 pages.

Weinberg, Gerald; *Quality Software Management: Volume 2, First-Order Measurement*; Dorset House Press, New York; 1993; ISBN 0-932633-24-2; 360 pages.

Wiegers, Karl A.; *Creating a Software Engineering Culture*; Dorset House Press, New York; 1996; ISBN 0-932633-33-1; 358 pages.

Wiegers, Karl E.; *Peer Reviews in Software: A Practical Guide*; Addison Wesley Longman, Boston, MA; 2002; ISBN 0-201-73485-0; 232 pages.

Yourdon, Ed; *Death March: The Complete Software Developer's Guide to Surviving "Mission Impossible" Projects*; Prentice Hall PTR, Upper Saddle River, NJ; 1997; ISBN 0-13-748310-4; 218 pages.

Additional Data Sources from Software Benchmark Organizations

In addition to published books and journal articles, current data on productivity and quality results of various methodologies are available from the major software benchmark organizations. It is suggested that to find actual productivity and quality data, the professional software benchmark companies are probably the best source. Obviously, popular methodologies such as Agile will have more benchmarks than older or obscure methodologies.

Software Benchmark Companies and Associations in 2016		
1	4SUM Partners	www.4sumpartners.com
2	Bureau of Labor Statistics, Department of Commerce	www.bls.gov
3	Capers Jones (Namcook Analytics LLC)	www.namcook.com
4	CAST Software	www.castsoftware.com
5	Congressional Cyber Security Caucus	cybercaucus.langevin.house.gov

(Continued)

6	Construx	www.construx.com
7	COSMIC function points	www.cosmicon.com
8	Cyber Security and Information Systems	https://s2cpat.thecsiac.com/s2cpat/
9	David Consulting Group	www.davidconsultinggroup.com
10	Economic Research Center (Japan)	www.zai-keicho.or.jp
11	Forrester Research	www.forrester.com
12	Galorath Incorporated	www.galorath.com
13	Gartner Group	www.gartner.com
14	German Computer Society	http://metrics.cs.uni-magdeburg.de/
15	Hoovers Guides to Business	www.hoovers.com
16	IDC	www.IDC.com
17	ISBSG Limited	www.isbsg.org
18	ITMPI	www.itmpi.org
19	Jerry Luftman (Stevens Institute)	http://howe.stevens.edu/index.php?id=14
20	Level 4 Ventures	www.level4ventures.com
21	Metri Group, Amsterdam	www.metrigroup.com
22	Namcook Analytics LLC	www.namcook.com
23	Price Systems	www.pricesystems.com
24	Process Fusion	www.process-fusion.net
25	QuantiMetrics	www.quantimetrics.net
26	Quantitative Software Management (QSM)	www.qsm.com
27	Q/P Management Group	www.qpmg.com
28	RBCS Inc.	www.rbcs-us.com
29	Reifer Consultants LLC	www.reifer.com

(*Continued*)

30	Howard Rubin	www.rubinworldwide.com
31	SANS Institute	www.sabs,org
32	Software Benchmarking Organization (SBO)	www.sw-benchmark.org
33	Software Engineering Institute (SEI)	www.sei.cmu.edu
34	Software Improvement Group (SIG)	www.sig,eu
35	Software Productivity Research	www.SPR.com
36	Standish Group	www.standishgroup.com
37	Strassmann, Paul	www.strassmann.com
38	System Verification Associates LLC	http://sysverif.com
39	Test Maturity Model Integrated	www.experimentus.com

Index

D

M

N

O

P

T

U

V

W

X